Electronic Portable Instruments

Design and Applications

Halit Eren

Electronic Portable Instruments

Design and Applications

CRC PRESS

Boca Raton London New York Washington, D.C.

TK
7878.4
.E74
2004

Library of Congress Cataloging-in-Publication Data

Eren, Halit.
    Electronic portable instruments: design and applications / Halit Eren.
        p. cm.
    Includes bibliographical references and index.
    ISBN 0-8493-1998-6 (alk. paper)
    1. Portable computerized instruments—Design and construction. 2. Detectors. I. Title.

TK7878.4.E74 2003
681'.2—dc21   2003055082

Visit the CRC Press Web site at www.crcpress.com

© 2004 by CRC Press LLC

No claim to original U.S. Government works
International Standard Book Number 0-8493-1998-6
Library of Congress Card Number 2003055082
Printed in the United States of America 1 2 3 4 5 6 7 8 9 0
Printed on acid-free paper

Dedication

To my mother, Mediha, and to my family: Gulden, Emre and Pelin

Preface

Today, in broad terms, portable electronic devices are very common. They are used in a large range of applications: science, engineering, research, and industry. They are not confined to only these areas, but find many applications and use by ordinary people in their everyday lives. In fact, these portable devices constitute a large part of consumer electronics, including portable tapes, radios, CD players, TVs, VCRs, cameras, power tools, calculators, games, toys, cellular telephones, clocks, watches, pacemakers, thermometers, microphones, shavers, notebooks, palm computers, electronic diaries, etc. Hence, the list can go on and on.

An important section of portable electronic devices is the electronic portable instruments that are used for measuring and monitoring purposes. Particularly in recent years, due to progress in the electronics technology, many instruments conventionally known as bench-top instruments are being replaced by their portable counterparts. As the electronic components get smaller in the form of microelectronics, the physical dimensions of conventional instruments become smaller too. Therefore, miniaturization of many features of the larger and bulkier electronic instruments opens up new avenues of applications in many areas. They find applications that include laboratory instruments, medicine, environmental monitoring, space explorations, optics, chemistry, biology, petroleum industry, aerospace, manufacturing, pollution control, military devices, navigation, personal security, hobby devices, process control, and so forth. The applications of portable instruments do not end in the areas mentioned above, but are found almost everywhere. They are used to measure variables such as size, light, color, temperature, pressure, radioactivity, currents, voltages, bacterial and viral presence, water and other chemical analyses, latitude, displacements, changes in position, chemical composition of solids, liquids, and gases, sound analysis, breath analysis, human or machine orientation, acceleration, homogeneity, humidity, quantity, density, etc.

Measurement is essential for observing and testing scientific and technological investigations. In fact, it is so important and fundamental that the whole science can be said to be dependent on it. Instruments are developed for monitoring the conditions of physical variables and converting these conditions into symbolic output forms. Therefore, instruments are man-made devices to enhance human interaction with nature. They help in the understanding of natural phenomena.

Since instruments are so vital in the progress of mankind, there have been many research efforts resulting in many publications on the topic over the past decades. There is a large reservoir of books written on measurements, instruments, instrumentation systems, sensors, and transducers. Almost exclusively all those books deal with conventional or bench-top instruments and instrumentation systems. As yet, there are virtually no books available on electronic portable instruments. One wonders how and why such an important topic has escaped the attention of many researchers, writers, and publishers. The answer may primarily be hidden in the philosophy that portable instruments are essentially a subsection of normal instruments and instrumentation systems. But not any more — with the availability and use of today's advanced technology on digital systems and communications, portable instruments are emerging as an important discipline in their own right. The emergence of satellite communications and advances in other means of communication such as optical, microwave, and radio frequency (RF), as in the case of mobile telephones, are greatly impacting the development and use of portable instruments.

Portable instruments are no longer small devices to take simple measurements of physical variables; they can communicate with other devices or digital systems while they are in service. They are used under all conditions, including in remote environments where conventional instruments cannot operate. In parallel to the recent developments in communications, the progress in digital systems, intelligent sensors, microsensors, virtual instruments, and electronic portable instruments is significant, offering challenges to researchers, designers, manufacturers, and consumers.

This book on electronic portable instruments is among the first of its kind. Due to the reasons given above, we should expect a great deal of attention on the topic in the near future, leading to the emergence of many other publications. This book is written primarily for three types of readers: (1) practicing scientists, engineers, and technologists who use electronic portable instruments in their work environment; (2) engineers and scientists who are involved in hardware and software design, construction, and manufacturing of electronic portable instruments with some degree of complexity to suit specific applications; and (3) those with backgrounds other than science and engineering who use portable instruments.

Although the topic of portable electronic devices is a fascinating and largely untouched area, this book concentrates on portable measurement instruments rather than portable electronic devices. Needless to say, portable electronic devices and portable electronic instruments have many features in common as far as design, construction, components, power requirements, and operations are concerned.

In the past decades, apart from a few low-power-demanding instruments, traditionally most measurement and monitoring instruments were first introduced as bench-top devices, with their performance being the primary concern and with no regard for portability. Only later, as the technology progressed, and as the need existed, instruments migrated to smaller and lighter formats. In recent years, this migration is accelerated due to rapid progress in digital and communications technologies. However, there were few instruments uniquely built to be portable, such as the classical AVO meters for current, voltage, and resistance measurements, and radiation and sound and light intensity measurement devices. Early electronic portable instruments, especially handheld types, were relatively simple and all analog devices. Today, due to their simplicity, robustness, and low costs, such instruments still exist and are widely used. However, the range of existing portable instrumentation now extends far beyond simple, low-accuracy portable devices to highly complex, network-integrated and high-performance measurement systems.

Design of electronic portable instruments is centered on the trade-offs between the instrument cost, performance, limited size and weight, power consumption, user interface options, ruggedness, and ability to operate in a range of environments. The operational requirements of many portable instruments are likely to vary widely from scientific laboratories to complex industrial plants to field use in the outback, far from civilization. Most of the portable instruments must be able to survive accidental or deliberate abuses such as drops and knocks, rain, and water. The environment where they are required to operate may be unconventional and harsh, as in the case of chemical plants. They must be able to operate in the open sunshine or in the darkness, in extreme heat or in cold conditions. Some are designed to be used by humans and machines, whereas others may be attached to animals. Therefore, they must have finely tuned design features that balance performance against power consumption, weight against ruggedness, simple user interface characteristics against the convenience of operations. In addition, they must have reasonable costs. This, in many ways, presents a truly challenging problem to researchers, designers, and manufacturers. All the main parts of an electronic portable instrument — the display case, power supply, sensors, electronic components, and processors — must be selected, designed, and tested carefully.

In the past, the unique nature of most instrumentation hardware demanded customized software support and precluded the use of standard instrumentation platforms in electronic portable instruments. However, today's portable instruments are largely software driven and therefore can be programmed for the implementation of various measurement techniques. Their front-end and back-end hardware can be standardized and designed to address large classes of similar applications. Instrument design that is based on the use of a standard instrumentation platform has been successfully implemented on desktop computers and has also reached the domain of portable instrumentation. In the following sections, basic principles behind electronic portable instruments will be introduced; operational principles will be described; and their typical specifications and limitations will be discussed.

Today's electronic portable instruments can be divided into three main groups:

1. *Portable and handheld instruments* that are built for a specific application. They carry the standard features of normal instruments with

user interface, complex signal processing electronics, and communication capabilities.

2. *Intelligent sensor-based devices* that contain few components with
   limited features. A typical example is the *implantables* in medical
   applications.

3. *Portable data systems* that contain fixed sensors and supporting mechanisms. These devices are capable of communicating with other
   digital devices and computers via communication networks. A typical example is mobile weather stations.

In the ensuing sections, the necessary theoretical and practical information
will be provided for the understanding of operational principles, design,
construction, and use of portable instruments.

Chapter 1 is dedicated to measurements, instruments, and instrumentation in general terms, with specific references to electronic portable instruments. *Measurement* is defined as the process of gathering information from
a physical world and comparing this information with agreed standards.
Instruments are developed for measuring the conditions of physical variables and for converting them into symbolic output forms so that humans
can interpret and understand the nature of these physical variables. As in
the case of all instruments, portable instruments have sensors or transducers,
signal conditioners, inputs and outputs, or termination stages. This chapter
deals with the general concepts of all types of instruments and then concentrates on the features of portable instruments, as they deviate from conventional instruments in a number of ways. The basic features of the portable
instruments that differentiate them from the conventional instruments are
given. The response characteristics, errors and error control, uncertainty,
standards, design, testing, calibration, and use of instruments are also discussed in detail.

As in the case of conventional bench-top instruments, portable instruments
rely on sensors and transducers; this is explained in Chapter 2. The development of semiconductor devices, including operational amplifiers, had an
impressive impact on the measurement instruments. For example, by the
end of the 1970s it had become apparent that essentially planar processing
integrated circuit (IC) technology could be modified to fabricate three-
dimensional electromechanical structures by a micromachining process.
Accelerometers and pressure sensors were among the first IC sensors, and
today there are many different types extensively used in electronic portable
instruments. Detailed treatment of sensors and sensors technology on which
instrumentation and particularly portable instruments depend is given in
this chapter.

Chapter 3 discusses the key concept of digital instruments, microprocessors, and microcontrollers — the conversion from analog signals to digital
signals and vice versa (the supporting software and network aspects are all
explained in this chapter). In addition to the practical digital aspects, the

supporting theory on analog-to-digital and digital-to-analog conversions, signal processing, is explained. The theory ensures that digital instruments truly represent the physical processes.

A new dimension in electronic portable instruments is the data and information gathering about a physical process by using suitable communication systems. One form of information gathering is realized by the use of telemetry. In telemetry, data are transmitted from remote locations to convenient places using various techniques, such as optical, electrical, microwave, RF, the Internet, etc. Chapter 3 highlights the operation of telemetry that is particularly relevant to electronic portable instruments. Communication networks protocols and basic principles are discussed.

As indicated earlier, portable instruments are no longer stand-alone, simple, and on–off devices. They can be equipped with communication features and networked as part of a large instrumentation system. Today's technology allows not only electronic portable instruments to be networked, but also intelligent and sophisticated sensors to be networked and integrated into large systems, as explained in Chapters 3 and 4.

Chapter 4 is dedicated solely to portable instruments, from basic components to complex features. It will be iterated that the art of measurement and monitoring by electronic portable instruments can be achieved in two different ways: (1) by the use of analog instruments, which deal directly with analog signals, and (2) by converting the analog signals into digital forms. In modern applications, digital instruments are rapidly replacing analog counterparts. One crucial advantage of digital information is that it can be managed and communicated much more easily and reliably than analog information. The power requirements, digital hardware and software aspects, enclosures and casing, construction, and noise problems will be discussed in detail in this chapter.

Chapter 5 gives many examples of portable instruments that are available in the marketplace. Among virtually thousands of portable instruments, only a few selected examples are provided. Their electrical properties, features, characteristics, limitations, and capabilities are discussed with particular relevance to information provided in the earlier chapters.

Chapter 6 summarizes the future trends of electronic portable instruments.

Acknowledgments

I started writing this book with Boris Donskoy of DVP Inc., U.S.A. Unfortunately, because of work commitments, Boris could not continue, but he encouraged me to press forward to complete this book. I am grateful to Boris for his valuable encouragement and initial contribution. I express my special thanks to Bernard Goodwin for his continual encouragement to come up with an informative and high quality book. I also thank my colleagues Dr. Lance C.C. Fung and Clive Maynard for their valuable help in the digital section in Chapter 3.

Special thanks also go to my colleague, James Goh, for offering constructive comments whenever I needed help during development of this book. I am thankful to my colleagues at Gazi University, Department of Electrical and Computer Engineering, Ankara, for providing offices and computers at their premises during the final stages of this book. My warm acknowledgment is also extended to Professor Izzet Kale of the University of Westminster, London, for having patient and extended discussions on various topics about portable instruments when we met at the IMTC/IEEE conferences.

The Author

Halit Eren, Ph.D., received his undergraduate degree in 1973, a Master's degree in electrical engineering in 1975, and a Ph.D. degree in control engineering in 1978 from the University of Sheffield, U.K. He also earned an MBA from Curtin University of Technology, Perth, Western Australia in 1998 majoring in international management and leadership.

Dr. Eren has lectured at Curtin University of Technology since 1983, first in Kalgoorlie School of Mines and then in the School of Electrical and Computer Engineering, Perth. He also served as head of the Department of Electronics and Communications Engineering. His expertise lies in the areas of instruments, instrumentation systems and networks, electronic portable instruments, signal processing, and engineering mathematics. Dr. Eren has been researching portable instruments for more than 15 years, mainly in the area of electromagnetic, ultrasonic and infrared techniques applied in mobile robots, density measurements, fluid flow measurements, and moisture and humidity measurements. His recent interests include fieldbus technology, telemetry, control systems via internet, and telecontrolers. He serves as a consultant to many industrial and government organizations. Dr. Eren has contributed to more than 100 conferences, journals, and transactions and various books published by CRC Press and John Wiley & Sons.

Abbreviations

AC	alternating current
A/D	analog-to-digital (converter)
AI	artificial intelligence
ALU	arithmetic logic unit
AM	amplitude modulation
AMR	anisotropic magnetoresistor
ANSI	American National Standards Institute
ARQ	automatic repeat request
ASIC	application-specific chip
ASMFS	adapted synchronous model feedback systems
BCD	binary-coded decimal
BCS	British Calibration Services
BDLC	byte data link controller
BF_3	boron trifluoride
BIOS	basic input/output system
BJT	bipolar junction transistor
bps	bits per second
CAD	computer-aided design
CAM	computer-aided manufacturing
CAN	controller area network
CASE	computer-aided software engineering
CCD	charge-coupled device
CdS	cadmium sulfide
CdSe	cadmium selenide
CISC	complex instruction set computer
CMOS	complementary metal oxide semiconductor
CMRR	common mode rejection ratio
CPM	counts per minute
CPU	central processing unit
CRC	cyclic redundancy code
CRT	cathode ray tube
CSMA/CD	carrier sense multiple access with collision detection
D/A	digital-to-analog (converter)
DAQ	data acquisition card
DC	direct current
DCE	data communication equipment
DDS	direct digital synthesis
DE	differential equation
DETF	double-ended tuning fork

DFT	discrete Fourier transform
DIP	dual in-line package
DLL	dynamic link library
DM	data memory
DMA	direct memory access
DMM	digital multimeter
DoD	Department of Defense
DR	dynamic range
DSN	distributed sensor network
DSO	digital storage oscilloscope
DSP	digital signal processor
DSSS	direct sequence spread spectrum
DTE	data terminal equipment
DUT	device under test
DVM	digital voltmeter
EBI	external bus interface
ECG	electrocardiogram
ECL	emitter-coupled logic
EDAS	Ethernet data acquisition system
EISA	extended industry standard architecture
EMC	electromagnetic compatibility
emf	electromotive force
EMI	electromagnetic interference
EOB	electrical oscillator-based
EPA	Environmental Protection Agency
ETS	equivalent-time-sampling technique
FCC	Federal Communications Commission
FDM	frequency division multiplexing
FET	field-effect transistor
FFT	fast Fourier transform
FIR	finite impulse response
FM	frequency modulation
FS	full scale
FSK	frequency-shift keyed
Gbps	gigabits per second
GBW	gain-bandwidth product
GEMS	Global Environmental Monitoring System
GMR	giant magnetoresistor
GPS	global positioning system
GSO	geostationary satellite orbit
HIS	hospital information system
IC	integrated circuit
I²C bus	interintegrated circuit bus
IDE	integrated development environment
IEA	Electronics Industries Association
IEEE	Institute of Electrical and Electronics Engineers

IF	intermediate frequency
IIR	infinite impulse response
INS	inertial navigation system
I/O	input/output
IP	ingress progress
IPAD	integrated passive and active device
IR	infrared
ISA	industry standard architecture
ISFET	ion-selective field-effect transistor
ISO	International Organization for Standardization
ITU	International Telecommunication Union
LAN	local area network
LCD	liquid crystal display
LDR	light-dependent resistor
LED	light-emitting diode
LIS	laboratory information system
LSB	least significant bit
LVDT	linear variable-differential transformer
MAC	multiply-accumulate
Mb	megabit
Mbyte	megabyte
Mbps	megabits per second
MCM	multichip module
MCU	microcontroller unit
MEM	micro-electro-mechanical
MIL-STD	military standard
MIPS	mega instructions per second
mmf	magnetomotive force
MMS	manufacturing message service
MOS	metal oxide semiconductor
MOSFET	metal oxide semiconductor field-effect transistor
MSB	most significant bit
MTBF	mean time between failure
MTU	master terminal unit
NATA	National Association of Testing Authorities
NBS	National Bureau of Standards
NCAP	network capable application processor
NEMA	National Electrical Manufacturers Association
NI	noise index
NMR	nuclear magnetic resonance
NPL	National Physical Laboratory
NTC	negative temperature coefficient
NTP	normal temperature and pressure
OOP	object-oriented programming
op-amp	operational amplifier
OTP	one-time programmable

PAN	personal area network
PbS	lead sulfide
PbSe	lead selenide
PC	personal computer
PCB	printed circuit board
PCI	protocol control information
PCMCIA	Personal Computer Memory Card International Association
PCR	polymerase chain reaction
PDIP	plastic dual-in-line packaging
PEM	proton exchange membrane
pf	power factor
PI	pulse induction
PIC	power integrated circuit
PIR	passive infrared
PLA	programmable logic array
PLL	phase locked loop
PM	program memory
PMP	post metal programming
ppm	parts per million
PPS	Precise Positioning Service
PROM	programmable read only memory
PCS/PCN	personal wireless communication system/network
PSD	position-sensitive detector
PSoC	programmable system-on-chip
PTC	positive temperature coefficient
PVA	polyvinyl alcohol
PWM	pulse width modulation
PZT	lead zirconate titanate
RAM	random access memory
RE	reference electrode
RF	radio frequency
RFI	radio frequency interface
RH	relative humidity
RISC	reduced instruction set computer
rms	root mean square
ROM	read only memory
RSSI	received signal strength indicator
RTD	resistance temperature detector
RTS	real-time sampling
RTU	remote terminal unit
SA	spectrum analyzer
SAD	stochastic analog-to-digital (converter)
SAR	successive approximation register
SAW	surface acoustical wave
SBC	single-board computer
SCADA	System Control and Data Acquisition

SCI	serial communications interface
SCIM	single-chip integration module
SG	specific gravity
S/H	sample-and-hold
SHA	sample-hold amplifier
SI	Système International d'Unités
SINAD	signal to noise and distortion
SLM	sound level meter
SNR	signal-to-noise ratio
SoC	system-on-chip
SOIC	small outline integrated circuit
SPI	serial peripheral interface
SPL	sound pressure level
SPS	Standard Positioning Service
sps	samples per second
SQUID	superconducting quantum interface device
SRAM	standby RAM
STP	standard temperature and pressure
TC	timer counter
TCP/IP	transmission control protocol/Internet protocol
TDM	time division multiplexing
TDR	time domain reflectometry
TEDS	transducer electronic data sheet
TFT	thin film technology
T/H	track/hold
THD	total harmonic distortion
TPU	time processor unit
TTL	transistor–transistor logic
UART	universal asynchronous receiver transmitter
UFPA	uncooled focal plane arrays
UHF	ultrahigh frequency
USART	universal synchronous/asynchronous receiver transmitter
UV	ultraviolet
VHF	very high frequency
VLF	very low frequency
VLSI	very large scale integration
VME	VersaModule Eurocard
VOC	volatile organic compound
WE	working electrode
WLAN	wireless local area network
WPAN	wide personal area network

Contents

1

Measurements, Instrumentation, and Electronic Portable Instruments

1.1 Fundamentals of Measurements

Measurement is a process of gathering information from the physical world and comparing it with agreed standards. Measurement is carried out with instruments that are designed and manufactured to fulfill given specifications.

Instruments are man-made devices that are designed to maintain prescribed relationships between the parameters being measured and the physical variables under investigation. The physical parameter under investigation is known as the *measurand*. Sensors and transducers are the most basic and primary elements that respond to physical variations to produce an output. The relation between sensor signal and physical variations can be expressed in the form of *transfer functions*. The transfer function may be linear or nonlinear. A linear relationship is governed by the equation

$$y = a + bx \qquad (1.1)$$

where y is the electric signal from the sensor, x is the physical stimulus, a is the intercept that is the output signal for a zero input signal, and b is the slope, also known as the sensitivity.

The output y may be in amplitude, frequency, phase, or other properties, depending on the characteristics of a particular sensor. The nonlinear transfer functions can be logarithmic, exponential or in other forms, such as a power function. In many applications, a nonlinear sensor may function linearly over a limited range.

The energy output from the sensor is usually supplied to a transducer, which converts energy from one form to another. Therefore, a transducer is a device capable of transferring energy between two physical systems.

For a specific application, a diverse range of sensors and transducers may be available to meet the measurement requirements of a particular physical system. Correct sensors must always be selected, and appropriate signal processing must be employed to retrieve the required information, which

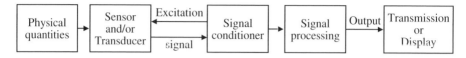

FIGURE 1.1
Essential components of an instrument.

can be directly related to the physical variable. Also, after having generated the signals by a sensor, the type of signal processing to be implemented depends on the information required from it.

Instruments are applied for information gathering about the physical variable. Once the information is obtained, it can be processed, interpreted, generalized, or reorganized. Appropriately designed tests and experiments may be necessary for information gathering. Tests can be done by humans or machines in an automatic or semiautomatic manner. Tests and experiments generally are conceived, planned, performed, and repeated until full confidence is acquired in the results. Applications of instruments range from laboratory conditions to arduous environments, such as inside nuclear reactors or remote locations, e.g., satellite systems and spaceships. Many manufacturers are producing a large range of instruments in order to meet diverse ranges of measurements with extensive degrees of complexity and broad application requirements.

The functionality of an instrument can be broken into smaller elements, as illustrated in Figure 1.1. All instruments have some or all of these functional blocks. In general, if the behavior of the physical system under investigation is known, its performance can be monitored and assessed by means of suitable methods of sensing, signal conditioning, and termination.

In the last 10 years or so, due to rapid progress in integrated circuit (IC) technology and availability of low-cost analog and digital components and microprocessors, considerable advancements have taken place in measurements, instruments, and instrumentation systems. This progress equally reflects on electronic portable instruments. Performance of electronic portable instruments has improved significantly by the availability of on-line and off-line backed analysis, enhanced signal processing techniques, agreed local and international standards, and, most importantly, progress in communications technology, which addresses particularly the needs of distributed instrumentation systems.

1.2   Units and Standards

Standards for the fundamental units, such as length, time, weight, temperature, and electrical quantities, were developed a long time ago. These standards enable consistency in measurements irrespective of time and place all

TABLE 1.1

Basic SI Units

Quantity	Unit	Symbol
Length	Meter	m
Mass	Kilogram	kg
Time	Second	sec
Electric current	Ampere	A
Temperature	Kelvin	K
Amount of substance	Mole	mol
Luminous intensity	Candela	cd
Plane angle	Radian	rad
Solid angle	Steradian	sr

over the world. The standard physical entities for length and weight (the meter and the kilogram) are kept in the International Bureau of Weights and Measures in Serves, France. Nevertheless, apart from the physically existing standard meter, in 1983, the meter was defined as the length of path traveled by light in vacuum in 1/299,792,458 of a second. Currently this is adopted as the standard meter.

The standard unit of time is the second, which is established in terms of known oscillation frequencies of certain devices, such as the radiation of the cesium-133 atom. The standards of electrical quantities are derived from mechanical units of force, mass, length, and time. Temperature standards are established as international scale by taking 11 primary fixed points of temperature. Historically, many different units are involved in different places of the world or even within the same country; the relationships between different units are defined in fixed terms. For example, 1 lb = 453.59237 g.

Based on these standards, primary international units, Système International d'Unités (SI), are established for mass, length, time, electric current, luminous intensity, and temperature, as illustrated in Table 1.1. From these units, SI units of all physical quantities can be derived, as exemplified in Table 1.2. The standard multiplier prefixes are illustrated in Table 1.3.

In addition to primary international standards of units, standard instruments are available. These instruments have stable and precisely defined characteristics that are used as a reference for other instruments performing the same function. Hence, the performance of an instrument can always be cross-checked against a known device. At the worldwide level, checking is done by using an international network of national and international laboratories, such as the National Bureau of Standards (NBS), the National Physical Laboratory (NPL), or the Physikalisch-Technische Bundesanstalt (PTB) of Germany. A treaty between the world's national laboratories regulates the international activity and coordinates development, acceptance, and intercomparisons. Standards are kept in four stages:

The *international standards* represent certain units of measurement with maximum accuracy possible within today's available technology.

TABLE 1.2

Fundamental, Supplementary, and Derived Units

Quantity	Symbol	Unit Name	Unit Symbol
Mechanical Units			
Acceleration	A	Meter/second^2	m/sec^2
Angular acceleration	a	Radian/second^2	rad/sec^2
Angular frequency	w	Radian/second	rad/sec
Angular velocity	w	Radian/second	rad/sec
Area	A	Square meter	m^2
Energy	E	Joule	J (kg·m^2/sec^2)
Force	F	Newton	N (kg·m/sec^2)
Frequency	F	Hertz	Hz
Gravitational field strength	G	Newton/kilogram	N/kg
Moment of force	M	Newton meter	N·m
Plane angle	$\alpha, \beta, \theta, \phi$	Radian	rad
Power	P	Watt	W (J/sec)
Pressure	P	Newton/meter^3	N/m^3
Solid angle	a	Steradian	sr
Torque	T	Newton meter	N·m
Velocity	V	Meter/second	m/sec
Volume	V	Cubic meter	m^3
Volume density	r	Kilogram/meter^3	kg/m^3
Wavelength	l	Meter	m
Weight	W	Newton	N
Weight density	g	Newton/cubic meter	N/m^3
Work	W	Joule	J
Electrical Units			
Admittance	Y	Siemens	S
Capacitance	C	Farad	F (A·sec/V)
Conductance	G	Siemens	S
Conductivity	g	Siemens/meter	S/m
Current density	J	Ampere/meter^2	A/m^2
Electric potential	V	Volt	V
Electric field intensity	E	Volt/meter	V/m
Electrical energy	W	Joule	J
Electrical power	P	Watt	W
Impedance	Z	Ohm	Ω
Permittivity of free space	e	Farad/meter	F/m
Quantity of electricity	Q	Coulomb	C (A·sec)
Reactance	X	Ohm	Ω
Resistance	R	Ohm	Ω
Resistivity	ρ	Ohm/meter	Ω·m
Magnetic Units			
Magnetic field intensity	H	Ampere/meter	A/m
Magnetic flux	ϕ	Tesla (weber/meter)	Wb
Magnetic flux density	B	Tesla (weber/meter^2)	T (Wb/m^2)
Magnetic permeability	m	Henry/meter	H/m
Mutual inductance	M	Henry	H

TABLE 1.2 (Continued)

Fundamental, Supplementary, and Derived Units

Quantity	Symbol	Unit Name	Unit Symbol
Permeability of free space	μ_o	Henry/meter	H/m
Permeance	P	Henry	H
Relative permeability	μ_r	—	—
Reluctance	R	Henry^{-1}	H^{-1}
Self-inductance	L	Henry	H
Optical Units			
Illumination	Lx	Lux	cd·sr/m^2
Luminous flux	Lm	Lumen	cd·sr
Luminance	Cd	Candela/meter^2	cd/m^2
Radiance	L_e	Watt/steradian·meter^3	W/sr·m^3
Radiant energy	W	Joule	J
Radiant flux	P	Watt	W
Radiant intensity	I_e	Watt/steradian	W/sr

TABLE 1.3

Decimal Multiples

Name	Symbol	Equivalent
Yotta	Y	10^{24}
Zetta	Z	10^{21}
Exa	E	10^{18}
Peta	P	10^{15}
Tera	T	10^{12}
Giga	G	10^9
Mega	M	10^6
Kilo	k	10^3
Hecta	h	10^2
Deca	da	10
Deci	d	10^{-1}
Centi	c	10^{-2}
Milli	m	10^{-3}
Micro	μ	10^{-6}
Nano	n	10^{-9}
Pico	p	10^{-12}
Femto	f	10^{-15}
Atto	a	10^{-18}
Zepto	z	10^{-21}
Yocto	y	10^{-24}

These standards are under the responsibility of an international advisory committee and are not available to ordinary users for comparison or calibration purposes.

The *primary standards* are the national standards maintained by national laboratories in different parts of the world for verification of secondary standards. These standards are independently calibrated by absolute measurements that are periodically made against the international standards. The primary standards are compared against each other.

The *secondary standards* are maintained in the laboratories of industry and other organizations. They are periodically checked against primary standards and certified.

The *working standards* are used to calibrate general laboratory and field instruments.

Another type of standards is published and maintained by organizations such as the Institution of Electrical and Electronics Engineers (IEEE), the International Organization for Standardization (ISO), etc. These standards cover test procedures, safety rules, definitions, nomenclature, various guidelines, and so on. The IEEE standards are adopted by many organizations around the world. Many nations also have their own standards for tests, instrument usage procedures, health and safety rules, and the like.

1.3   General Concepts on Instruments

Instruments are designed on the basis of existing knowledge about a physical process or from the structured understanding of the process. Ideas conceived about an instrument are translated into hardware or software that can perform well within the expected specifications and standards so that they can easily be accepted with confidence by the end users.

In the wake of a rapidly progressing technology, instruments are upgraded often to meet the changing and expanding demand of the market. As the conventional instruments are upgraded by the use of new technologies, more and more portable instruments are introduced that can fulfill similar or even better functions as the conventional instruments.

Usually, the design of an instrument requires many multidisciplinary activities. Depending on the complexity, it may take many years to produce an instrument for a relatively short commercial lifetime. In the design and production of instruments, one must consider factors such as simplicity, appearance, ease and flexibility of use, maintenance requirements, lower production costs, lead time to the product, and positioning strategy in the marketplace with respect to the competitors who can offer similar products with comparable performances. In this respect, electronic portable instruments are rapidly replacing many types of fixed or bench-top instruments.

While designing and producing instruments, firms consider many factors, including sound business plans, suitable infrastructure, plant setup, appropriate equipment for production, understanding of technological changes, skilled and trained personnel, adequate finances, marketing and distribution channels, and so on. A clear understanding and careful analysis of the worldwide trends in instrument and instrumentation systems help considerably in the successful acceptance of new products. It is important to choose the right products that are likely to be in demand in the years to come. This concern is particularly important for portable instruments since many of them are newly emerging in the wake of advanced technology. Here entrepreneurial management skills may play a very important role.

The design process of an instrument itself may follow well-ordered procedures, starting from conception of a feasible idea to successful marketing. In the design stages, engineers seek a global perspective solution to the problem in hand, evaluate numerous parameters, determine priorities, synthesize solutions, and determine the effective methods. The process may further be broken down into smaller tasks, such as identifying specifications, developing possible solutions for these specifications, modeling, prototyping, installing and testing, making modifications, manufacturing, planning, marketing and distribution, evaluating customer feedback, and making design and technological improvements as a result of this feedback. Figure 1.2 illustrates the basic steps for the design and marketing of an instrument. Each one of these steps can be viewed in detail in the form of subtasks. For example, many different specifications may be considered for a particular product. These specifications include but are not limited to operational requirements, functional and technological requirements, quality, installation and maintenance, documentation and servicing, and acceptance level determination by the customers.

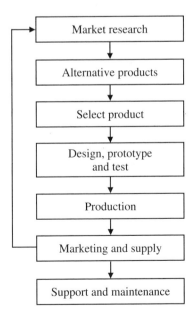

FIGURE 1.2
Simplified production process of portable instruments.

In the design and development of electronic portable instruments different approaches may be adapted. For example, for a handheld altimeter for skydivers one emphasizes reliability and precision, whereas for a bathroom scale durability may be the main issue. Nevertheless, all approaches must have good foundations of responsibility, ethics, integrity, and harmonious teamwork. This approach must be consistent during the product life cycle in that issues such as product functionality, safety, reliability, maintainability, and utility can continuously be improved from the experience gained.

In recent years, in the instrument manufacturing industry, computers have been used extensively in forms of computer-aided design (CAD), automated testing, simulation, circuit analysis and design, and many other applications. Computers enable rapid access to knowledge-based information and make design time considerably shorter. This speeds up the process so that manufacturers can meet rapidly changing demands. In some CAD systems, mechanical drafting software, electronic circuit design tools, control analysis tools, and mathematical and word processing tools operate together to assist the design process.

1.3.1   Types of Instruments

From the user's point of view, instruments can be portable, nonportable, or fixed. Portable instruments rely on on-board power sources and can be used in different places. Nonportable instruments have unlimited power from the power outlets (the mains) and are extensively used in a laboratory and an industrial environment. Nonportable instruments operate in relatively unchanging environments having steady temperature, humidity, and so on. They are designed to operate for the steady environmental conditions; therefore, their failure does not come as a surprise when they are used in unexpected places such as in hazardous conditions. On the other hand, the fixed instruments are part of a system, such as dashboards of vehicles, cockpits of airplanes, machinery in industry, buildings, and power stations, etc.

Irrespective of being portable, nonportable, or fixed, instruments can be classified to be analog or digital or a combination of the two. Nowadays, most instruments are digital because of the advantages that they offer. However, the front end of many instruments is still analog; that is, the majority of sensors and transducers generate analog signals. The signals of such instruments are conditioned initially by analog circuits before they are converted into digital forms for further processing. It is important to mention that in recent years, digital instruments operating purely on digital principles have been developing fast. For instance, today's smart sensors contain the complete signal condition circuits in a single chip integrated with the sensor itself. The output of smart sensors can be interfaced directly with other digital devices. This requires the designers to think in digital terms from the outset. Many new portable instruments are developed using advanced digitally-based sensors.

In the case of analog instruments, the changes in the current and voltage amplitudes, phases, and frequencies, or combinations of these, convey the useful information. These signals can be deterministic or nondeterministic. Also, in both analog and digital instruments, as in the case of all signal-bearing systems, there are useful signals that respond to the physical phenomena and unwanted signals resulting from various forms of noise. In the case of digital instruments, additional noise is generated in the process of analog-to-digital (A/D) conversion.

A detailed treatment of analog and digital aspects of electronic portable instruments will be given in the following chapters. Common features of the majority of analog instruments and digital instruments will be discussed next.

1.3.2 Analog Instruments

A purely analog system measures, transmits, displays, and stores data in analog form. Analog instruments generate continuous current and voltage waveforms in response to physical variations; therefore, they are characterized by their continuous signals. The signals are processed by using analog electronics; therefore, signal transmission between the instruments and other devices is also done in this form. In assembling analog instruments, the following design considerations are necessary:

- Signal transmission and conditioning
- Loading effects and buffering
- Shielding
- Inherent and imposed noises
- Ambient conditions
- Signal level compatibility
- Impedance matching
- User interface
- Display units
- Data storage

In analog instruments, signal conditioning is usually realized by integrating many functional blocks, such as bridges, amplifiers, filters, oscillators, modulators, offset circuits, level converters, buffers, and the like. Some of these functional blocks are illustrated in Figure 1.3. For the majority of instruments, in the initial stages, signals produced by sensors and transducers are conditioned mainly by analog electronics, even if they are configured as digital instruments later. Therefore, it is worth paying special attention to analog instruments, keeping in mind that much of the information given here also may be used at various stages of the digital instruments.

FIGURE 1.3
Block diagram of analog instruments.

As is true for all instruments when connecting electronic building blocks, it is necessary to minimize the loading effects of each block by ensuring that the signals are passed without attenuations, losses, or magnitude or phase alterations. It is also important to ensure maximum power transfer between blocks by appropriate impedance-matching techniques. Proper impedance matching plays a significant role in the robust operations of all instruments, but particularly at high frequencies of 1 MHz and above. As a rule of thumb, output impedance of the blocks is usually kept low and input impedance is kept high so that the loading effects can be minimized.

The sensor connection to the rest of the processing circuits bears a particular importance. The output impedance of the sensor, Z_o, must match the impedance of the follow-up circuit, Z_i. The connections of a voltage-generating sensor and a current-generating sensor are illustrated in Figure 1.4(a) and (b), respectively. The output impedance of the sensors and the input impedance of follow-up circuits generally include active and reactive components. In general, to minimize signal distortions from the sensors, the output impedance, Z_o, of the voltage-generating sensor should be low and the circuit impedance, Z_i, high. In contrast, the output impedance, Z_o, of a current-generating sensor should be high and the circuit impedance, Z_i, should be low.

Operational amplifiers and high-precision instrumentation amplifiers are the building blocks of all types of instruments, including electronic portable instruments. They are made as either integrated (monolithic) circuits or hybrid (combination of monolithic and discrete parts). An operational amplifier may be made from hundreds of transistor resistors and capacitors in the

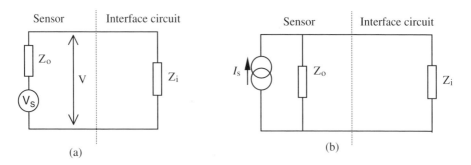

FIGURE 1.4
Sensor connection circuits: (a) voltage output; (b) current output.

single chip. They have two inputs, inverting (–) and noninverting (+), and a single output. A good operational amplifier has the following properties:

- High-input resistance, hundreds of megaohms or few gigaohms
- Low-output resistance, less than a few ohms or a fraction of 1 Ω
- Low-input offset voltage, few millivolts or in microohms
- Low-input bias current, few picoamps
- Very high open-loop gain, 10^4 to 10^6
- Low intrinsic noise
- High common mode rejection ratio (CMRR) (suppression of signals applied to both inputs at the same time)
- Low sensitivity to changes in power voltage
- A broad operating frequency range
- High environmental stability

Operational amplifiers form basic elements for the amplification of signals generated by sensors. They can be configured as inverting or noninverting amplifiers. In addition, by means of combination of suitable external components, they can be configured to perform many other functions, such as multipliers, adders, limiters, and filters.

Instrumentation amplifiers, essentially operational amplifiers, are high-performance differential amplifiers that consist of several closed-loop operational amplifiers within the chip. They are used in situations where operational amplifiers do not meet the requirements. The instrumentation amplifiers have improved common mode rejection ratios (CMRRs) (up to 160 dB), high-input impedances (e.g., 500 MΩ), low-output impedances, low offset currents and voltages, and better temperature characteristics.

Operational amplifiers also find extensive applications for processing analog signals that are nondeterministic. That is, for nondeterministic signals, the future state of the signal cannot be determined. If the signal varies in a probabilistic manner, its future can be determined only by stochastic and statistical methods. The mathematical and practical treatment of analog and digital signals having foreseen stochastic and nondeterministic properties is a very lengthy subject, and a vast body of information can be found in books specifically written on the topic (e.g., Valkenburg, 1982); therefore, it will not be treated here.

1.3.3   Digital Instruments

In modern instruments, the original data acquired from the physical variables are usually in analog form. However, in digital instruments, analog signals are converted to digital forms before they pass on to other parts of the system. For conversion purposes, A/D converters are used together with

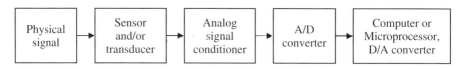

FIGURE 1.5
Components of a typical digital instrument.

appropriate sample-and-hold devices. In addition, analog multiplexers enable the connection of many sensors and transducers to the same signal processing media. The typical components of a digital instrument are illustrated in Figure 1.5. The digital systems are particularly useful in performing mathematical operations and storing and transmitting data.

Typically, a digital instrument consists of three main subsystems — the general-purpose computer core with display, mass storage and communication peripherals, and the application-specific input/output (I/O) signal conditioning hardware, together with power supply. In the case of electronic portable instruments, on-board batteries supply the power. All of these components may not be present in some of the simpler types. However, in all digital or semidigital instruments, the sensor signals are amplified or attenuated, filtered by the signal conditioning circuitry and converted to their digital representations by an A/D converter. Digital signals are then processed in accordance with the measurement algorithms, and the results are displayed, stored, or sent for remote processing or storage media. Some signals may be sufficiently high quality to be interfaced directly to a digital system, as in the case of optical measurements of rotation and linear motions. These signals eliminate the need for analog processing and hence require minimum additional preprocessing.

The analog signal generation from the digital world may be part of the measurement process or may be required for process control, or sensor stimulation purposes. Digital signals are converted to analog waveforms by digital-to-analog (D/A) converters and then filtered and amplified before being applied to the appropriate output transducers. However, some new types of instrumentation now deal exclusively with digital signals. These include logic and computer network analyzers, logic pattern generators, and multimedia and digital audio devices. Such instruments usually include high-speed digital interfaces to move data in and out of the instrument at high speeds.

Analog-to-digital conversion involves three stages: sampling, quantization, and encoding. Detailed treatment of these stages is given in Chapter 3. In all digital instruments, the Nyquist sampling theorem must be observed; that is, the number of samples per second must be at least twice the highest frequency present in the continuous signal. As a rule of thumb, depending on the significance of the high frequencies, the sampling must be about 5 to 10 times the highest frequency component in the signal. The next stage is quantization, which determines the resolution of the sampled signals. Quantization involves errors that decrease as the number of bits increase. In the

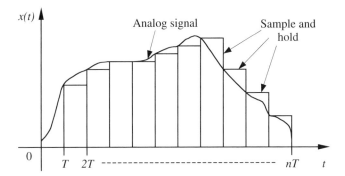

FIGURE 1.6
Process of converting an analog signal to digital form.

encoding stage, the quantized values are converted to binary numbers to be processed digitally. Figure 1.6 illustrates a typical A/D conversion process of an analog signal.

After the signals are converted into the digital forms, the data can further be processed by employing various techniques, such as fast Fourier transform (FFT) analysis, digital filtering, sequential or logical decision making, correlation methods, spectrum analysis, and so on.

Generally, digital circuits yield cheaper, more reliable, and more flexible systems. Since digital hardware allows programmable operations, it is possible to modify the functions of an instrument through the software. Thus, digital hardware and associated software provide greater flexibility in system design than does an equivalent counterpart. However, it is worth mentioning that some signals with extremely wide bandwidths require intensive real-time processing. For such signals, analog signal processing may be the only solution.

In the case of electronic portable instruments, analog signal processing is still preferred at the front end when cost, size, and power consumption are critical or whenever digital hardware cannot meet the required level of performance. However, most portable instruments are digital or, at the very least, display results digitally.

1.4 Introduction to Portable Instruments

Most bench-top or fixed instruments have virtually unlimited power availability. Also, unlike the portable instruments, the nonportable instruments usually operate in comparatively stable environments without being shifted around all that often. Therefore, in the design and manufacturing of nonportable instruments, the consideration of usage under different environmental conditions bears little importance. However, in contrast, the design and construction of portable instruments require additional considerations, such as:

- Available power is limited.

- Voltage level to supply the electronic components is low, usually less than 12 V. Although the voltage levels may be raised or lowered, it is better to stick to the voltage levels of the supply battery.

- Under normal operation conditions, power supply voltage levels may vary considerably due to diminishing stored power of the batteries as they are used. Even if the portable instruments are kept on the shelves without being used, the power of the batteries diminishes due to leakage.

- Conditions of use and the environment within which the portable instruments operate may vary often. Many portable instruments are designed for and can be used outdoors. Naturally, the environmental conditions outdoors change much more dramatically than stable indoor conditions. The possibility of portable instruments being subjected to severe variations in temperatures, moisture, and intense conditions of electromagnetic noise increases. Also, they are very likely to be subjected to accidental or deliberate damaging mechanical shocks. We all remember dropping or harshly treating our mobile telephones occasionally.

- Likelihood of the portable instruments being operated by the unskilled and untrained users increases compared to nonportable instruments.

- The human machine interface of electronic portable instruments may have to be very different from that of nonportable counterparts. The designer no longer has the luxury of ample space, size, and power supply that he or she does with nonportables.

- The weight and size of the portable instruments may have to be as minimum as possible.

1.4.1 Design Consideration of Electronic Portable Instruments

In the design and construction of electronic portable instruments, environmental conditions such as temperature, humidity and moisture, corrosive fumes and aerosols, and electromagnetic interference need to be considered carefully. This is because many test and measurement tasks are performed in environments where use of stationary or lab-grade instruments is either inconvenient or not feasible. Electronic portable instruments are likely to be used in different locations; therefore, they are relatively small, lighter, and self-powered. Typical sizes of such instruments range from the dimensions of a small suitcase and the size of a wristwatch to microcircuits as in the case of implantables. Traditionally, the term *portable* had been reserved for self-powered devices; however, the term nowadays is changing its meaning, since there are many instruments that are powerful enough in their functions to replace many bench-top or fixed instruments, and yet they are portable and much smaller in size.

The size of portable instruments varies widely and reflects a wide variety of measurement environments. For example, the existing oscilloscopes range from high-performance suitcase-size devices to medium-performance hand-held devices to low-performance PC cards and pods to be used as peripherals with portable PCs. Devices such as PC cards and pods may be powered by the unit to which they are attached.

Generally, one can divide portable electronic instruments by their size and power consumption. Examples of such instruments are the luggable, externally powered "notebooks" (5 to 15 W), handheld (0.1 to 5 W) instruments, and the micropower devices (<100 mW). This division reflects different design trade-offs. On one extreme, notebooks tend to sacrifice the duration of battery-powered operation for maximum performance, while micropower devices realize the opposite goal, with the handheld being in the middle.

The reduction of power consumption is perhaps the most significant problem in the design of electronic portable instruments. As a rule, better performance demands more power, while the desire to keep instruments small and light precludes the use of larger battery packs. Additionally, the desire to operate longer without the need for battery recharge or replacement puts further pressure on power consumption. Still, advancements in battery technology, although less significant than the impact of the new generation of semiconductor devices, lead to the development of many different types of light and compact electronic portable instruments.

For the power source, both primary and rechargeable batteries can be used, although primary batteries maintain an edge in energy and weight per volume. Solar and wind power are viable alternatives to chemical batteries, especially in long-term monitoring instrumentation in remote areas. Computer-based instruments usually implement some control features that slow down or altogether turn off parts of the instrument to save power. For example, if no operator intervention is detected for a set period, the instrument may slow down the measurement rate and eventually "put itself to sleep," until "wakened" by an operator request. This approach in some cases can result in a reduction of power consumption by a factor of 5 to 10 times. Another simple way to save battery power is to allow operation from an external power source, usually a small wall transformer, whenever an alternating current (AC) power source is available.

Size limitations of electronic portable instruments can also lead to an inconvenient user interface. While the large front panels of bench-top instruments can accommodate multiple signal connectors, large displays, keyboards, and switches, these large components may not be suitable for portable devices. New user interface devices such as small high-resolution flat-panel graphic displays, touch screens, and touch pads ease many of the user interface difficulties. It is possible that the introduction of speech recognition technology can largely solve most of the user interface problems in the near future.

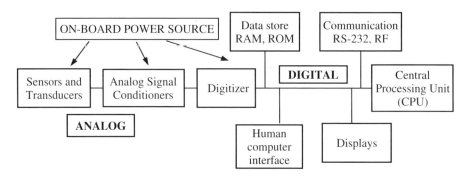

FIGURE 1.7
A typical computer-based portable instrument.

Figure 1.7 illustrates the block diagram of a typical computer-based portable digital instrument. Existing digital-based electronic portable instruments can implement very complex and sophisticated measurement techniques, store the results, display them graphically, and also operate automatically or remotely by using advanced communication techniques. They can communicate with other devices at high speeds and therefore can easily be combined in complex measurement systems. Digital hardware is much less sensitive to environmental conditions such as ambient temperature and electromagnetic interference.

With the easy availability of many components, an instrument designer can now choose from a full spectrum of digital signal processors, memories, peripherals, and other devices with varying degrees of performance, power consumption, and size. Similar developments in analog signal processing devices contributed to the improvements in operational bandwidths, sensitivity, and accuracy of measurements. Nowadays, the trend is toward the predominantly digital designs. Faster and more sensitive A/D and D/A converters are available with reasonable costs. The introduction of novel converter architectures that incorporate internal digital anti-aliasing filters and delta-sigma (Δ/Σ) converters allows designers to shrink analog hardware to a reasonable minimum size, thereby extending benefits of the digital approach to a greater portion of the measurement devices. Digital instrumentation has also benefited from general development of portable computing, which led to the development of user-friendly operating systems and high productivity software development tools.

In recent years, the progress in electronic portable instruments has made a major turnaround due to the availability of micro and intelligent (or smart) sensors. Smart sensors have intelligence due to integration of complex digital processors in the same chip. A general scheme of a smart sensor is illustrated in Figure 1.8. In this particular example, the transducer is under a microprocessor control. Excitation is produced and modified depending on the required operational range. The processor may contain full details of the

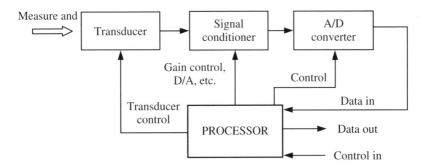

FIGURE 1.8
The block diagram of a smart sensor.

transducer characteristics in read only memory (ROM), enabling the correct excitation, gain, and so on.

Many measurement techniques that operate at the low end of the frequency spectrum bandwidth (e.g., 0 Hz to 100 kHz) now can routinely be configured as instruments similar to the features shown in Figure 1.8. Applications of such instruments include general electronic instrumentation, biomedical diagnostic and monitor devices, vibration and sound analysis, telecommunications, environmental measurements, and nondestructive testing devices using ultrasonic or eddy current techniques. Many portable instruments combine digital hardware with specialized analog and nonelectrical components and sensors in the hybrid instrumentation systems. Examples include high-voltage test instruments, spectrometers, fiber-optic instruments, time domain reflectometry (TDR) and optical TDR measurements, radio frequency (RF) and microwave devices, mobile communication devices, global positioning system (GPS)-based instruments, and so on.

Recent technological progress in electronic telemetry is making a significant impact on the development and use of portable electronic instruments. In some applications, they can transmit the data acquired from remote locations to base stations. A basic measurement system with telemetry is illustrated in Figure 1.9. The desired signal is acquired by the transducers and encoded into a signal suitable for the transmitter. The signal can then be used to modulate a high-frequency radio signal that is fed to the transmission system. In many applications, signals are multiplexed and sent via the same link. The transmission link can be very high frequency (VHF), microwave, radio, satellite, infrared, laser, and so on.

Examples of advanced electronic portable instruments have a wide range of design and application possibilities, such as:

- Personal safety equipment
- Model aircraft, boats, and land vehicles
- Patient ambulatory monitors

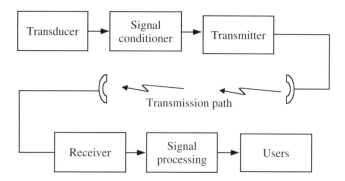

FIGURE 1.9
Basic elements of telemetry.

- Musical instruments
- Land, geophysical, and hydrographic surveying equipment
- Scientific data recorders to monitor weather, seismic activity, animal movements, pollution levels, etc.
- Recording instruments for sporting events
- Instruments in mining and chemical industries
- Radioactivity and toxic gas detectors and analyzers
- Photosynthesis systems
- Appliances for the handicapped
- Complex simulation games and toys

In addition to these, many others will be explained in this book.

1.5 Response and Drift

The following discussions are applicable to all types of instruments, including electronic portable instruments. The physical entity or process under investigation can be *static* or *dynamic*. At the same time, instruments themselves inherently exhibit *static* and *dynamic* characteristics. It is important that instruments' characteristics match with the process behavior. In static measurements, the process (e.g., fixed dimension or weight) does not change in time; therefore, its measurement is relatively easier. If the process changes in time, as often is the case, the measurement is said to be dynamic. In this case, the steady-state and transient behaviors of the physical variable must be analyzed so that it can be matched with the dynamic behavior of the instrument. Nevertheless, the total response of an instrument is still the combination of both dynamic and static responses.

Performance of an instrument largely depends on the static and dynamic characteristics of the physical variable, and this performance may be indicated by its *accuracy*. Accuracy may be described as the closeness of the measured values to the real values of the physical variable. If the signals generated by the variable are changing rapidly, then the dynamic properties of the instrument become important. In general, the dynamic errors may be neglected for slow-varying systems.

The performance of an instrument may also be decided by other factors, such as magnitudes of errors; the *repeatability*, which indicates the closeness of sets of measurements made in the short term; and the *reproducibility* of the instrument. The reproducibility is the closeness of measurements when repeated in similar conditions over a longer period.

An ideal or perfect instrument has perfect sensitivity, good reliability, and consistent repeatability without any spread of values. It also performs within the expected standards. However, in many measurements, because of internal and external factors of instruments, we can expect imprecise and inaccurate results. The departure from the expected perfection is called the *error*. Often, sensitivity analyses are conducted to evaluate the effect of individual components that are causing these errors. Sensitivity of an instrument can be obtained by conducting measurements with respect to the effecting parameter and keeping the others constant.

Instruments responding to physical phenomenon are observed from the generating signals. Depending on the type of instrument used and the physical phenomenon, the signals may be either slow or fast and may also contain transients. The response of the instruments to the signals can be analyzed by establishing static and dynamic performance characteristics as explained next.

1.5.1   Static Response

Instruments are often described by their dynamic range and full-scale (FS) deflection, also known as *span* of an instrument. The *dynamic range* of an instrument indicates the largest and smallest quantities that can be measured. The *full-scale deflection* refers to the maximum permissible value of the input quoted in the units of the particular quantity to be measured. There are two other terms often used to describe the performance of an instrument, the *input span* and *output span*. The *input span* indicates the dynamic range of stimuli that may be converted by a sensor. It represents the highest possible input that can be applied to a sensor resulting in an output within acceptable accuracy. The *output span*, also termed the *full-scale output*, indicates the algebraic difference of the output in response to minimum and maximum input physical applied stimuli.

In instruments, the change in the amplitude of an output resulting from a change in the input amplitude is called the *sensitivity*. The total system sensitivity often is a function of environmental conditions, such as temperature and humidity. The relative ratio of the output signal to the input signal is the *gain*. Both, the gain and sensitivity are dependent on signal amplitude and frequency, both of which will be discussed in the next section.

1.5.2 Dynamic Response

The dynamic response of an instrument is characterized by its natural frequency, amplitude, frequency response, phase shift, linearity and distortions, rise and settling times, and slew rates. These characteristics are common themes in many instrumentation, control, and electronics books (e.g., Doeblin, 1990; Holman, 1989). Although sufficient analysis will be given here, a comprehensive treatment of this topic can be very lengthy and complex; therefore, it will not be detailed here, as the full treatment of this topic is not within the scope of this book.

The dynamic response of an instrument can be linear or nonlinear. Fortunately, most instruments exhibit linear characteristics that yield in the simple mathematical models obtained from the differential equations (DEs) that describe the behavior. In simple cases, the DE is as follows:

$$a_n \frac{d^n y}{dt^n} + a_{n-1} \frac{d^{n-1} y}{dt^{n-1}} + \cdots + a_0 y = x(t) \tag{1.2}$$

where x is the input variable or the forcing function, y is the output variable, and a_n, a_{n-1}, and a_0 are the coefficients or the constants of the system, depending on the characteristics of the instrument and the physical variable.

The dynamic response of many instruments can be categorized as the zero-order, first-order, or second-order responses. Although, higher-order instruments may exist, their behavior can be understood adequately by reducing them to a second-order system. From Equation 1.2,

$$a_0 y = x(t) \qquad \text{Zero-order} \tag{1.3}$$

$$a_1 \frac{dy}{dt} + a_0 y = x(t) \qquad \text{First-order} \tag{1.4}$$

$$a_2 \frac{d^2 y}{dt^2} + a_1 \frac{dy}{dt} + a_0 y = x(t) \quad \text{Second-order} \tag{1.5}$$

Equation 1.3, Equation 1.4, and Equation 1.5 can be written in Laplace transform, thus enabling analysis in the frequency domain:

$$\frac{Y(s)}{X(s)} = \frac{1}{a_0} \tag{1.6}$$

$$\frac{Y(s)}{X(s)} = \frac{1}{(\tau_1 s + 1)} \tag{1.7}$$

$$\frac{Y(s)}{X(s)} = \frac{1}{(\tau_1 s + 1)(\tau_2 s + 1)} \tag{1.8}$$

where s is the Laplace operator and τ_1 and τ_2 are the coefficients, also called the time constants.

In zero-order instruments, there is no frequency dependence between the input and output. Therefore, alteration in the amplitude is uniform across the whole range of frequency spectrum. In practice, such instruments are difficult to obtain, except in a very limited range of operations.

In first-order instruments, the relation between the input and the output is dependent on frequency. Figure 1.10 illustrates the response of a first-order instrument for a unit step input. The output in the time domain may be written as

$$y(t) = K\,e^{-t/\tau} \tag{1.9}$$

where K and τ are constants determined by the system parameters.

The second-order systems exhibit the laws of simple harmonic motion, which can be described by linear wave equations. Equation 1.8 may be rearranged in Laplace as

$$\frac{Y(s)}{X(s)} = \frac{1/a_0}{s^2/\omega_n^2 + 2\zeta s/\omega_n + 1)} \tag{1.10}$$

where ω_n is the natural or undamped frequency (rad/sec) and ζ is the damping ratio.

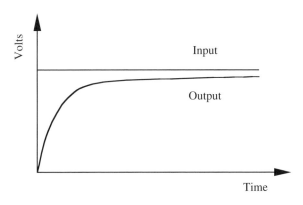

FIGURE 1.10
Time response of a first-order instrument.

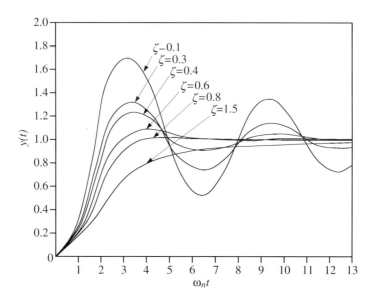

FIGURE 1.11

Response of a second-order system to a unit step input.

As can be seen, the performance of instruments becomes a function of natural frequency and the damping ratio of the system. Both the natural frequency and damping ratios are related to the behavior of the electronic components and other physical parameters of the device, such as mass and dimensions. The physical parameters may be selected, tested, and modified to obtain a desired response from the instruments during the early design stages.

The typical time response of a second-order system to a unit step input is illustrated in Figure 1.11. The response indicates that a second-order system can either resonate or be unstable. Furthermore, we can deduce that since the response of a second-order system is dependent on time, the wrong readings can be taken depending on the time the results are taken. Clearly, recording the output when the instrument is still under transient conditions will result in wrong readings and will give an inadequate representation of the physical variable. The frequency compensation, selection of appropriate damping, acceptable time responses, and rise and settling times of instruments may need careful attention during both design and application.

In many cases, the input signal may be a complex rather than a simple step input. In the analysis, we need to multiply the transfer function (second member of Equation 1.9 to Equation 1.10) by the Laplace transform of the input signal and then transform it back to the time domain. This operation can give good insight on the nature of the transient and steady-state responses for that particular input. Also, in some cases, the first-order systems may be cascaded. In this case, the relative magnitudes of the time constants become important; while some may be dominant, others may be neglected.

FIGURE 1.12
Drift of an instrument in time.

Two important concepts in instruments are the *response time* and *bandwidth*. The response time can be described as the time required for an instrument to respond to a change in the input signal. The response time is important, particularly in digital instruments and automatic measurements, since it is an indication of how many readings can be taken per second. Response time is affected by many factors, such as A/D conversion time, settling time, delays in the electronic components, and delays generated in the sensors. The bandwidth, on the other hand, is the frequencies over which the gain is reasonably constant. Usually half-power point (3-dB point), which is 70.7% of the constant gain against frequency, is taken as the bandwidth.

1.5.3 Drift

During the design stages or manufacturing, there might be small differences between the desired input and desired output, which are called *offsets*. Basically, when the input is zero, the output is not zero, and vice versa. The signal output of an instrument for the same input may change in time, which is known as the *drift*. Drift can happen for many reasons, including temperature, aging, and oxidation of components. Fortunately, drift usually happens in a predictable manner. A typical drift curve of an instrument against time is illustrated in Figure 1.12.

In many practical applications, readings taken from an instrument may not be repeatable even if the conditions of the tests are kept exactly the same. In this case, a repeatability test may be conducted and statistical techniques may be employed to evaluate the repeatability of the instrument within some error budgets and confidence intervals.

1.6 Errors and Uncertainty

It is essential to understand how errors arise in instruments. There may be many sources of errors; therefore, it is important to identify these sources and draw up an error budget. In the error budget, there may be many factors, such as (1) imperfections in electrical and mechanical components (e.g., high tolerances and noise or offset voltages), (2) changes in component performances (e.g., shift in gains, changes in chemistry, aging, and drifts in offsets), (3) external and ambient influences (e.g., temperature, pressure, and humidity), and (4) inherent physical fundamental laws (e.g., thermal and other electrical noises, Brownian motion in materials, and radiation).

Errors can broadly be classified to be systematic, random, or gross errors.

1.6.1 Systematic Errors

Systematic errors remain constant with repeated measurements. They can be divided into two basic groups: instrumental errors and environmental errors. Instrumental errors are inherent within the instrument, arising because of mechanical structures, electronic designs, improper adjustments, wrong applications, and so on. They can also be subclassified as loading, scale, zero, and response time errors. Environmental errors are caused by environmental factors such as temperature and humidity. Systematic errors can also be viewed as static or dynamic errors.

Systematic errors can be quantified either mathematically or graphically. They can be a result of the nonlinear response of an instrument to different inputs due to hysteresis. They also emerge from wrong biasing, wearing and aging of components, and other factors such as electromagnetic interference. Typical systematic error curves are illustrated in Figure 1.13.

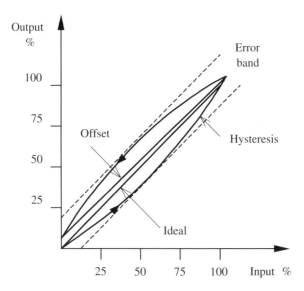

FIGURE 1.13
Systematic errors that include linearity, hysteresis, and offset errors.

Because of the predictability of systematic errors in most cases, deterministic mathematics can be employed. In the simplest form, the error of a measurement may be expressed as

$$\Delta x(t) = x_m(t) - x_r(t) \tag{1.11}$$

where $\Delta x(t)$ is the absolute error, $x_r(t)$ is the correct reference value, and $x_m(t)$ is the measured value.

The relative error, $r_e(t)$, may be calculated as

$$r_e(t) = \frac{\Delta x(t)}{x_r(t)} \tag{1.12}$$

However, in complex situations, correction curves, which are obtained either empirically or theoretically, may be used. Manufacturers usually supply correction curves, especially if their products embrace wide-ranging and different types of applications.

In many applications, the measurement system is made up of many components that have errors in their own right. In these cases, the deterministic approach may be adapted to calculate the overall propagated error of the system, as

$$y = f(x_1, x_2, x_3, \dots, x_n) \tag{1.13}$$

where y is the overall output and $x_1, x_2, \dots$ are the components that contribute to the error of the output.

Each variable affecting the output will have its own absolute error of Δx_i. That is, the term Δx_i indicates the error of each component under specified operating conditions. The absolute error of Δx_i can be determined mathematically or experimentally. The overall performance of the entire system with the errors may be expressed as

$$y \pm \Delta y = f(x_1 \pm \Delta x_1, x_2 \pm \Delta x_2, \dots, x_n \pm \Delta x_n) \tag{1.14}$$

For an approximate solution, the Taylor series may be applied to Equation 1.14. By neglecting the higher-order terms of the series, the total absolute error Δy of the system may be written as

$$\Delta y = \left| \Delta x_1 \frac{\delta y}{\delta x_1} \right| + \left| \Delta x_2 \frac{\delta y}{\delta x_2} \right| + \cdots + \left| \Delta x_n \frac{\delta y}{\delta x_n} \right| \tag{1.15}$$

The absolute error is predicted by measuring or calculating the contribution of each variable.

1.6.1.1 Uncertainty

Slight modification of Equation 1.15 leads to uncertainty analysis, as

$$w_y = [(w_1 \, \delta y / \delta x_1)^2 + (w_2 \, \delta y / \delta x_2)^2 + \dots + (w_n \, \delta y / \delta x_n)^2]^{1/2} \qquad (1.16)$$

where w_y is the uncertainty of the overall system and $w_1, w_2, \dots, w_n$ are the uncertainties of affecting the component.

Uncertainty differs from error in that it involves such human judgment factors as estimating the possible values of errors. For example, tolerance values of components lead to uncertainty analysis. This is because the absolute error of each component is not known, but the possibility of error is expressed within a band.

In measurement systems, apart from the uncertainties imposed by the instruments, experimental uncertainties also exist. In evaluating the total uncertainty, several alternative measuring techniques should be considered and assessed. The estimated accuracy must be worked out with care.

1.6.1.2 Calibration Error in Sensors

Some of the errors are generated by sensors. When sensors are calibrated in the factory, they still can contain some inaccuracies, known as *sensor calibration errors*. Systematic in nature, the sensor calibration error can differ from one sensor to the next. Also, it may not be uniform over the entire range. In sensitive applications of sensors, this point must be considered carefully, particularly when designing and constructing sensitive and accurate portable instruments. The errors quoted by the manufacturers must not be taken at face value, but mathematical and experimental analyses must be conducted to assess the error budget of sensors.

1.6.2 Random Errors

Random errors appear as a result of noise and interference, backlash and ambient influences, and so on. In experiments, random errors may vary by small amounts around a mean value. Therefore, the future value of any individual measurement cannot be predicted in a deterministic manner. Since the random errors may not be easily offset electronically, stochastic approaches are adapted for the analysis and compensation of these errors. This can be realized by using the laws of probability.

Depending on the system, random error analysis may be made by applying different probability distribution models. But most instrumentation systems obey normal distribution laws; therefore, the Gaussian model can broadly be applied, thus enabling the determination of the mean values, standard deviations, confidence intervals, and the like, depending on the number of samples taken. A typical example of a Gaussian curve is given in Figure 1.14.

The mean value $\bar{x}$ and the standard deviation σ may be found by

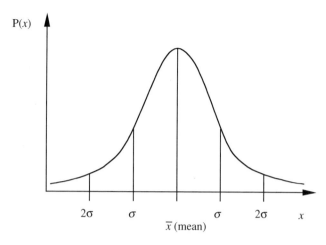

FIGURE 1.14
Random errors described in the normal distribution.

$$\bar{x} = \frac{\sum x_i}{n} \qquad (1.17)$$

and

$$\sigma = \frac{\sum (x_i - \bar{x})^2}{n-1} \qquad (1.18)$$

Discussions relating to the application of stochastic theory in error analysis are very lengthy and will not be repeated here. Interested readers should refer to the sources given in the Bibliography section of this book (e.g., Holman, 1989).

1.6.3 Gross Errors

Gross errors are the result of human mistakes, equipment fault, etc. Human errors may occur in the process of observations or during the recording and interpretation of experimental results. A large number of errors can be attributed to carelessness, the improper adjustment of instruments, and the lack of knowledge about the instrument and process. These errors cannot be treated mathematically and eliminated completely, but they can be minimized by having different observers repeat the same experiment.

1.6.4 Error Reduction Techniques

Controlling errors is an essential part of instruments and instrumentation systems. Various techniques are available to achieve this objective. The error control begins in the design stages by choosing the appropriate components, filtering, and bandwidth selection, and by reducing noise. Minimizing the errors generated by the individual subunits of the complete system will result in lowering the errors of the instrument. In a good design, the errors of a subunit may be compensated adequately by the ensuing subunit.

The accuracy of instruments can be increased by postmeasurement corrections. Various calibration methods may be employed to alter parameters slightly to give the correct results. In many cases, calibration graphs, mathematical equations, tables, and the experience of the operators can be utilized to reduce measurement errors. In recent years, with the application of digital techniques and intelligent instruments, error corrections are made automatically by computers, or within the devices and sensors themselves.

In many instrumentation systems, the application of compensation strategy is used to increase static and dynamic performances. In the case of static characteristics, compensations can be made by many methods, including the introduction of opposing nonlinear elements into the system, by using isolation and zero environmental sensitivity conditions, and the introduction of opposing compensating environmental inputs, by using differential and feedback systems. At the same time, dynamic compensation can be achieved by applying these techniques as well as by taking measures to reduce harmonics, using filters, adjusting bandwidths, using feedback compensation techniques, and so on.

Open-loop and closed-loop dynamic compensations are popular methods employed in both static and dynamic error corrections. For example, using a high-gain negative feedback can reduce the nonlinearity that may be generated by a system. A recent and fast-developing trend is the use of digital techniques for estimating measured values and providing compensation during the operations if and when any deviation occurs from the expected values.

1.7   Calibration and Testing of Instruments

The calibration of all instruments is essential for checking their performances against known standards. This provides consistency in readings and reduces errors, thus validating the measurements universally. The calibration procedure involves comparison of the instrument against primary or secondary standards. In some cases, it may be sufficient to calibrate a device against another one with a known accuracy. After the calibration of an instrument, future operation is deemed to be error-bound for a given period for similar operational conditions.

Many nations and organizations maintain laboratories with the primary functions of calibrating instruments and field measuring systems that are used in everyday operations. Examples of these laboratories are the National

Association of Testing Authorities (NATA) of Australia and the British Calibration Services (BCS).

Calibrations may be made under static or dynamic conditions. During calibration, the input is varied in increments in increasing and decreasing directions over a specified range. The observed output then becomes a function of that single input. For better results, this procedure may be repeated by varying the input, thus developing a family of relationships between the inputs and outputs. The input/output relationship usually demonstrates statistical characteristics. From these characteristics, appropriate calibration curves can be obtained, and then statistical techniques can be applied for analysis.

Most instrument manufacturers supply calibrated instruments together with fairly reliable information about their products. But their claims of accuracy and reliability must be taken at face value since your specific application may not exactly match their calibration conditions. Therefore, in many cases, application-specific calibrations must be made periodically within the recommended intervals. Usually, manufacturers supply calibration programs. In the absence of such programs, it is advisable to conduct frequent calibrations in the early stages of service of the instrument and then lengthen the period between calibrations as the confidence builds on the satisfactory performance.

Nowadays, with the wide applications of digital systems, many intelligent instruments can make self-calibrations. In these cases, postmeasurement corrections are made, and the magnitudes of various errors are stored in the memory to be recalled and used in the laboratory and field applications. A new trend is that calibrations can be conducted over the Internet by entering the appropriate sights of manufacturers of calibration authorities.

1.7.1 Testing of Instruments

After the instrument is designed and prototyped, various evaluation tests are conducted. These tests may be performed under reference conditions or under simulated environmental conditions. Some examples of reference condition tests are the accuracy, response time, drift, and warm-up time. Simulated environmental tests may be compulsory, thus being regulated by the governments and other authorities. Some simulated environment tests include climatic test, drop test, dust test, insulation resistance test, vibration test, electromagnetic compatibility tests, safety and health hazard tests, and ingress progress (IP) tests against water and dust. The IP tests are particularly important for electronic portable instruments since they are most likely to be used during their operations in unsteady environments containing rain, water, humidity, and dust. Many of these tests are strictly regulated by national and international standards. Detailed information on this topic is given in Chapter 4.

Adequate testing and proper use of instruments is important to achieve the best results from them. When the instruments are commissioned for use, regular testing is necessary to ensure the consistency of the performance over the period of operation. Incorrect measurements can cost a considerable amount of money and time, or even result in the loss of lives.

For maximum efficiency, an appropriate instrument for the measurement must be selected. Users should be fully aware of their application requirements, since instruments that do not fit their purposes will deliver false data, resulting in wasted time and effort. For a particular application, users must carefully study the documents about all the possible instruments to be used and make comparisons among all the options. While selecting the instrument, users must evaluate many factors, such as accuracy, frequency response, electrical and physical loading effects, sensitivity, response time, calibration intervals, power supply needs, spare parts, technology, and maintenance requirements. They must ensure compatibility of new instruments with the existing equipment.

Also, when selecting and implementing instruments, quality becomes an important issue from both quantitative and qualitative perspectives. The quality of an instrument may be viewed differently depending on the people involved. For example, quality in the eyes of the designer may be an instrument designed on sound physical principles, whereas from the user's point of view, it may be reliability, maintainability, cost, and availability.

In many applications, the reliability of the instrument must be assessed, and the performance of it must be checked regularly. The *reliability* of the system may be defined as the probability that it will operate at an agreed level of performance for a specified period. The reliability of instruments follows a bathtub shape against time, as illustrated in Figure 1.15. Instruments tend to be unreliable in the early stages of use mainly due to burn-in of components. They are also unreliable in the later stages of their lives due to wear and tear, aging, drift, and so on. During normal operations, if the process conditions change, calibrations must be conducted to avoid possible performance deterioration of the instrument. Therefore, the correct operations of the instruments must be assured at all times throughout the lifetime of the device.

Once the instruments are commissioned properly, they may be expected to operate reliably. They may be communicating with other devices, and

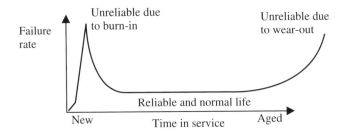

FIGURE 1.15
The bathtub-shaped reliability curve of instruments.

their performance may affect the performance of the rest of the system. In some applications, the instruments may be part of a large instrumentation system, taking a critical role in monitoring and controlling the process and operations. However, in many applications, as in the case of the majority of electronic portable instruments, they are used on a stand-alone basis for laboratory and experimental work. In these cases, the experiments must be designed and conducted carefully by identifying the primary variables, controlling and selecting the correct instruments, assessing the relative performances, validating the results, and using the data effectively by employing comprehensive data analysis techniques. Set procedures for experimental designs can be found in various sources given in the Bibliography (e.g., Sydenham et al., 1989).

After having performed the experiments, the data must be analyzed appropriately. This can be done in various stages by examining the consistency of the data, performing appropriate statistical analyses, estimating the uncertainties of the results, relating the results to the theory, and correlating the data. Details of statistical data analysis can be found in many books; also, many computer software programs are available for the purpose of analysis, including common packages such as Microsoft Excel.

1.8   Controlling and Networking of Instruments

Instruments and electronic portable instruments can be manual, semiautomatic, or fully automatic. Manual instruments need human intervention for adjustment, parameter setting, and interpretation of the readings. Semiautomatic instruments require limited intervention, such as the selection of operating conditions. In fully automatic instruments, however, the variables are measured either periodically or continuously without human intervention. The information gathered by the instrument may be either stored or transmitted to other devices automatically.

In many processes, it is often necessary to measure many parameters by using two or more instruments. Portable instruments are finding applications in these processes due to two main reasons: (1) their networked communication capabilities by means of remote communication methods such as RF, microwave, Internet, and optical techniques, and (2) availability of on-board memory for data storage. The resulting arrangement for performing the overall measurement function is called the *measurement system*. In measurement systems, instruments operate autonomously but in a coordinated manner. The information generated by each device is communicated between the instruments themselves, or between the instrument and other devices, such as recorders, display units, some base stations, or a host computer. The coordination of instruments can be done in three ways: analog to analog, analog-to-digital, and digital-to-digital. Digital networking of instruments is discussed in detail in Chapter 4. Thanks to recent progresses in communication and telemetry technologies, unlike the situation five or six years ago, portable instruments can easily be networked and can act as an active part of a networked system.

Offset and level conversion is used to convert the output signal of an instrument from one level to another to make it compatible with the transmission medium selected. In digital instrumentation systems, analog data are converted and transmitted in the digital form. The transmission of data between digital devices can be done relatively easily, by using serial or parallel transmission techniques. However, as the measurement system becomes large by the inclusion of many instruments, communication becomes complex. To avoid this complexity, message interchange standards are used for digital signal transmission, e.g., RS-232, USB, EIA-485, and IEEE-488. IEEE-488 is primarily used for instrument control in test and measurement and finds applications in many portable instruments too.

Today many portable instruments contain an RS-232 interface. Also, there are many companies offering radio frequency RS-232 systems for remote data transmission. The RS-232 was first issued by the Electronic Industries Association (IEA). It uses serial binary data interchange and applies specifically to the interconnection of data communication equipment (DCE) and data terminal equipment (DTE). DCE may include modems, which are the devices that convert digital signals suitable for transmission through telephone lines. The RS-232 uses standard DB-25 connectors. In DB-25 connectors, although 25 pins are assigned, a complete data transmission is possible by using only three pins: 2, 3, and 7. The transmission speed can be set to certain baud rates such as 1200, 2400, 4800, 9600, 19,200, and 38,400 bits/sec or higher. The RS-232 can be used for synchronous or nonsynchronous communication purposes. The signal voltage levels are very flexible, with any voltage between −3 and −25 V representing logic 1 and any voltage between +3 and +25 V representing logic 0. The RS-232 standard dates back to the 1960s; therefore, it has evolved into a number of diverse configurations that need to be understood and configured for the system in hand.

Many portable instruments also contain parallel ports since they are faster than serial counterparts. Peripheral devices usually operate on parallel I/O ports. The architecture of a parallel bus defines the width of data paths, transfer rates, protocols, cable lengths, and connector configurations. IEEE-488 is a common parallel bus that is used in variety of instruments and instrumentation systems.

When data are transmitted over long distances, modems, microwave, or radio frequency (RF) transmissions are used on the receiving end for further transmission. In these cases, various signal modulation techniques are used to convert digital signals to suitable formats. For example, most modems, with medium-speed asynchronous data transmission, use frequency-shift keyed (FSK) modulation. The digital interface with modems uses various protocols such as MIL-STD-188C to transmit signals in simplex, half-duplex, or full-duplex forms, depending on the directions of the data flow. The simplex interface transmits data in one direction, whereas full duplex transmits it in two directions simultaneously.

In industrial applications, several standards for digital data transmission are available. These are commonly known as *fieldbuses* in engineering literature. Among many others, some of these standards are widely accepted and used, such as WorldFIP, Profibus, Foundation Fieldbus, etc. The fieldbuses are supported by hardware and software (e.g., National Instruments chips and boards) that allow increases in the data rates suitable with high-speed protocols.

As far as design software is concerned, there are a number of important tools that help implement control (application) software for automatic measuring equipment, such as LabWindows and LabView from National Instruments.

1.9   Signals and Signal Conditioning

A *signal* is defined as "any physical quantity that varies with time, space, or any other independent variable or variables." Mathematically, a signal can be described as a function of one or more independent variables. In some cases, signals can be described by specifying the functional dependence on an independent variable or set of variables, whereas in other cases, the functional relationship is unknown or highly complicated to be of any practical value. This indicates that different types of signals may require different processing techniques. And indeed, there are techniques that can apply to only specific families of signals.

1.9.1   Types of Signals

The mathematical analysis and processing of signals require the availability of a mathematical description of the signal. As far as mathematical description is concerned, the signals generated by sensors can be classified in a number of ways, such as multichannel and multidimensional, continuous time or discrete, deterministic or random (stationary, nonstationary), or transient.

1.9.1.1   Multichannel and Multidimensional Signals

The signal is described by a function of one or more independent variables. The independent variable can be real-valued scalar or complex-valued quantity. For example,

$$x(t) = A\, e^{j\pi t} = A \sin(\pi t) + jA\cos(\pi t) \tag{1.19}$$

is a complex valued signal.

In some applications, multiple sources or multiple sensors generate signals. These signals can be represented in vectors. Generally, if the signal is a function of a single independent variable, it is called a *one-dimensional* signal. Similarly, a signal is called *m-dimensional* if its value is a function of *m* independent variables (e.g., earthquake signals picked up by accelerometers). Vectors in matrix form can represent such signals.

1.9.1.2   Continuous or Discrete Signals

Signals may be classified according to characteristics of independent (e.g., time) variables and the values they take.

1.9.1.2.1   Continuous Signals

Continuous signals, also known as analog signals, are defined for every value of time from $-\infty$ to $+\infty$. They can be periodic or nonperiodic.

In periodic signals, the signal repeats itself. A good example of continuous signal is the sinusoidal signal, which can be expressed as:

$$x(t) = X_m \sin(\omega t) \tag{1.20}$$

where $x(t)$ is the time-dependent signal, $\omega = 2\pi f t$ is the angular frequency, and X_m is the maximum value.

The signals can be periodic but not necessarily sinusoidal, such as triangular, sawtooth, rectangular, etc. If they are periodic but nonsinusoidal, they can be expressed as a combination of a number of pure sinusoidal waveforms as

$$x(t) = X_0 + X_1 \sin(\omega_1 t + \phi_1) + X_2\sin(\omega_2 t + \phi_2) + \ldots + X_n\sin(\omega_n t + \phi_n) \tag{1.21}$$

where $\omega_1, \omega_2, \ldots, \omega_n$ are the frequencies (rad/sec), $X_0, X_1, \ldots, X_n$ are the maximum amplitudes of respective frequencies, and $\phi_1, \phi_2, \ldots, \phi_n$ are the phase angles.

In Equation 1.21, the number of terms may be infinite, and the higher the number of elements, the better the approximation. These elements constitute the frequency spectrum. The signals can be represented in the time domain or frequency domain, both of which are extremely useful in the analysis.

1.9.1.2.2   Discrete-Time Signals

Discrete-time signals are defined only at discrete values of time. These time instants may not be equidistant, but in practice, for computational convenience, they are taken as equally spaced. For example, Equation 1.20 is represented in discrete form as

$$x(nT) = X_m\sin(\omega nT) \qquad (1.22)$$

where $t = nT$, $n = 0, \pm1, \pm2, \pm3. \ldots$

Discrete representation of signals is used in *sampling* theory and *quantization*, as is explained in greater detail in Chapter 3.

1.9.1.3   Deterministic and Random Signals

Any signals that can be expressed by an explicit mathematical expression, or by some well-defined rules, are called *deterministic signals*. That is, the past, present, and future values of the signals are known precisely without uncertainty.

In practical applications, there are signals that evolve *randomly* in time in an unpredictable manner. They cannot deterministically be described to any reasonable degree of accuracy, or description is too complicated for practical use. The random signals are met often in nature, where they constitute irregular cycles that never repeat themselves exactly. Theoretically, an infinitely longtime record is necessary to obtain a complete description of these signals. However, statistical methods and probability theory can be used for the analysis by taking representative samples. Mathematical tools such as probability distributions, probability densities, frequency spectra, cross-correlations, auto-correlations, digital Fourier transforms (DFTs), FFTs, and autospectral analysis, root mean square (rms) values, and digital filter analysis are some of the techniques that can be employed.

If the statistical properties of signals do not vary in time, the signal is called *stationary random signal*. However, in many cases, the statistical properties of signals do vary in time, thus giving *nonstationary random signals*. In this case, methods such as time averaging and other statistical techniques can be employed.

1.9.1.4   Transient Signals

Often, short duration and sudden occurrence may have to be sensed and processed as in the case of random transients and shocks. Statistical methods and Fourier transforms can be used for their analysis.

1.9.2   Signal Classification

Signals can be classified according to their amplitude levels, or relationships of their source terminals and ground, or bandwidths, and also according to their output impedance properties. Generally, signals lower than 100 mV are considered to be low-level signals, and they need amplification. Nevertheless, large signals may also need amplification depending on the input range of the receiver.

1.9.2.1   Single-Ended and Differential Signals

A *single-ended* signal comes from a source that has one of its two output terminals at a constant voltage. Usually one of the terminals is grounded. A source of a *differential signal*, on the other hand, has two output terminals whose voltages change simultaneously by the same magnitudes but in opposite directions. The output voltages of bridges are typical examples of differential signals.

1.9.2.2   Narrowband and Broadband Signals

A narrowband signal has a very small frequency range relative to its central frequency. A narrowband signal can be near direct current (DC), or almost static, resulting in low frequencies (such as thermocouples), or AC (such as those obtained from AC-driven modulating sensors).

Broadband signals have a large frequency range relative to their central frequency. The value of central frequency is important; for example, a signal from 1 to 10 Hz may be considered a broadband instrumentation signal, whereas two 20-kHz sideband signals around 2 MHz may be regarded as a narrowband signal.

1.9.2.3   Low- and High-Output Impedance Signals

The output impedance of a signal determines the requirements of the input impedance of the conditioning circuits to avoid loading effects. This is commonly known as *impedance matching*.

At low frequencies, it is relatively easy to achieve large-input impedances, such as those from piezoelectric sensors. However, at high frequencies, stray input capacitances play important roles and can make impedance matching difficult. This may not be an important problem for narrowband signals, but in broadband signals, if the impedance is frequency dependent, each frequency signal goes through different attenuations, thus making impedance matching very difficult.

Signals with very high-output impedances are usually modeled and treated as current sources. Special techniques are applied (e.g., charge amplifiers) to process such signals.

1.9.2.4   Signal Conditioning at Sensor Level

Signals from sensors are not usually suitable for displaying, recording, and transmitting. For instance, the amplitudes, power levels, and bandwidths may be very small, or they may carry excessive noise and superimposed interference that mask the desired information. Signal conditioners are used to adapt sensor signals to the acceptable requirements for processing and display purposes.

As mentioned previously, the signals coming from sensors can be analog or digital. Digital signals come from devices such as position encoders, switches, or oscillator-based sensors. Digital signals from the sensors must be compatible with the receiving devices. Analog signals, on the other hand, can be self-generated or modulated. Self-generated sensors usually yield a voltage (thermocouples, electrochemical sensors) or current (piezo- and pyroelectric sensors). Modulating signals yield a variation in resistance, capacitance, self-inductance, or other electrical quantities.

Modulating sensors need to be excited externally in order to provide an output voltage or current. Impedance variation-based signals are normally placed in voltage dividers or in a suitable bridge network. Current signals can be converted into equivalent voltage forms by inserting a series resistor into the circuit or can be processed by suitable amplifiers such as charge amplifiers. Further discussions on signal conditioning will be given in the ensuing sections.

1.10 Noise and Interference

1.10.1 Noise

Noise is an important concept in all types of electronic devices. Noise can be described as either the unwanted signals generated internally by the device itself or the signals that are externally imposed upon it. Therefore, the noise can be *inherent*, arising within the circuit, or *interference*, picked up from outside the circuit. There are many different types of inherent and interference noise. Depending on the device, some noise may be significant for that device, whereas others may not be all that significant. Since noise is a concern for all devices, a general discussion on noise will be given here, and specific concerns of noise related to electronic portable instruments will be discussed in Chapter 4.

There are many different types of inherent noise, such as thermal noise, shot noise, excess noise, burst noise, partition noise, generation–recombination noise, spot noise, and total noise. They will briefly be discussed next.

1.10.1.1 Thermal Noise

Thermal noise, also known as Johnson noise, is generated by the random collision of charge carriers with a lattice under conditions of thermal equilibrium. Thermal noise can be modeled by a series voltage source, v_t, or a parallel current source, i_t, having the mean-square values

$$\overline{v_t^2} = 4kTR\Delta f \quad \text{and} \quad \overline{i_t^2} = \frac{4kT\Delta f}{R} \tag{1.23}$$

where T is the temperature in Kelvin, R is the resistance in ohms, and Δf is the bandwidth in hertz over which the noise is measured. The equation for

$\overline{v_t^2}$ is known as the Nyquist formula. The amplitude distribution of the thermal noise can be modeled by a Gaussian probability density function.

1.10.1.2   Shot Noise

Shot noise is caused by the random emission of electrons and by random passage charge carriers across potential barriers. The shot noise generated in a device is modeled by a parallel noise current source. The mean-square shot noise current in the frequency band Δf is given by

$$\overline{i_{sh}^2} = 2qI\Delta f \tag{1.24}$$

where I is the DC through the device. This equation is commonly referred to as the Schottky formula. Like thermal noise, shot noise is white noise with a uniform and Gaussian distribution.

1.10.1.3   Excess Noise

In resistors, excess noise is caused by variable contact between particles of the resistive material. It is known that carbon composition resistors generate the most excess noise, while the metal-film resistors generate the least. The mean-square excess noise current can be written as

$$\overline{i_{ex}^2} = \frac{10^{NI/10}}{10^{12}\ln 10} \times \frac{I^2 \Delta f}{f} \tag{1.25}$$

where I is the DC through the resistor and NI is the noise index.
    The noise index (NI) is defined as the number of microamperes of excess noise current in each decade of frequency per ampere of the DC through the resistor.

1.10.1.4   Burst Noise

Burst noise, also known as popcorn noise, is caused by a metallic impurity in pn junctions. When amplified and reproduced by a loud speaker, it sounds like corn popping. On oscilloscopes, it appears as fixed amplitude pulses of random varying width and repetition rate. The rate can vary from 1 pulse/ sec to several hundred pulses per second.

1.10.1.5   Partition Noise

Partition noise occurs when the charge carriers in a current have a possibility of dividing between two or more paths. The noise is generated in the resulting components of the current by the statistical process of partition. Partition noise occurs in bipolar junction transistors (BJTs) where the current flowing from the emitter into the base can take one or two paths.

1.10.1.6   *Generation–Recombination Noise*

Generation–recombination noise is generated in semiconductors by the random fluctuation or free carrier densities caused by spontaneous fluctuations in the generation, recombination, and trapping rates. This noise appears both in BJTs and field-effect transistors (FETs), but not in metal oxide semiconductor field-effect transistors (MOSFETs).

1.10.1.7   *Spot Noise*

Spot noise is the rms in a band divided by the square root of the noise bandwidth. For a noise voltage and current, it has the units of $V / \sqrt{Hz}$ and $A / \sqrt{Hz}$, respectively. For white noise, the spot noise in any band is equal to the square root of the spectral density. The spot noise voltage of a device output is given by $\sqrt{v_{no}^2 / B_n}$. Often the noise bandwidth is expressed in the context of a second-order bandpass filter, given by $B_n = \pi B_3 / 2$, where B_3 is the −3 dB bandwidth. In the case of bandpass filters with two poles at frequencies f_1 and f_2, the noise bandwidth is given by $B_n = \pi(f_1 + f_2)/2$.

1.10.1.8   *Total Noise*

The total noise is the combination of all noise currents and noise voltages. In total noise, the instantaneous voltage ($v = v_n + i_n R$, where v_n is the noise voltage and i_n is the noise current) plays an important role. The mean-square voltage of the total noise can be expressed by

$$\overline{v^2} = \overline{(v_n + i_n R)^2} = \overline{v_n^2} + 2\rho\sqrt{\overline{v_n^2}}\sqrt{\overline{i_n^2}}R + \overline{i_n^2}R^2 \tag{1.26}$$

where ρ is the correlation coefficient defined by

$$\rho = \frac{\overline{v_n i_n}}{\sqrt{\overline{v_n^2}}\sqrt{\overline{i_n^2}}} \tag{1.27}$$

For the case ρ = 0, the sources that generate the noise are said to be uncorrelated or independent.

In many circuit analyses, noise signals are represented by phasors in the form of complex impedances. Also, the noise generated by electronic devices, such as amplifiers, can be modeled by referring all internal noise sources to the input.

1.10.1.9   Noise Bandwidth

Noise bandwidth is the bandwidth of a device having a constant passband gain, which passes the same rms noise voltage when the input signal is a white noise. White noise has the same amplitude for the given frequency range. The noise bandwidth is given by

$$B_n = \frac{1}{A_{vo}^2} \int_0^\infty \left| A_v(f) \right|^2 df \qquad (1.28)$$

where A_{vo} is the maximum value $\left| A_v(f) \right|$. For a white noise input voltage with a spectral density $S_v(f)$, the mean-square noise voltage of the filter output is $v_{no}^2 = A_{vo}^2 s_v(f) B_n$.

The noise bandwidth can be measured with a white noise source with a known voltage spectral density. If the spectral density of the source is not known, the noise bandwidth can be determined by comparing it with another source with a known noise bandwidth. The noise bandwidth of the unknown source can be expressed as

$$B_{n2} = B_{n1} \frac{\overline{v_{o1}^2}}{\overline{v_{o2}^2}} \left(\frac{A_{vo1}}{A_{vo2}} \right)^2 \qquad (1.29)$$

The white noise source must have low-output impedance so that the loading effect does not change the spectral density of the source.

1.10.1.10   Spectral Density

The spectral density of a noise signal is defined as the mean-square value of noise per unit bandwidth. For example, for thermal noise generated by a resistor, the voltage spectral density may be expressed by

$$S_v(f) = 4kTR \qquad (1.30)$$

As can be seen from Equation 1.30, since the spectral density is independent of the frequency, the thermal noise is known to have a flat or uniform distribution; therefore, it is called white noise.

1.10.2   Interference

Noise can also be transmitted to electronic circuits due to interference of signals generated by external sources. There are several classifications of transmitted noise, depending on how it affects the output signals and how it enters the electronic circuits. They can be magnetic, capacitive, or electromagnetic. The transmitted noise can be *additive* or *multiplicative* with respect to the output signals.

In additive noise, the noise magnitude is independent of the magnitude of the useful signal. Clearly, if there is no signal from the device, only the noise can be observed. An additive noise can be expressed as

$$v_{out} = v_{signal} + v_{noise} \qquad (1.31)$$

In multiplicative noise, the signal becomes modulated by the transmitted noise. It grows together with the magnitude of the signal. The multiplicative noise can be expressed as

$$v_{out} = v_{signal} + v_{signal} \times v_{noise} \qquad (1.32)$$

The transmitted noise can be periodic or random, depending on where and how it is generated. It can be inductive due to magnetic pickups, capacitive due to electrostatic pickups or electric charges, and, at high frequencies, electromagnetic; various combinations of all three are possible.

The offending external sources can be identified from the noise picked up by the instrument. For example, 50/60 Hz power supplies usually generate about 100-pA signals whereas 120/100 Hz supply ripples can generate only a few microvolts. Radio broadcasting stations can induce noise voltages up to 1.0 mV. Mechanical vibrations can generate currents from 10 to 100 pA (Fowler, 1996).

Identification and elimination of noise is an important topic for electronic portable instruments; therefore, further discussions are given in Chapter 4.

2

Sensors, Transducers, and Electronic Portable Instruments

Introduction

Electronic portable instruments interface with a physical phenomenon by means of sensors and transducers in the input and with actuators at the output. In literature, there is extensive information on sensors, transducers and actuators available (Webster, 1999; Fraden, 1993; Dyer, 2001); therefore, it will not be repeated here in great detail. However, some of the most common sensors and transducers used in electronic portable instruments will be explained in detail whenever necessary.

Sensors and transducers produce outputs in response to physical variations. These outputs are processed suitably by analog and digital methods to have desired information about the physical process. Similar to all other instruments, electronic portable instruments are developed for measuring and controlling some aspects of the physical environment. Most measurements are conducted by sensors that convert physical energy into electrical forms. An actuator, on the other hand, converts an electrical signal into an external action.

The criteria of selection of appropriate sensors and transducers for electronic portable instruments depend on many factors, such as power consumption, operating voltages, environmental sensitivity, noise resistance, weight, size, and reliability. It is the designer's choice to select the best sensor for a particular physical application. The best choice can only be made after all characteristics of the physical phenomenon are considered.

There are many different approaches that may be adopted to discuss the topic of sensors and transducers. In this chapter, a general approach will be adopted with a particular emphasis on recent developments in the sensors and transducers. Those that are applicable to electronic portable instruments will be elaborated.

As in the case of all instruments, for portable instruments, sensors and transducers can be categorized in a number of ways depending on factors such as energy input and output, input variables, sensing elements, and electrical or physical principles. For example, from the energy input and output point of view, there are three fundamental types of transducers: modifiers, self-generators, and modulators.

In *modifiers*, a particular form of energy is modified rather than converted; therefore, the same form of energy exists in both the input and output stages. In *self-generators*, electrical signals are produced from nonelectric inputs without the application of external energy. These transducers produce very small signals, which may need additional signal conditioning. Typical examples are piezoelectric transducers and photovoltaic cells. *Modulators*, on the other hand, produce electric outputs from nonelectric inputs, but they require an external source of energy. Strain gauges are typical examples of such devices.

In this book, the sensors and transducers for portable instruments are viewed from an applications point of view and classified in 16 main headings:

1. Voltage and current sensors
2. Magnetic and electromagnetic sensors
3. Capacitive and charge sensors
4. Semiconductor and intelligent sensors
5. Acoustic sensors
6. Temperature and heat sensors
7. Light sensors
8. Radiation sensors
9. Chemical sensors
10. Gas sensors
11. Biomedical and biological sensors
12. Environmental sensors
13. Pollution sensors
14. Distance and rotation sensors
15. Navigational sensors
16. Mechanical variables sensors

The choice of grouping is arbitrary, as the sensors for electronic portable instruments can be grouped in many other different ways.

2.1 Voltage and Current Sensors

The electrical, magnetic, and electromagnetic sensors are the most basic sensors, as their principles apply to many others.

2.1.1 Voltage Sensors

Voltage sensing is an important concept in all types of instruments, since the signals generated by most of the sensors are in the voltage form. Laboratory

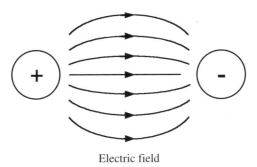

Electric field

FIGURE 2.1
Electric potential difference between two charged bodies related to electric.

type portable voltmeters (together with ammeters and ohmmeters) have been well established for many decades. The experiences gained from these instruments led to the development of many different portable instruments in other areas. Today, the majority of voltage, current, and resistance measurements are largely done by portable devices.

Since the basic principles of voltages are applicable to many other sensors in one way or another, it is worth discussing these principles in detail.

The voltage measurement involves the determination of the electric potential difference between two points. In the Système International d'Unités (SI), the measurement unit for the voltage is the volt, named so in honor of Italian physicist Alessandro Volta (1745–1827). The term *voltage* is also used to indicate the electric potential or the electromotive force. The volt is one of the fundamental electrical units.

In order to perceive how the difference in electric potential may arise, consider two charged bodies as illustrated in Figure 2.1. An attractive or repulsive force is generated between the two bodies, which represents an electric field. From here the potential may be defined as follows:

The potential is the amount of work needed to move a unit charge from a reference point to a specific point against an electric field. The work done in moving the charge is always relative to another one; therefore, the potential difference is always relative to some reference point. Typically, the reference point is the Earth, although other references can be used. Nevertheless, in many applications, the relativity of potential is omitted and the reference point is usually taken to be at the ground. The ground is the body of the Earth, which can accept or supply charge without changing its characteristics.

Since the potential difference is taken as the work per unit charge, the unit of potential, the volt, is related to the unit of work (joule, J) and the unit of charge (coulomb, C) by

$$1 \text{ V} = 1 \text{ J/C} \tag{2.1}$$

Although the concept of electric potential is useful in understanding electrical phenomena, it is worth noting that only the differences in potential energy are measurable. In general, it is commonly understood that the term *voltage* refers to potential difference.

Humans can sense many physical quantities such as temperature, light, and acoustics quite well. Contrast to that, the electric quantities can interfere with our life systems, and an attempt of sensing them directly may be severely harmful to our health. Therefore, we always relied on instruments that are able to convert the electric quantities into observable forms by humans. Instruments that measure voltages are called voltmeters, and instruments that measure currents are called the ammeters. Voltmeters and ammeters constitute the bulk of portable instruments.

Voltage can be measured by a diverse range of instruments operating on different principles. Some instruments rely on the relationship between currents and magnetic fields. In many cases, these relationships together with the suitable mechanical constructions have been applied to generate mechanical torque proportional to voltages or squared voltage under investigation; whereas other instruments rely on the thermal effects of currents flowing in conductors. Today, electronic voltmeters are mostly based on semiconductors and operate digitally. Depending on the operating principles, there are four distinct types of voltmeters:

1. *Electromechanical voltmeters* are based on the mechanical interaction between currents and magnetic fields. This interaction generates a mechanical torque proportional to the voltage or the squared voltage to be measured (in the voltmeters), or proportional to the current or the squared current to be measured (in the ammeters). This torque is balanced by a restraining torque, usually generated by a spring. The spring causes the instrument pointer to be displaced by an angle proportional to the driving torque, and hence to the quantity, or squared quantity, to be measured. The value of the input voltage or current is therefore given by the reading of the pointer displacement on a graduated scale. Nowadays, these instruments have been replaced by electronic counterparts and find only few applications, such as in vehicles, the transportation industry, and panel indicators where visual and scaled displays are preferred instead of digital numbers.

2. *Thermal type voltmeters* are based on the thermal effects of a current flowing into a conductor, and their reading is proportional to the squared input voltage or current. These instruments are not as widely used as in the case of electromechanical ones.

3. *Electronic voltmeters* constitute a wide range of electronic portable voltmeters and ammeters. They are based on purely electronic circuits and attain the required measurement by processing the input signal by means of semiconductor devices. The method employed to process the input signal can be either analog or digital: in the first case, analog electronic instruments are obtained, while in the second

TABLE 2.1

Classification of the Voltage and Current Meters

Class	Operating Principle	Subclass	Application Field
Electromagnetic	Interaction between currents and magnetic fields	Moving magnet	DC, current and voltage
		Moving coil	DC, current and voltage
		Moving iron	DC and AC, current and voltage
Electrodynamic	Interactions between currents	—	DC and AC, current and voltage
Electrostatic	Electrostatic interactions	—	DC and AC voltage
Thermal	Current's thermal effects	Direct action	DC and AC, current and voltage
		Indirect action	DC and AC, current and voltage
Induction	Magnetic induction	—	AC, current and voltage
Electronic	Signal processing	Analog	DC and AC, current and voltage
		Digital	DC and AC, current and voltage

case, digital electronic instruments are obtained. A peculiar characteristic of the electronic instruments is that their input should always be in voltage form. The currents can only be measured after they have been converted into voltages.

4. *Oscilloscopes* are basically voltmeters. In the case of electronic portable instruments, the main characteristic of oscilloscopes is to allow a graphic representation of measurements on a liquid crystal display (LCD). The desired measurements can be performed directly on the displayed signal.

In many applications, there are two basic types of voltage sensors commonly used: inductive sensors based on the characteristics of magnetic fields and capacitive sensors based on the characteristics of electric fields. Inductive voltage sensors generally use voltage transformers, alternating current (AC) voltage inductive sensors, eddy current sensors, and so on. Capacitive voltage sensors detect voltages by different methods, such as electrostatic force, Kerr or Pockels effects, the Josephson effect, and the change of reflective index of the optic fibers. Capacitive voltage sensors are generally used in low-frequency, high-voltage measurements. In high-voltage applications, the capacitive dividers are used to reduce the voltages to low levels. Many of these sensors will be explained in ensuing sections. Table 2.1 shows a rough classification of the most commonly employed voltmeters and ammeters in accordance to their operational principles and typical application fields.

The development of the semiconductor devices, including operational amplifiers, had an impressive impact on the measurement instruments, including voltmeters and ammeters. Both voltmeters and ammeters have steadily moved from the conventional electromechanical architectures to the electronic ones. Although all types of portable voltmeters and ammeters are commonly available, this book will briefly discuss the electronic types.

2.1.1.1   *Electronic Voltmeters*

Electronic voltmeters are designed to measure high- or low-frequency AC and direct current (DC) voltages. They are basically three port elements with an input port, an output port, and a power supply. The input port has a high impedance to eliminate the loading effect of the input signal. The output port provides the measurement result in analog or digital form. The power port supplies power to internal circuits.

Modern electronic voltmeters can be classified as analog or digital. The main features of the analog voltmeters are high-input impedance, high gain, and a wide bandwidth for AC measurements. They are primarily based on an electronic amplifier for processing of the input signal. An ammeter-like device displays the output signal of the amplifier. The operating principle consists of the generation of an output DC from the amplifier proportional to the input signal to be measured. This DC flows into the DC ammeter, thus forming an arrangement like a moving-coil milliammeter.

Digital voltmeters display the results in digital forms. They are characterized by speed, automatic operation, and programmability. The basic structure of a digital voltmeter is shown in Figure 2.2. The first stage of the device takes care of conditioning the input signal and adapting it to the requirements of the analog-to-digital (A/D) converter. The converter section is responsible for sampling of the input signals and converting each sampled value into digital forms. The sequence of digital values is obtained after the sampling and conversion operations, the data are stored into the memory of a digital signal processor (DSP), and they are processed to attain the desired measured values.

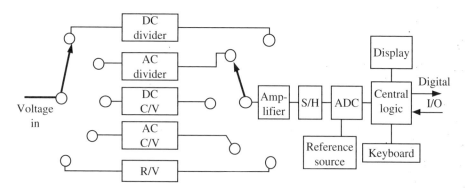

FIGURE 2.2
Basic structure of a digital meter.

2.1.2 Current Sensors

A charged body in an electric field is subject to a force. If the body is not restrained, it will start moving in the electric field. The result of the movement of charged bodies from one point to another point in the space is the electric current. If metallic conductors are used in electric circuits, the charged bodies that can flow through the conductors are the electrons. A basic property of an electron is its charge (1.603×10^{-19} C). The electric current is the rate of charge flowing in a cross section of the conducting element in 1 sec, and its measurement unit in the SI is the ampere (A), named so in honor of French physicist André-Marie Ampère (1775–1836).

Since the current is taken as the charge flow per unit time, its unit is related to the unit of charge and the unit of time (second, sec); therefore,

$$1 \text{ A} = 1 \text{ C/sec} \tag{2.2}$$

Current flowing through a circuit element can also be related to the properties of the circuit element and the voltage across its terminals by the well-known Ohm's law as

$$V = R\,I \tag{2.3}$$

where R is a quantity representing the electric behavior of the circuit element. Under DC conditions, this quantity is called resistance and its measurement unit in the SI is the ohm (Ω), named so in honor of German physicist George Simon Ohm (1787–1854).

The knowledge of the voltage located at the terminal of each circuit element as well as the current flowing in each element gives full knowledge of the circuit behavior. The measurement of voltages and currents is hence of utmost importance and extensive use in all branches of electrical and electronic engineering, and in instrumentation engineering. Moreover, when electronic devices for signal processing are involved, such as those used in telecommunication systems, control systems, and informatics, the majority of the signals are in voltage form. Therefore, the voltage and current measurements constitute an important area in laboratory and industrial measurements, sensors, and other systems.

Fundamentally, the properties of magnetic fields, electric fields, and heat can be used for currents sensing. In addition, any magnetic field sensor can be configured for current measurements. As an example, the Rogowski coil is illustrated in Figure 2.3. The Rogowski coil is a solenoid air core winding of a small cross section looped around a conductor carrying the current. The voltage induced across the terminals of the coil is proportional to the derivative of the current.

Nevertheless, one of the simplest methods of current sensing is the current-to-voltage conversion by means of resistors. In current measurements,

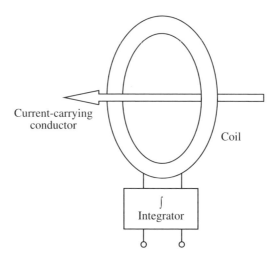

FIGURE 2.3
Rogowski coil for current sensing.

the resistor is called the shunt resistor, despite its connection in series with the load. The shunt produces a voltage output that can be presented by a variety of secondary meters, such as analog meters, digital meters, oscilloscopes, etc. In addition, there is a range of noncontact sensors that are available, such as Hall-effect probes and current transformers, which can be used in wrap-around form so they can be installed without disconnecting the power. The wrapping process involves breaking a magnetic path and sensing the magnetic intensity proportional to the current that a conductor is carrying.

2.1.2.1 Power and Power Factor Measurements

Ammeters and voltmeters exist as single units equipped with selectable switches. One application of combined current and a voltage measurement device is the wattmeter. With slight design modifications, it can be used for power measurements as well as power factor and harmonics measurements. In order to measure the power factors, the voltages and currents must be sensed separately, and the time relation between the two must be worked out to obtain the phase difference. Other AC quantities such as power and power factor are derived mathematically from these measured quantities.

There are many different types of handheld instruments for harmonic and power factor measurement. Many devices use some form of Fourier calculations or digital filtering to determine power values within the accepted definitions. They sense currents and voltages by using current probes and voltage leads. The current probes can be purchased to measure currents from a few amperes to several thousand amperes.

2.2 Magnetic Sensors

Magnetic sensors find many applications in everyday life and industry. They provide convenient, noncontact, simple, rugged, and reliable operational devices compared to many other sensors. The technology to produce magnetic sensors involves many different disciplines such as physics, metallurgy, chemistry, and electronics.

Magnetic sensors are based generally on sensing the properties of magnetic materials, which can be done in many ways. For example, magnetization, which is the magnetic moment per volume of materials, is used in some measurement systems by sensing force, induction, field methods, and superconductivity. However, the majority of sensors make use of the relationship between magnetic and electric phenomena. A typical application of these phenomena is in the computer memory requiring the reading of the contents of a disc without making any contact between the sensor and the device. In other applications, the position of ferromagnetic objects that are sensitive to magnetic fields (e.g., metals buried underground) can be sensed magnetically.

The magnetic elements in sensors are used in a wide range of forms: toroids, rods, films, substrates, and coatings. Some elements are essentially freestanding, whereas others are an integral part of more complex devices. In order to obtain maximum material response in magnetic sensors, the relative orientation and coupling between sensors and measurand is important. The magnetic properties of materials used must be considered carefully, and the orientation and coupling between materials must be optimized at the design stages.

Many different types of magnetic sensors are available, but they can broadly be classified as the *primary sensors* or the *secondary sensors*. In the primary sensors, also known as the magnetometers, the physical parameters to be measured are the external magnetic fields. They are largely used in biological, geophysical, and extraterrestrial measurements. In the secondary sensors, the external parameters are made from other physical variables such as force and displacement.

Magnetic sensors are very commonly used in electronic portable instruments; therefore, detailed attention will be paid to them in this section. Both the primary and secondary sensors will be introduced in detail.

2.2.1 Primary Magnetic Sensors

Magnetometers are based on magnetic sensors for measuring of external magnetic fields. They find applications in many areas from biological and geophysical measurements to the determination of characteristics of extraterrestrial objects and stars. They also find applications where sensitivity is of

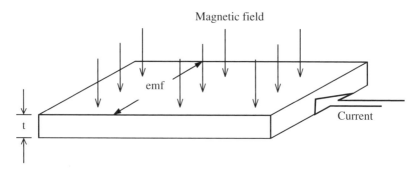

FIGURE 2.4
A Hall-effect sensor.

utmost importance, as in the case of devices for diagnosing and curing human illnesses. The superconducting quantum interface devices (SQUIDs) and nuclear magnetic resonance (NMR) imaging are examples of such devices.

There are many different types of magnetometers that are used in electronic portable instruments:

- Hall-effect sensors
- Magnetodiode and magnetotransistors
- Magnetoresistive sensors
- Magneto-optical sensors
- Integrated magnetic field sensors
- Magnetic thin films
- Fluxgate magnetometers
- Search coil magnetometers

Some of these magnetic sensors will be explained next. Fluxgate and search coil magnetometers will be explained in the section allocated on environmental sensors.

2.2.1.1   Hall-Effect Sensors

Hall-effect sensors operate on the principles that the voltage difference across a thin conductor carrying a current depend on the intensity of magnetic field applied perpendicular to the direction of the current flow. This is illustrated in Figure 2.4. Electrons moving through a magnetic field experience Lorentz force perpendicular to both the direction of motion and the direction of the field. The response of electrons to Lorentz force creates a voltage known as the Hall voltage. If a current I flows through the sensor, the Hall voltage can mathematically be expressed by

$$V = R_\mathrm{H}\, I\, B/t \tag{2.4}$$

where R_H is the Hall coefficient (m^3/°C), B is the flux density (T), and t is the thickness of the sensor (m).

Hall-effect sensors can be made by using metals or silicon, but they are generally made from semiconductors with high electron mobility such as indium antimonide. They are usually manufactured in the form of probes with sensitivity down to 100 μT. Silicon Hall-effect sensors can respond to magnetic flux having an operational frequency from DC to 1 MHz, within the ranges of 1 mT to 0.1 T. They have good temperature characteristics, varying from 200°C to near absolute zero.

2.2.1.2   Magnetodiodes and Magnetotransistor Sensors

Magnetodiodes and magnetoresistor sensors are made from silicon substrates with undoped areas that contain the sensor between *n*-doped and *p*-doped regions, forming *pn*, *npn*, or *pnp* junctions. In the case of magnetotransistors, there are two collectors, as shown in Figure 2.5.

Depending on the direction, an external magnetic field deflects electron flow between the emitter and collector in favor of one of the collectors. The two collector voltages are sensed and related to the applied magnetic field. Basically, an external magnetic field perpendicular to the flow of charges deflects the holes and electrons in opposite directions, resulting in variations of the resistance of the undoped silicon layers.

2.2.1.3   Magnetoresistive Sensors

The resistivity of some current-carrying material changes in the presence of a magnetic field, mainly due to the inhomogeneous structure of the material. For example, the resistance of bismuth can change by a factor of 10^6. Most conductors have a positive magnetoresistivity. Magnetoresistive sensors are largely fabricated from permalloy stripes positioned on a silicon substrate. Each strip is arranged to form one arm of a Wheatson bridge so that the output of the bridge can directly be related to the magnetic field strength.

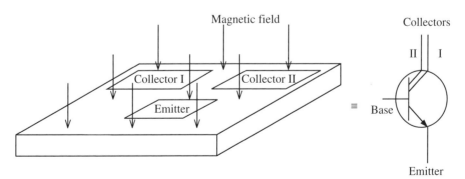

FIGURE 2.5
A semiconductor magnetotransistor.

As in the case of Hall-effect sensors, the basic cause of magnetoresistivity is the Lorentz force, which causes electrons to move in a curved path between collisions. One advantage is that for small values of magnetic field, the change in resistance is proportional to the square of the magnetic field strength, thus giving better sensitivity.

2.2.1.4   Magneto-Optical Sensors

Magneto-optical sensors constitute an important component of magnetometers. In recent years, highly sensitive magneto-optical sensors have been developed. These sensors are based on various technologies, such as fiber optics, polarization of light, the Moire effect, the Zeeman effect, etc. These types of sensors lead to highly sensitive devices and are used in applications requiring good resolution, such as human brain function mapping, magnetic anomaly detection, and so on.

2.2.1.5   Integrated Magnetic Field Sensors

Integrated magnetic field sensors constitute a new class of sensors; they are termed as semiconductor magnetic microsensors. This technology uses either high-permeability (ferromagnetic) or low-permeability (paramagnetic) materials. The integrated circuit magnetic techniques support many sensors, such as magnetometers, optoelectronics, Hall-effect sensors, magnetic semiconductor sensors, superconductive sensors, etc. At present, silicon offers an advantage of inexpensive batch fabrication. Most of the integrate circuit magnetic sensors are manufactured by following the design rules of standard chip manufacturing. For example, metal oxide semiconductor (MOS) and complementary MOS (CMOS) technologies are used to manufacture highly sensitive Hall-effect sensors, magnetotransistors, and other semiconductor sensors.

2.2.1.6   Magnetic Thin Films

Magnetic thin films are an important part of superconducting instrumentation, sensors, and electronics in which active devices are made from deposited films. The thin films are usually made from amorphous alloys, amorphous gallium, and the like. As an example, a thin-film Josephson junction is given in Figure 2.6. The deposition of thin films can be done by several methods, such as thermal evaporation, electroplating, sputter deposition, or by some chemical means. The choice of technology depends on the characteristics of the sensors. For example, thin-film superconductors require low-temperature operations, whereas common semiconductors normally operate well at room temperatures.

Magnetic thin films find extensive applications in memory devices where high density and good sensitivities are required. In such applications, the magnetic properties of the coating are determined by the magnetic properties

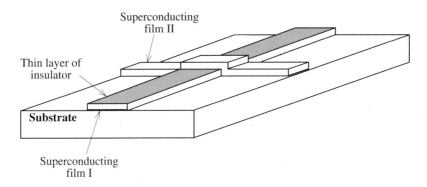

FIGURE 2.6
A magnetic thin-film Josephson junction.

of the particles that can be controlled before coating. The choice of available materials for this purpose is extremely large. Thin-film technology is also developed in magneto-optics applications where erasable optical media for high-density magnetic storage is possible. Miniature magnetoresistive sensors for magnetic recording and pickup heads, Hall-effect sensors, and other magnetic semiconductors make use of thin-film technology extensively.

2.2.2 Secondary Magnetic Sensors

The secondary magnetic sensors are basically inductive sensors, which make use of the principles of magnetic circuits. They can be classified as *passive sensors* or *active sensors* (self-generating). The passive sensors require an external power source; hence, the action of the sensor is restricted to the modulation of the excitation signal in relation to external stimuli. On the other hand, self-generating sensors generate signals by utilizing the electrical generation principle based on Faraday's law of induction: when there is a relative motion between a conductor and a magnetic field, a voltage is induced in the conductor. Or a varying magnetic field linking a stationary conductor produces voltage in the conductor. This relationship can be expressed as

$$e = -d\Phi/dt \tag{2.5}$$

where Φ is the magnetic flux and t is time.

In Equation 2.5, the time dependence of the induced voltage should be carefully noted. In instrumentation applications, the magnetic field under investigation may be varying in time irregularly or with some frequency. In addition, the conductor sensing the magnetic field may be moving in space too. Both the time variations of the magnetic field and the relative motion of the conductor affects the voltage induced; therefore, during the measurements, the results obtained by the sensors must be carefully analyzed.

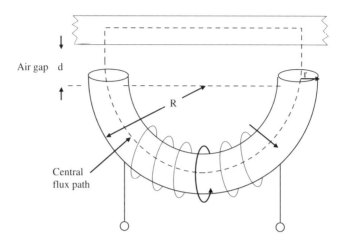

FIGURE 2.7
A basic inductive sensor.

In order to explain the operation of the basic principles of inductive sensors, a simple magnetic circuit is shown in Figure 2.7.

The magnetic circuit consists of a core, made from a ferromagnetic material, and a coil with n number of turns wound on it. The coil acts as a source of magnetomotive force (mmf), which drives the flux Φ through the magnetic circuit. If we assume that the air gap is zero, the equation for the magnetic circuit is expressed as

$$\text{mmf} = \text{flux} \times \text{reluctance} = \Phi \times \Re \text{ (A-turns)} \qquad (2.6)$$

such that the reluctance, $\Re$, limits the flux in a magnetic circuit just as resistance limits the current in an electric circuit. By writing the magnetomotive force in terms of current, the magnetic flux may be expressed in webers, after William Eduard Weber (1804–1891), as

$$\Phi = ni/\Re \text{ (weber)} \qquad (2.7)$$

In Figure 2.7, the flux linking a single turn is expressed by Equation 2.7. But the total flux linking by the entire n number of turns of the coil is

$$\Psi = n\Phi = n^2 i/\Re \text{ (weber)} \qquad (2.8)$$

Equation 2.8 leads to self-inductance L of the coil. Inductance in henry, after Joseph Henry (1797–1878), is described as the total flux per unit current for a particular coil. That is,

$$L = \Psi/I = n^2/\Re \text{ (henry)} \qquad (2.9)$$

This indicates that the self-inductance of an inductive element can be calculated from magnetic circuit properties. Expressing $\mathfrak{R}$ in terms of dimensions,

$$\mathfrak{R} = l/\mu\mu_0 A \text{ (A-turns/weber)} \tag{2.10}$$

where l is the total length of the flux path (meters), μ is the relative permeability of the magnetic circuit material, μ_0 is the permeability of free space ($= 4\pi \times 10^{-7}$ H/m), and A is the cross-sectional area of the flux path.

If the air gap is allowed to vary, the arrangement becomes a basic inductive sensor. In this case, the ferromagnetic core is separated in two parts by the air gap. The total reluctance of the circuit now is the addition of the reluctance of the core and the reluctance of the air gap. The relative permeability of air is close to unity, and the relative permeability of the ferromagnetic material may be on the order of a few thousand, indicating that the presence of the air gap causes a large increase in circuit reluctance and a corresponding decrease in the flux. Hence, a small variation in the air gap causes a measurable change in inductance.

There are many different types of inductive sensors, such as linear and rotary variable-reluctance sensors, sychros, microsyn, linear variable inductors, induction potentiometers, linear variable-differential transformers, and rotary variable-differential transformers. Some of these will briefly be discussed next.

2.2.2.1 *Linear and Rotary Variable-Reluctance Sensors*

Linear and rotary variable-reluctance sensors are based on change in the reluctance of a magnetic flux path. These sensors find applications particularly in acceleration and force measurements. However, they can suitably be constructed for sensing displacements as well as velocities. There are quite a few different types of variable-reluctance sensors, e.g., single-coil linear variable-reluctance sensors and variable-reluctance tachogenerators.

A *single-coil linear variable-reluctance sensor* consists of three elements: a ferromagnetic core in the shape of a semicircular ring, a variable air gap, and a ferromagnetic plate. The sensor's performance depends on the core geometry and permeability of the materials; therefore, it can be highly nonlinear. Despite this nonlinearity, these sensors find applications in many areas, such as measurements of force. The coil usually forms one of the components of an *LC* oscillator circuit whose output frequency varies with the applied force. Hence, the coil modulates the frequency of the local oscillator.

In a *variable-differential reluctance sensor*, the problem of the nonlinearity may be overcome by modifying the single-coil system into variable-differential reluctance sensors (also known as push–pull sensors). This sensor consists of an armature moving between two identical cores separated by a fixed distance. In a typical commercially available variable-differential sensor, the iron core is located halfway between the two E-shaped frames. The flux generated by primary coils depends on the reluctance of the magnetic

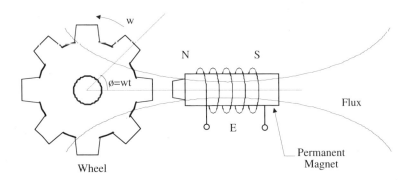

FIGURE 2.8
A variable-reluctance tachogenerator.

path, the main reluctance being the air gap. Any motion of the core increases the air gap on one side and decreases it on the other side. Consequently, the reluctance changes in accordance with the principles explained previously, thus inducing more voltage on one of the coils than on the other. A motion in the other direction reverses the action with a 180° phase shift from the null position. The output voltage can be modified depending on the requirements in signal processing by means of rectification, demodulation, or filtering. In these instruments, full-scale motion may be extremely small, on the order of a few thousandths of a centimeter.

The *variable-reluctance tachogenerator* is another example of the variable-reluctance sensor shown in Figure 2.8. These sensors are based on Faraday's law of electromagnetic induction; therefore, they may also be referred to as electromagnetic sensors. Basically, the induced electromagnetic force (emf) in the sensor depends on the linear or angular velocity of the motion.

The variable-reluctance tachogenerator consists of a ferromagnetic toothed wheel attached to the rotating shaft, and a coil wound onto a permanent magnet, extended by a soft iron pole piece. The wheel moves in close proximity to the pole piece, causing the flux linked by the coil to change, thus inducing an electromotive force in the coil. The reluctance of the circuit depends on the width of the air gap between the rotating wheel and the pole piece. When the tooth is close to the pole piece, the reluctance is at a minimum; it increases as the tooth moves away from the pole. If the wheel rotates with a velocity ω, the flux may mathematically be expressed as

$$\Psi(\theta) = \Psi_m + \Psi_f \cos m\theta \tag{2.11}$$

where Ψ_m is the mean flux, Ψ_f is the amplitude of the flux variation, and m is the number of teeth.

The induced emf is given by

$$E = \frac{-d\psi(\theta)}{dt} = \frac{-d\psi(\theta)}{d\theta} \times \frac{d(\theta)}{dt} \tag{2.12}$$

or

$$E = \Psi_f \, m\omega \sin n\omega t \qquad (2.13)$$

Both the amplitude and frequency of the generated voltage at the coil are proportional to the angular velocity of the wheel. In principle, the angular velocity ω can be found from either the amplitude or the frequency of the signal. In practice, the loading effects and electrical interference may influence the measured amplitudes. In signal processing, the frequency is the preferred option because it can be easily converted into digital signals.

2.2.2.2 Linear Variable-Differential Transformer

The linear variable-differential transformer (LVDT) makes use of the principles of transformer action; that is, magnetic flux created by one coil links with the other coil to induce voltages. The LVDT is a passive inductive transducer that has found many applications. It consists of a single primary winding positioned between two identical secondary windings wound on a tubular ferromagnetic former. The primary winding is energized by high-frequency (50 Hz to 20 kHz) AC voltages. The two secondaries are made identical by having an equal number of turns. They are often connected in series opposition so that the induced output voltages oppose each other.

One important advantage of the LVDTs is that there is no physical contact between the core and the coil form; hence, there is no friction or wear. They find a variety of applications, which include jet engine controls that are in close proximity to exhaust gases and controls that measure roll positions in the thickness of materials in hot-slab steel mills. After some mechanical conversions, LVDTs may also be used for force and pressure measurements.

2.3 Capacitive and Charge Sensors

Many sensors are based on the characteristics of electrical charges and the associated electric field properties. The electrical charge and field are related by capacitance. Capacitive sensors are extensively used in electronic portable instruments. These sensors are based on changes in capacitances in response to physical variations. The changes in the capacitance can occur as a response to variations in physical dimensions, such as the area or the distance between the plates, or the variations in the dielectric properties of the material between the plates. In electronic portable instruments, capacitive sensors find many diverse applications from humidity and moisture sensing to displacement measurements. In this section, the basic principles of capacitive sensors will be introduced, and in the ensuing sections, more discussions on specific sensors will be given.

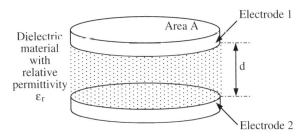

FIGURE 2.9
A typical capacitor.

Capacitors are made from two charged electrodes separated by a dielectric material, as shown in Figure 2.9. The capacitance, C, of this system is equal to the ratio of the absolute value of the charge, Q, to the absolute value of the voltage, V, between charged bodies:

$$C = Q/V \qquad (2.14)$$

where C is in farads (F), Q is in coulombs (C), and V is in volts.

The unit of capacitance, farad, is named after Michael Faraday (1791–1867). The farad is a large unit; therefore, practical capacitors have the capacitances in microfarads (10^{-6} F = 1 µF), nanofarads (10^{-9} F = 1 nF), or picofarads (10^{-12} F = 1 pF).

In general, the capacitance can be determined by using Laplace's equation: $\nabla^2 V(x, y, z) = 0$ with the appropriate boundary conditions. The boundary conditions specify the electrode voltages V_1 and V_2 of the plates. For two electrodes, Laplace's equation yields to the voltage $V(x, y, z)$ and the electric field $E(x, y, z) = -\nabla V(x, y, z)$. The charge of each electrode can also be obtained by integration of the electric flux density over each electrode surface, as

$$Q = \int \varepsilon(x, y, z)\, E(x, y, z)\, dA \qquad (2.15)$$

In Figure 2.9, if the edge effects are ignored, capacitance in terms of dimensions may be expressed by

$$C = \varepsilon\, A/d = \varepsilon_r \varepsilon_0\, A/d \qquad (2.16)$$

where ε is the dielectric constant or permittivity, ε_r is the relative dielectric constant (in air, $\varepsilon_r = 1$), ε_0 is the dielectric constant of vacuum (8.854188×10^{-12} F/m or $10^{-9}/36\pi$ F/m), d is the distance of the plates in meters, and A is the effective area of the plates in square meters.

The capacitance, C, depends on the size and shape of charged bodies and their positioning relative to each other. In many electrical and electronic systems, it is necessary to deal with the useful capacitances as well as the stray capacitances. Stray capacitances may be introduced externally or inter-

TABLE 2.2

Permittivity (Dielectric Constants) of
Materials Used in Capacitors

Material	Permittivity
Vacuum	1.0
Air	1.0006
Polythene, etc.	2.0–3.0
Silicone resins	3.8
Impregnated paper	4.0–6.0
Glass and mica	4.0–7.0
Ceramic (low K)	1–20.0
Ceramic (medium K)	80.0–100.0
Ceramic (high K)	1000.0 up
Thick-film capacitor compounds	300.0–5000.0

nally whenever two charged bodies exist side by side. The cables and other external components introduce additional capacitances that need to be taken care of for a desirable performance of the system. More information on capacitors and stray capacitances can be found in Chapter 4.

As it can be seen in Equation 2.16, the value of the capacitance largely depends on the permittivity of the dielectric material. In electronic portable instruments, there are many sensors that are based on the properties of dielectric materials. The typical values of permittivity of commonly used materials (at 1.0 KHz) are listed in Table 2.2.

The dielectrics of capacitors can be made from polar or nonpolar materials. Polar materials have dipolar characteristics. That is, they consist of molecules whose ends are oppositely charged. This polarization causes oscillations at certain frequencies, resulting in high losses at those frequencies.

2.3.1 Integrated Circuit and Computable Capacitors

Integrated circuit capacitors find wide applications in electronic portable instruments. They are made as MOS integrated circuits, and the monolayer versions contain tantalum or other suitable deposits. Two heavily doped polysilicon layers formed on a thick layer of oxide generally form the plates of capacitors of integrated circuits. The dielectric is usually made from a thin layer of silicon oxide. Important parameters for integrated circuit (IC) capacitors are the tolerances, voltage coefficients, temperature coefficients, and capacitance values. These capacitors are temperature stable with about 20 ppm/°C temperature coefficients. The voltage coefficients are usually less than 50 ppm/V. Integrated circuit capacitive sensors are achieved by incorporating a dielectric material sensitive to physical variables.

Integrated circuit capacitors are adapted for use in microelectronic circuits. They include some miniature ceramic capacitors, tantalum oxide solid capacitors, and tantalum electrolyte solid capacitors. The ceramic and tantalum oxide ICs are not encapsulated, but they are fitted with end caps for direct surface-mounting purposes onto circuit boards. Ceramic IC capacitors are most suitable for radio frequency (RF) applications. Different manufacturers offer many different types of ceramic capacitors with high Q multilayer. They are available from 0.5 to 1000 pF, with voltage ratings from 100 to 500 V DC, depending on capacitance value.

Another version of IC capacitors is the thin-film integrated circuit computable capacitor. In a single IC, there may be many capacitors with typical values, up to 1, 2, 4, 8, and 16 pF. Binary ratios starting at 0.25 pF are also available. In the design of these capacitors, a back metallization layer forms one plate of the capacitor; individual contacts form the other plate; silicon dioxide dielectric features high Q over a wide temperature range. These capacitors are commonly used in circuits involving active filters, such as Chebyshev or Butterworth, for high-pass, low-pass, bandstop, or bandpass applications.

2.4 Semiconductor and Intelligent Sensors

2.4.1 Semiconductor Sensors

Silicon is the basic element of most semiconductors. Properties of silicon are well studied and most suitable in developing sensors (Fraden, 1993). Silicon-based sensors find many applications such as:

Radiation measurements: Photovoltaic effect, photoelectric effect, photoconductivity, photomagneto effects

Mechanical measurements: Piezoresistivity, photoelectric and photovoltaic effects

Thermal measurements: Seebeck effect, temperature, Nernst effect

Magnetic measurements: Hall effect, magnetoresistance, Suhi effect

Chemical measurements: Ion sensitivity

Developments of silicon-based sensors are enhanced considerably due to advances in embedded controllers, microelectronics, and micromachining. Micromachining technology combined with semiconductor processing technology provides sensors for mechanical, optical, magnetic, chemical, biological, and other phenomena, as listed above. Digital logic provided by either logic circuits, dual in-line packages (DIPs), and microcontrollers plays a vital role in the performance, design, and construction of semiconductor sensors. Some of these sensors will be explained in this section.

Batch processing techniques are ideally suited for making high-volume, low-cost sensors. Because of mass production, the characteristics of individual sensors may have some deviations from ideal values; therefore, most specifications are given in terms of full-scale values. In some cases, typical specifications are given, which indicate a certain degree of spread in the performance from one sensor to another. Nevertheless, the advances in digital technology and cost-effective manufacturing techniques of IC sensors reflect positively on the development of electronic portable instruments. Today there are many new portable instruments that are based on microsensors and microprocessors. This trend is expected to continue in the near future, thus revolutionizing the portable instrumentation technology.

Semiconductor sensors provide easier-to-interface, lower-cost, and more reliable inputs to electronic control systems. Before the availability of microelectronics, the sensors and transducers were directly coupled with the readout devices. However, with the advent of microelectronics technology, sensors and transducers are developed in such a way that many processing components are integrated with the sensor itself. Most micro and semiconductor sensors interface to a microcontroller unit (MCU) directly without requiring any A/D conversion. This can be achieved either by inherent digital output sensors or by the integration of on-chip processing electronics within the sensing unit.

The sensing technique used for a measurement can vary considerably depending on the range of the measurand, accuracy required, and environmental considerations. Characteristics of a measurement system can introduce constraints for signal conditioning, signal transmission, data display, calibrations, input and output impedances, supply voltage, frequency response, etc.

Among many others, typical examples of IC sensors are the micro-electromechanical accelerometers, such as ADXL150 and ADXL250 manufactured by Analog Devices. The ADXL150 is capable of sensing acceleration in a single axis, whereas the ADXL250 senses acceleration in two axes. These sensors include transducer elements and the necessary signal conditioning electronics integrated together on a single IC. Both ADXL150 and ADXL250 offer low noise (1 mg /Hz) and a good signal-to-noise ratio. The data obtained from each sensor can be acquired by suitable microcontrollers such as PIC16F874, which has a 10-bit internal A/D converter.

Microsensors and semiconductor sensors are dealt with in detail in appropriate sections. In this section, a general brief description of piezoelectric, capacitive, chemical, and oscillatory and resonance sensors is given.

2.4.1.1 Piezoelectric Sensors

A piezoelectric sensor produces an electric charge when a force is applied to it. This effect exists in natural crystals such as quartz (SiO_2) or is manmade from polarized ceramics and polymers. A piezoelectric sensor can be viewed to have a crystal structure with a dielectric material. When this dielectric material is arranged like a capacitor with electrodes, it behaves like an electric charge generator when stress is applied to it. Therefore, a voltage, V, across the capacitor appears as a result of stress. The piezoelectric effect is a reversible physical phenomenon. That is, applying a voltage across the terminals, the crystal produces a mechanical strain. The magnitude of the piezoelectric effect can be expressed by a vector of polarization as

$$P = P_{xx} + P_{yy} + P_{zz} \qquad (2.17)$$

where x, y, and z refer to a conventional orthogonal system related to the crystal axis. In terms of stress, σ, the piezoelectric effects, P, in the x, y, and z directions may be written as

$$P_{xx} = d_{11}\,\sigma_{xx} + d_{12}\,\sigma_{yy} + d_{13}\,\sigma_{zz}$$

$$P_{yy} = d_{21}\,\sigma_{xx} + d_{22}\,\sigma_{yy} + d_{23}\,\sigma_{zz} \qquad (2.18)$$

$$P_{zz} = d_{31}\,\sigma_{xx} + d_{32}\,\sigma_{yy} + d_{133}\,\sigma_{zz}$$

where constants d_{mn} are the piezoelectric coefficients (in coulombs per Newton) along the orthogonal axis of the crystal.

In most modern piezoelectric sensors is the lead zirconate titanate (PZT). This ceramic-based sensor is used to construct biomorph sensors for force, sound, motion, vibration, and acceleration sensing. The biomorph consists of two layers of different PZT formulations that are bonded together in rectangular strips. Rectangular shape structure can conveniently be mounted as a cantilever on structures, and the motions perpendicular to the surface of the sensor generate signals of measurements.

Other IC sensors are the *piezofilm* sensors, which are made from polymer films that generate a charge when they are deformed. They also exhibit mechanical motions when a charge is applied. Also, surface machining techniques are combined with piezoelectric thin-film materials (e.g., zinc oxide) to produce semiconductor pressure sensors particularly useful in low-level acoustic applications.

A variation in piezoelectric technology is the piezoresistive sensors, which are the most common micromachined sensors. Their output signals are highly uniform and predictable; therefore, these signals can easily be conditioned.

2.4.1.2 Capacitive IC Sensors

Capacitive IC sensors have typically one plate fixed and one that moves as a result of the applied measurand. Most microelectronic capacitive sensors are based on silicon technology.

Silicon technology combined with the surface micromachining allows the capacitance between interdigitated fingers to measure acceleration as well as other inputs. The value of nominal capacitance is 100 fF to 1 pF, and the variations in capacitance are in the order of femtofarads. Usually, the signal conditioning circuits integrate either in the same chip or in the same package that provides the necessary output to the external digital system. The micro-

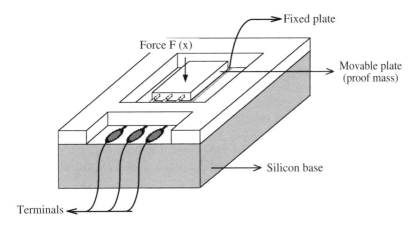

FIGURE 2.10
A typical micromachined sensor.

machining technology of capacitive sensors forms the basis of many sensors, and a typical example is given in Figure 2.10. This device is a force transducer and makes use of changes in the distance of plates when force is exerted.

2.4.1.3   Chemical Sensors

There are many different types of IC chemical sensors. A commonly used sensor is based on the resistivity changes of the metal oxides in response to reducing or oxidizing properties of the chemical environment around the sensor. Further details on this topic are provided in Section 2.9, which is allocated to chemical sensors.

2.4.1.4   Resonance Sensors

Some sensors are based on mechanical resonant structures. For example, a micromachined silicon can act as a resonant element if it is designed with a membrane or tuning fork. Electric activation is achieved by using some piezolectric films such as ZnO. Most resonant structures are implemented in pairs, one of which acts as a reference; therefore, it is not interfaced to the input. Comparison of the output frequency of the sensing element to the reference element reduces the influence of the unwanted parameters. This technique is often used in chemical sensors using surface acoustical wave (SAW) delay-line oscillators.

2.4.1.5   Electrical Oscillators

Electrical oscillators are another form of IC sensors that make use of electrical oscillator-based (EOB) technology and the stochastic analog-to-digital (SAD) converters. An EOB sensor generates a periodic voltage or current when it is subjected to a measurand. The voltage-to-frequency or current-to-frequency converters are used to convert the signals into digital forms. Some

EOB sensors make use of ring oscillators based on integrated-injected logic gates. SAD converters usually employ flip-flops and comparators.

2.4.2   Embedded Sensing

Power ICs (PICs) are typical examples of electronic devices that are already equipped with embedded or internal sensors. They are produced by combining bipolar and MOS circuitry with power metal oxide semiconductor field-effect transistor (MOSFET) technology. They are capable of providing direct interface between microcontroller units (MCUs) and system loads, such as solenoids, lamps, and motors. These devices also provide increased functionality, as well as sophisticated diagnostics and protection circuitry. Sensing of current levels and junction temperatures is one of the key aspects for control strategy during normal operations and for detecting several fault modes. Sensors within small PICs detect fault and threshold conditions, thus allowing the implementation of control strategies for situations such as over-temperatures, overcurrents, and overvoltages.

The approach to PICs is based on the consolidation of a number of circuit elements into one single device. These devices would have normally been discrete components, or a combination of standard or custom ICs with some discrete output device backups. In this type of single chip, some circuit elements (e.g., operational amplifiers, comparators, regulators) are best implemented by the bipolar IC technology. MOS circuitry handles logic, active filters, and time delays. Some circuits, such as A/D converters and power amplifiers, can be implemented by either bipolar or MOS technology.

2.4.3   Sensing Arrays

In many cases, more than one sensor is required to provide sufficient infor-mation. There is considerable effort to integrate many sensors on the same silicon wafer with the necessary signal conditioning and computational capa-bilities. Sensing arrays include a number of sensors for different measurands, such as pressure, flow, temperature, and vibration. Multiple sensors are used for many reasons, such as increasing the range, providing redundancy, or capturing information at different spatial points. A good example of sensor arrays is in chemical applications, where the same chip can measure different chemicals. Multielement gas analyzers are developed by using thin-film detectors. For example, a four-element gas analyzer can be achieved by multiple-chip hybrid configurations.

Photodiode arrays are another example of such devices, and they prove to be highly cost-effective in many applications. The popular silicon charge-coupled device (CCD) with many sensing elements (e.g., 2048 elements) is typical of such devices. A complete spectral measurement can be made by a single unit. Some array sensors use pyroelectric and piezoelectric effects in zinc oxide thin films for gas flow, chemical reactions, and cantilever beam acceleration, tactile, and SAW vapor sensing.

CMOS technology is also used in sensing arrays. This technology allows integration of many sensors in a single chip, and they are used in ion detectors, moisture sensors, electrostatic discharge sensors, strain gauges, edge damage detectors, corrosion detectors, and so on.

2.4.4 Amorphous Material Sensors

Amorphous materials are manufactured in the form of powders, ribbons, or flakes by using methods such as evaporation and vapor deposition, ion mixing or ion implantation, and liquid quenching. The vapor deposition technique is used in semiconductor type sensors. They can be classified as amorphous alloys (Fe, Co, Ni), amorphous rare earths, and amorphous superconductors. Amorphous alloys have good soft magnetic material properties and are extensively used in magnetic heads as sensors. They have high-saturation magnetization, high permeability, and flat high-frequency dependence. Amorphous alloys are produced in the form of 20- to 50-μm-thick ribbons using rapid solidification methods. The thermal stability of these alloys is a major drawback, thus preventing wider applications of these materials.

On the other hand, amorphous superconductors are a class of superconducting materials such as bismuth and gallium. These materials are used in high-field magnets, memory devices, and some computer applications. Many of the amorphous metallic films are based on rare earth three-dimensional transition metal alloys, such as Gd_xCo_x They are extensively used in bubble domain devices.

2.4.5 Intelligent Sensors

Intelligent sensors are mainly based on silicon technology. They contain sensors, signal processors, and intelligence capabilities in a single chip. They are appearing in the marketplace as pressure sensors and accelerometers, biosensors, chemical sensors, optical sensors, magnetic sensors, etc. Some of these sensors are manufactured with the neural network and other intelligence techniques on board the chip. Intelligent vision systems and parallel processors-based sensors are typical examples of such devices.

The NC3002 is an example of such a sensor, which is based on the digital very large scale integration (VLSI) parallel processing technique. These types of sensors find applications in machine leaning and image recognition supported by artificial neural networks in quality determination and inspection applications. The architecture of NC3002 is structured in a way to implement the Reactive Tabu Search learning algorithm, a competitive alternative to back-propagation. This algorithm does not require derivatives of the transfer functions. The chip is suitable to act as a fast parallel number-crunching engine intended for operation with a standard central processing unit (CPU) in single-chip or multiple-chip configurations.

Another example is the intelligent image sensors. These sensors are based on monolithic CMOS technology and contain on-chip A/D converters and appropriate microprocessor interface circuits. These types of sensors constitute an important part of digital cameras. They incorporate sensors, analog signal conditioning circuits, and memory elements on-chip. In digital camera applications, they are designed to operate in direct connection with a microprocessor bus, eliminating the need for a frame grabber. As an example of such image sensors, the NC1002 has the following specifications:

- A 256 × 256 pixel resolution, square pixels, a noninterlaced operation, and an array size of 5.12 × 5.12 mm
- On-chip charge amplifiers and successive approximation A/D converters with 8-bit resolution
- Control logic and registers, general-purpose digital interface to microprocessor bus
- Single-line digital memory for synchronization with processor clock
- Sensitivity adjustment circuits to enable operation under diverse illumination conditions
- Acquisition rate of up to 25 frames per second
- Single 5-V power supply, low power consumption, and <10 mA at 5-V 44-pin package

Naturally, many intelligent sensors are finding wide applications in electronic portable instruments, such as the digital cameras in the consumer market. Every day, new and novel portable instruments are appearing in the marketplace. In addition to the availability of technology, these types of portable devices are gaining wider acceptance by consumers, which encourages the manufacturers to supply with a degree of confidence.

A new emerging technology that is expected to find wide applications in electronic portable instruments is the system-on-chip (SoC) technology. Integrated circuit technology allows the design of complex systems on a single chip that incorporates high-performance analog subsystems such as operational amplifiers (op-amps) and data converters on the same die with the digital circuits.

Further information on the use of intelligent sensors is given in Section 3.5, "IEEE-1451 Standards for Sensors and Actuators: Smart Transducer Interface."

2.5 Acoustic Sensors

Vibrations in air, liquid, or solids are generally measured by accelerometers. Some of the accelerometers make use of piezoelectric sensors. A common accelerometer used in electronic portable instruments is the piezoelectric type. A piezoelectric sensor is a ceramic or quartz crystal sandwiched by two electrodes in either side. Slight deformation on the crystal due to vibration causes a signal to appear at the terminals. Their response is restricted for high frequencies, and the output of the sensor is best near the resonant frequency.

Sound is a form of vibration. Like the other devices, sound vibrations are commonly sensed by piezoelectric sensors. However, there are many other acoustic sensors based on capacitive or inductive properties. For example, moving-coil inductive sensors give good responses at low frequencies compared to piezoelectric types. Apart from the vibration measurements, sound has unique properties that are explained below.

Sound is defined as the vibrations of a solid, liquid, or gaseous medium in the frequency range of 20 Hz to 20 kHz, which can be detected by the human ear. Sound travels in a media by obeying the laws of shear and longitudinal forces. In contrast to solids, liquid or gaseous media cannot transmit shear forces; therefore, in these media, the sound waves are always longitudinal (that is, the particles move in the direction of the propagation of the wave). As the sound waves travel in medium, the medium compresses or expands; hence, its volume changes from V to $V - \Delta V$. As a result, the *bulk modulus* of elasticity of a medium, which is the ratio of change in pressure Δp relative to change in volume, can be written as

$$B = \frac{\Delta p}{\Delta V / V} = \rho_0 v^2 \qquad (2.19)$$

where ρ_0 is the density outside the compression zone and v is the speed of sound in the medium.

In practice, pressure variations of the sound waves carry useful properties and find wider applications. In this case, sound waves can be viewed as pressure waves. Traveling pressure waves can be described by

$$p = (k\rho_0 v^2 y_m)\sin(kx - \omega t) \qquad (2.20)$$

where $k = 2\pi / \lambda$ is a wave number, ω is the angular frequency, and y_m is the amplitude of the wave.

As the pressure waves travel, the pressure at any given point changes constantly. The difference between instantaneous pressure and average pressure is called *acoustic pressure*, P. The power transferred per unit area due to acoustic pressure can be expressed as

$$I = \frac{P^2}{Z} \qquad (2.21)$$

where Z is the acoustic impedance that can be determined from acoustic pressure and instantaneous velocity.

TABLE 2.3

Speed of Sound in Materials

Material	Speed of Sound (m/sec)
Aluminum	5100
Glass	4500
Copper	3700
Seawater	1510
Helium	965
Air (20°C)	343

Pressure levels are often expressed in decibels, as

$$P_L = 20 \log_{10} \frac{p}{p_0} \qquad (2.22)$$

where p_0 is the lowest pressure levels that the human ear can detect (2×10^{-5} N/m^2).

In sensing of sounds, there are three important parameters: velocity, frequency, and wavelength. The frequency is determined by the source, but the velocity and the wavelength are determined by the medium of propagation. The velocity of sound depends on the density and elastic constants of the material. Therefore, velocity is high in the dense solids and relatively lower in the gaseous environment. Table 2.3 shows the speed of sound in different materials.

Due to lack of response or overresponse of the human ear to time-varying sound signals and impulses, the development of sound level meters has been necessary. Sound meters have four distinct parts: microphones and preamplifiers, an A-type weighting filter, root mean square (rms) detectors, and displays. There are a number of different weighting filters available, such as B-weighting, C-weighting, and D-weighting filters. These are specialized equipment that give better correlations with subjective responses in specific applications, such as determination of aircraft noise.

In practice, the sound pressure is almost always the only parameter measured directly. All other parameters, such as sound power, particle velocity, reverberation time, directivity, etc., are derived from the pressure measurements. The sound pressure measurements are performed by microphones in gaseous media and hydrophones in liquid media. There are many types of microphones that can be listed for gaseous media application, for example:

- Capacitor microphones
- Piezoelectric and electret microphones
- Fiber-optic microphones
- Carbon microphones

- Moving-iron (variable-reluctance) microphones
- Moving-coil microphones

The characteristics of a microphone are analyzed for both acoustical and electrical properties. The overall sensitivity is expressed in millivolts or microvolts per unit intensity of the sound wave. Microphones can be directional or omni-directional. If a microphone responds to the sounds coming from any direction, it is said to be omni-directional.

The most commonly used microphones for sound measurements are capacitor, piezoelectric, electret, and fiber-optic microphones. They tend to have good stability, be sensitive and precise, and possess excellent temperature properties. Other types (carbon, moving-iron, and moving-coil) still find some limited applications, but they will not be discussed in this book.

2.5.1 Capacitor Microphones

A capacitor microphone, also known as a condenser, shown in Figure 2.11, consists of five elements: diaphragm, backplate, insulator, protection grid, and casing. The diaphragm and the backplate form the parallel plates of the capacitor. The capacitor is polarized with a charge from an external power supply. When the sound pressure fluctuates, the distance between the diaphragm and the backplate changes in response to pressure variations on the diaphragm, thus changing the capacitance. The acoustic performance of the capacitor microphone is determined by the physical dimensions of the capacitor arrangement, the stiffness and mass of the diaphragm, and the internal volume of the microphone casing.

Capacitor type microphones are linear devices. They can provide good-quality audio signals without the need for elaborate constructional techniques. Some capacitor microphones use a light film of pyroelectric material

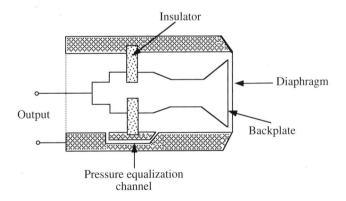

FIGURE 2.11
A capacitance type microphone.

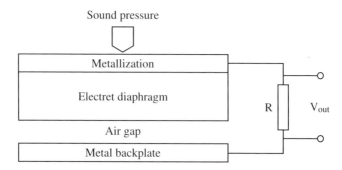

FIGURE 2.12
An electret microphone.

whose polarization changes with strain. Using pyroelectric material for polarization creates high-quality performance. The metallization on one side of the film leads to simple and cost-effective microphones.

Today, many capacitive microphones are fabricated with a silicon diaphragm for converting acoustic pressure to displacement and acting as the moving plate of the capacitor. In these microphones, a large bias voltage (20 to 200 V) provides large static deflection of the diaphragm, hence giving good sensitivity.

2.5.2   Piezoelectric and Electret Microphones

Many microphones use piezoelectric materials such as lead zirconate titanate. They have very high impedance levels and very broad operating frequencies. They are suitable for ultrasonic applications.

The operation principles of electret microphones are similar to those of piezoelectric microphones. The principles of electrets also are similar to those of piezoelectric and pyroelectric materials. Electret is permanently polarized crystalline dielectric material. An electret microphone consists of a metallized diaphragm and backplane separated from the diaphragm by an air gap, as shown in Figure 2.12.

Electret microphones do not require a DC bias voltage. The tension of the mechanical membrane is kept low to enable speedy restoration of the membrane. The membrane is usually fabricated by Teflon®, which has been charged permanently for electret properties. These microphones have an excellent frequency range, from 10^{-3} Hz to hundreds of megahertz. They have low harmonic distortion, good impulse response, and excellent sensitivity.

2.5.3   Fiber-Optic Microphones

Fiber-optic microphones make use of the reflection of laser beams from a vibrating diaphragm. The diaphragm vibrates in response to sound pressure, and the interference between incident light and the reflected light beams is sensed by a suitable interferometer (e.g., Michelson interferometer).

Optical microphones are suitable to be used in hostile environments such as monitoring the noise in rocket and jet engines. In some applications, extreme measures may need to be taken for good sensitivity by cooling the temperature-sensitive components by water or other suitable fluids.

2.5.4 Ultrasonic Sensors

Ultrasonic refers to acoustic waves of frequencies higher than 20,000 Hz, and they find many medical and industrial applications. Ultrasonic is generated naturally or by transducers, which are available over a wide range of frequencies, sizes, and shapes. An ultrasonic detection system is illustrated in Figure 2.13. A common method of transduction is piezoelectric material, but in some applications magnetorestrictive materials can be used too. Both piezolectric and magnetorestrictive transducers can function as transmitters and receivers at the same time. They can transmit and receive signals through air, solids, or liquids.

Some ultrasonic sensors are made from magnetoresistive materials that exhibit magnetorestriction properties. Magnetorestriction is the change of dimensions of magnetic materials as they are magnetized and demagnetized. They consist of a magnetorestrictive metal core inside a coil. The electric waveform is applied to the coil that causes the core to vibrate due to magnetic induction. Several types of nickel alloys or alfenol are strongly magnetorestrictive and have been used in transducers for the lower ultrasonic frequencies, in the range of 20 to 100 kHz, particularly in underwater applications.

Piezoelectric transducers have a much wider range of applications. Piezoelectric materials exhibit a mechanical strain in the form of relative deformation by the presence of an electric field. Conversely, they generate an electric field when subjected to a mechanical stress.

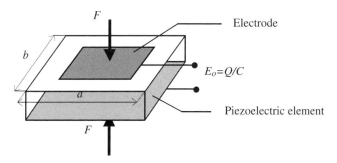

FIGURE 2.13
A piezoelectric ultrasonic detection system.

Ultrasonic sensors in acoustic applications are commonly used in the following areas:

- Detection of defects
- Fluid and gas flow measurements
- Ultrasonic cleaner
- Security devices
- Elastic properties of solids
- Porosity and size estimation
- Acoustic microscopy
- Medical applications
- Motion, displacement, and range measurements
- Thickness measurements in process control, etc.

2.6   Temperature and Heat Sensors

2.6.1   Temperature Sensors

Temperature measuring devices constitute an important part of electronic portable instruments. Temperature is a measure of intensity of heat. The simplest and most widely used method of measuring temperature is with thermometers which make use of thermal expansion of materials such as liquid in glass. In electronic temperature measurements, there are many different methods and sensors available, such as resistive sensors, thermoelectric sensors, thermocouples, thermostats, IC temperature sensors, piezoelectric sensors, and fusible links.

The basic principle of temperature measurement is heat detection. Heat is a form of energy (thermal energy), and it is measured in joules, named after James Prescott Joule (1818–1889). The quantity of heat contained in an object cannot easily be measured, but the changes in the heat content can be measured as the temperature. In temperature measurements, a small portion of thermal energy of the object is transmitted to the sensors for conversion of that energy to an electrical signal. The operational principles of temperature sensors depend on the changes that take place in the materials containing the heat. The method of sensing can be conductive, convective, or radiative.

The techniques involved in measurement of thermal energy can be subclassified as:

- Temperature measurements
  - Bimetallic thermometers

- Liquid or gas expansion methods
- Resistive temperature sensors
- Thermistors
- Thermocouples
- Semiconductor junction devices
- Infrared sensors
- Pyroelectric sensors
- Optical temperature sensors
  - Fiber-optic thermometers
  - Fluoroptic sensors
  - Interferometric sensors
- Acoustic temperature sensors
- Thermal conductivity measurements
- Heat flux measurements
- Calorimetric measurements

Temperature measurements involve two basic types, *contact* and *noncontact*. In contact measurements, the sensing element makes contact with the object whose measurement is made. The measurement is regarded to be complete once there is no thermal gradient between the object and the sensor. The amount of transferred heat is proportional to a temperature gradient between the instantaneous temperature, T, of the sensor and the temperature of the object, T_o. The heat transfer or the absorbed heat, dQ, in a small duration of time, dt, can be expressed as

$$dQ = aA(T_o - T)dt \qquad (2.23)$$

where a is the thermal conductivity of the sensor–object interface and A is the surface where that heat transmission is taking place.

Ignoring the heat loss of the sensor to the environment and support structure, the heat absorbed, dQ, by the sensor may be written as

$$mcdT = aA(T_o - T)dt \qquad (2.24)$$

where c is the specific heat of the sensor and m is the mass that absorbs the heat.

The differential equation (Equation 2.24) can be expressed as

$$T = T_o - Ke^{-t/\tau} \qquad (2.25)$$

where K is a constant and τ is the thermal time constant ($\tau = mc/aA$).

Noncontact temperature sensors make use of thermal radiation. Noncontact sensors are also known as thermal detectors such as thermopiles, pyroelectric sensors, active infrared sensors, and so on.

In portable instruments, the temperature sensors are used for two different purposes. The first is for sensing of the temperature in various parts of the instrument for protection purposes, and the second is for measuring the temperature of a physical process. Temperature measurements can involve mechanical and electrical methods. The topic of temperature measurements is vast in literature, and due to space limitations in this book, only the electronic sensors will be discussed.

2.6.1.1 *Thermocouples*

Thermocouples are contact type sensors. They are made from bonding of two different metals in the form of two symmetrical junctions, as shown in Figure 2.14. One of the junctions is kept at a constant temperature (e.g., room temperature) to act as a reference point. The output voltage of a thermocouple pair is in the order of tens of millivolts and usually fed into a FET amplifier. The most popular thermocouples used in portable instruments are copper–constantan (–190 to 400°C), iron–constantan (–190 to 870°C), chromel–alumel (–200 to 1000°C), and platinum–rhodium (–0 to 1760°C).

In general, an electric potential is always produced at the junctions of two dissimilar metals bonded together. The junction potential usually exhibits temperature dependence with magnitudes varying in accordance to the characteristics of metals selected. The potential difference between the junctions, known as the *Thomson effect*, is described by

$$E_j = \alpha_1(T_1 - T_2) + \alpha_2(T_1^2 - T_2^2) \qquad (2.26)$$

where E_j is the voltage (in μV) across the junctions, α_1 is the *Seebeck coefficient*, α_2 is the *Thomson coefficient*, and T_1 and T_2 are the absolute junction temperatures in Kelvin.

The coefficients α_1 and α_2 are dependent on thermoelectric and bulk properties of the materials. They also are independent of homogeneous length

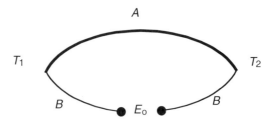

FIGURE 2.14
A thermocouple arrangement.

TABLE 2.4

Types of Thermocouples

Type	Range, °C	ρ (μΩ-cm)	Typical emf, mV	Common Name
B (BX, BP, BN)	870–1700	16–34	0.78	Pt30Rh, Pt6Rh
E (EX, EP, EN)	−200 to 870	46–127	29.0	Chromel–constantan
J (JX, JP, JN)	0–760	10–46	22.0	Iron–constantan
K (KX, KP, KN)	−200 to 1260	31–112	16.4	Chromel–alumel
N (NX, NP, NN)	0–1260		13.0	Nisil–nicrosil
R (RX, RP, RN)	0–1480	10–29	3.4	Pt13Rh, Pt
S (SX, SP, SN)	0–1480	10–30	3.3	Pt10Rh, Pt
T (TX, TP, TN)	−200 to 370	2–48	20.8	Copper–constantan

and shapes of materials that are forming the thermocouple; hence, they have constant values. Depending on the materials used, Equation 2.26 can be expanded as a third-order or higher-order equation as a function of temperature gradient of the two dissimilar materials.

The component of the voltage proportional to junction temperature [$\alpha_1 (T_1 - T_2)$] is known as the *Peltier effect*. The Peltier effect is considered to be localized at the junctions. The component of the voltage proportional to the square of the junction temperature [$\alpha_2(T_1^2 - T_2^2)$] is the *Thomson effect*, and it is considered to be distributed along each conductor between the junctions.

For a single junction, the contact potential is not measurable. A voltage can be detected when junctions are subjected to different temperatures. A two-junction arrangement is illustrated in Figure 2.14. By some suitable physical arrangements and proper electronic compensation techniques, the temperature of the object can be measured linearly.

There are many different types of thermocouples that can be made from different materials, as shown in Table 2.4. Thermocouples made from copper–constantan are used mainly for lower-range temperatures, while the platinum–rhodium types are used in the higher ranges. Conditioning of the output signals requires considerable attention, as the signals generated can be extremely small and noisy. Usually, chopper amplifiers are used to amplify the signals associated with thermocouples. Their lag time is small, and they respond rapidly to temperature changes. In some applications, cold junction compensations and zero settings may be necessary. An advantage of thermocouples is that the sensing elements are small, allowing them to be used in confined spaces.

2.6.1.2 *Thermoresistive Sensors*

Resistance thermometers are based on the changes in electrical resistance with respect to temperature. Platinum is a commonly used element because of its good corrosive resistance properties. Generally, thermoresistive sensors require simple signal processing circuits. They have good sensitivity and stability. These sensors can be divided into three main groups:

1. Resistance temperature detectors (RTDs)
2. Silicon resistive sensors, also known as *pn*-junction sensors
3. Thermistors

RTDs are fabricated either in the form of wires or in thin films. The resistivities of virtually all metals are dependent on temperature, while some metals and alloys are most exclusive because of their predictable responses, stability, and durability. Platinum and its alloys are typical examples of such metals. Another popular metal is tungsten, which is used at high temperatures, over 600°C.

In general, all metallic conductors exhibit changes in their resistivity when subjected to variations in temperatures. The resulting change in the resistance of conductors can be expressed by

$$R_\theta = R_0 \left(1 + \alpha\, \theta\right) \tag{2.27}$$

where R_θ is resistance at temperature θ 0°C, R_0 is resistance at 0°C, and α is the temperature coefficient of resistance at 0°C.

For most metals, the coefficient α is close to 0.00366. Nickel or nickel alloys are selected in low-temperature measurements of up to 400°C due to their greater resistance to oxidation. Their variations in the resistances exhibit linear characteristics against temperature. Nevertheless, the platinum resistance thermometers exhibit many advantages compared to other alternatives. They are suitable in high-temperature applications and have very low drifts. They are noncorrosive and have excellent aging properties. Also, their resistance–temperature relationship is linear for much wider temperature ranges.

In many instruments, the RTDs are configured usually as arm(s) of a suitable bridge circuit.

2.6.1.3   Silicon Resistive Sensors

Silicon resistive sensors make use of conductive properties of bulk silicon. Pure silicon intrinsically has a negative temperature coefficient (NTC). That is, resistance decreases as temperature increases. However, when doped with *n*-type impurity, its temperature coefficient becomes positive at a certain temperature range. Therefore, as one would expect, resistance vs. temperature characteristics of silicon resistive sensors is highly nonlinear. However, in some ranges, the characteristics transfer function can be approximated by a second-order polynomial as

$$R_T = R_0 \left[1 + A(T - T_0) + B(T - T_0)^2\right] \tag{2.28}$$

where R_0 and T_0 are resistance and temperature (°K), respectively, at a reference point. Silicon resistive sensors are used inside IC chips to detect internal temperatures.

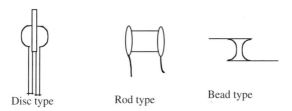

Disc type   Rod type   Bead type

FIGURE 2.15
Various types of thermistors.

2.6.1.4 Thermistors

Thermistors are based on semiconductors and often used as a part of transistor switches or an operational amplifier inside the chips. Thermistors enjoy wide applications in electronic portable instruments.

Thermistors are a mixture of semimetal oxides (rare earths), and they can be classified in three major groups: *bead type* thermistors, *chip* thermistors, and *semiconductive* thermistors. Each one of these types is manufactured in various sizes and shapes, as illustrated in Figure 2.15.

Termistors can be made from NTC materials or positive temperature coefficient (PTC) materials, where resistance increases as temperature increases. Most metal oxide thermistors have a negative temperature coefficient. The physical dimensions and the resistivity of the material determine the value of resistance. The relation between temperature and resistance can be highly nonlinear, such as an exponential function. This relationship can be expressed as

$$R_T = R_0 e^{\beta\left(\frac{1}{T}-\frac{1}{T_0}\right)} \tag{2.29}$$

where R_T is the resistance at temperature T in $°K$ and β is the characteristic temperature coefficient of the thermistor.

Positive temperature coefficient thermistors are generally made from polycrystalline ceramic materials such as barium titanate and strontium titanate. PTC thermistors are capable of maintaining their temperature coefficients constant.

Thermistors are a good choice to measure temperatures particularly in applications at near-normal ambient temperatures. They are available in different sensitivities. The resistance values of thermistors are usually much higher than those of platinum thermometers. Although the output of a thermistor is nonlinear, the sensitivity can be very high and its response rapid.

2.6.1.5 Semiconductor Temperature Sensors

The *pn*-junction diode and bipolar transistors exhibit a strong thermal dependence. For example, the current-to-voltage equation of a diode can be expressed as

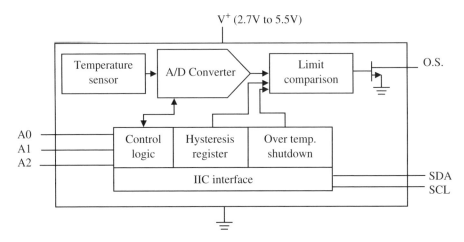

FIGURE 2.16
A typical IC temperature sensor.

$$I = I_0 e^{\left(\frac{qV1}{2kT}\right)} \qquad (2.30)$$

where I_0 is the saturation current, q is the charge of an electron, k is the Boltzman constant, and T is the temperature in °K.

In principle, the transistors themselves can be used as sensing elements, since they exhibit exponential negative temperature characteristics. Indeed, inherent amplification properties of transistors make them very attractive for temperature measurements. In particular, the use of temperature sensitivity of the base-emitter voltage becomes a good proposal since the output can be taken in amplified form from the collector. However, this needs extensive compensation due to the high degree of nonlinearity. Therefore, the use of transistors as temperature sensors is confined mainly to ICs. A typical transistor-based IC temperature sensor is illustrated in Figure 2.16.

In electronic portable instruments, semiconductors in the IC form with good temperature properties are often used. Most of these sensors come with the associated signal processing circuits that are integrated in the same IC chip. Some of these sensors, such as the LX5600, can be used to measure temperatures as well as other physical properties, such as air speed, liquid level, and so on.

2.6.2 Heat Sensors

Heat can be viewed as a form of radiated energy. Radiant energy can be in many forms, such as heat, light, radio waves, or ionizing radiation. Therefore, the operation principles and manufacturing techniques of sensors for radiant energy are different for different types of radiations. For example, *blackbody radiation* is sensed by bolometers; light is sensed by photometers; radio waves are sensed by antenna; and particles are measured by various types of radiation detectors.

Any object whose temperature is greater than absolute zero radiates energy, termed *blackbody radiation*. The spectrum of radiated energy at each detectable frequency is dependent on the temperature of the object. The radiation usually takes place over a wide band of frequencies, with a peak at some frequency. At low temperatures, radiation is predominantly in the far infrared region; as the temperature rises, the spectrum shifts toward higher frequencies and the absolute amount of energy at each frequency increases.

Thermal detectors are used for noncontact heat and temperature sensing. Thermal detectors known as *pyrometers* are based on the conversion of thermal radiation into heat, and then converting the heat into electrical signals. All thermal radiation detectors can be divided into two groups: *passive* detectors and *active* detectors. Passive detectors absorb incoming radiation and convert it into heat. Active detectors, on the other hand, emit thermal radiation directed toward the object under investigation.

Four different types of thermal sensors will be discussed here:

1. Pyroelectric sensors
2. Thermopile sensors
3. Active infrared sensors
4. Bolometers

2.6.2.1 Pyroelectric Sensors

Pyroelectric sensors are made from pyroelectric materials. These are crystals that are capable of generating charge in response to heat flow. Pyroelectric sensors consist of thin slices or films with electrodes deposited on the opposite sides to collect the thermally induced charges. Most common pyroelectric materials are lead zirconate titanate, triglycine sulfate, lithium tantalite, and polyvinyl fluoride. Many commercial pyroelectric sensors are made from single crystals, such as triglycine sulfate, PZT ceramics, or $LiTaO_3$. The polyvinyl fluoride-based sensors exhibit high-speed responses and good lateral resolutions.

Physically, there are several mechanisms that change in heat flow (e.g., temperature), resulting in pyroelectricity. Heat flow affects the randomness and shapes of dipoles in the crystal structure, known as the *primary pyorelectricity*. The *secondary* effect is the piezoelectric properties; that is, the structure is strained due to thermal expansion of the material. This can be viewed as thermally induced stress or heat absorption. Taking the second option, the thermal absorption can be expressed as

$$\Delta Q = A\mu \, (T, \Delta W) \tag{2.31}$$

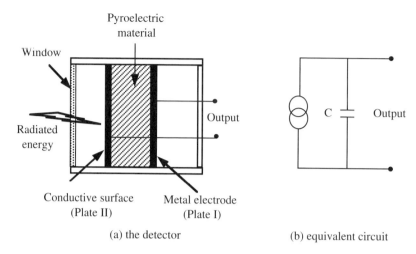

FIGURE 2.17
(a) A pyroelectric detector and (b) its equivalent circuit.

where μ is the dipole moment per unit volume, A is the sensor area, T is the temperature, and ΔW is the incremental thermal energy absorbed by the crystal.

Pyroelectric sensors are manufactured by having metallized surfaces and a protective window, as shown in Figure 2.17(a). In the case of infrared (IR) detectors, the thin-slice films are made from the dielectric materials whose surfaces are electrically charged by the incoming IR radiation. Therefore, these detectors are primarily capacitive types with very high impedances. The equivalent circuit of an IR detector is illustrated in Figure 2.17(b). Although many plastic films are suitable, in most applications, a lithium tantalate film forms the basic material of passive IR detectors. These detectors are constructed just like capacitors. A metal plate forms one of the electrodes. The other side of the pyroelectric material contains a conductive material, thus acting as the second plate. The voltages across the plates alter since the IR radiated energy alters the charge of the pyroelectric material. In practical applications, most pyroelectric detectors and signal processing circuits are integrated in a MOSFET structure.

2.6.2.2   Thermopile Sensors

Thermopiles are passive infrared (PIR) sensors. They operate on the same principles as thermocouples. That is, they have a number of serially connected hot junctions, shown in Figure 2.18, with improved absorptivity and cold junction that give the temperature difference between the two. The sensor is hermetically sealed in a metal can with a hard infrared transparent window such as silicon, germanium, or zinc selenide. The operating frequency is determined by the thermal capacity, and thermal conductivity of the membrane is sandwiched between the hot junctions and cold junction. These sensors have excellent noise properties.

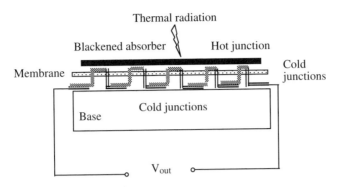

FIGURE 2.18
A thermopile.

2.6.2.3 Active Infrared Sensors

The temperature properties of active infrared sensors are dependent on the ambient temperature and temperature of the object. Therefore, their surface temperatures are to be maintained constant during the sensing process. The temperature control is maintained either by a cooling process or by a heating process using an external electric power. In both cases, a feedback mechanism is necessary to monitor the surface temperatures. In many cases, self-balancing bridge circuits with voltage-controlled current sinks are used. In the operations, an active infrared sensor acts as an infinite heat source in response to the temperature of the object. A heat flux is generated due to the difference in the ambient temperature and the temperature of the object. This heat flux is sensed and processed suitably.

2.6.2.4 Bolometers

The blackbody radiations at low frequencies are sensed by bolometers. Basically, they are blackened materials that absorb the radiation quite well. As a result, the temperature of the materials rises. Modern bolometers are almost exclusively made from semiconductors. The surface of the semiconductor arrangement is blackened for the maximum absorption of the radiated energy that falls onto it.

2.7 Light Sensors

Light sensors can be divided into two major groups: *quantum* sensors and *thermal* sensors. Within each group, there are many different types, some of which are commonly used in portable instruments. Photoconductive sensors and photovoltaic sensors operate on quantum principles. Phototransistors, charge-coupled sensors, and photoemissive sensors are a few others of many that operate on thermal principles. These sensors are available in many different configurations, some built with powerful lenses and high-speed responses to monochromatic light, and others having different color sensitivities to chromatic light.

Light, basically, is an electromagnetic radiation that consists of an electric field and a magnetic field component. Compared to radio waves, light has a short wavelength, and hence a very high frequency. Many different properties of light are used for instrumentation and measurement purposes. Applications vary from photographic imaging to high-speed data transmission via fibers. Once the light is generated and propagated from a source, it can be expanded, condensed, collimated, reflected, polarized, filtered, diffused, absorbed, refracted, and scattered to develop sensors and measurement systems. Some of these properties of light are manipulated on purpose to serve a particular application need, and sometimes manipulations are not necessary since they happen naturally due to optical properties of the media and physical characteristics of light.

One of the properties of light is polarization, as in the case of any other radiated wave. Normal unpolarized light consists of waves whose direction of oscillation can be in any plane at right angles to the direction of the motion. When such light is passed through a polarizing material, the light can become plane polarized. Materials that polarize light in this way are generally crystals, which contain lines of atoms aligned at critical spacing.

Like the other waves in the electromagnetic range, light has a velocity, c, in space (3×10^8 m/sec) so that the relationship between frequency, f, and wavelength, λ, can be written as

$$c = \lambda \times f \tag{2.32}$$

Also, the energy is carried by photons. The energy of a single photon is given by

$$E = h \times f \tag{2.33}$$

where h is Planck's constant ($h = 6.63 \times 10^{-34}$ J or 4.13×10^{-15} eV·sec).

When a photon strikes a surface of a conductor, part of the photon energy, E, is used to generate some free electrons in the conductor. This is known as the photoelectric effect and can be described by

$$h \times f = \varphi + K_m \tag{2.34}$$

where φ is called the work function of the emitting surface and K_m is the maximum kinetic energy attained by the electron.

When light travels through transparent materials, the velocity is reduced by a factor known as the refractive index. For visible light, values of the refractive index range just above unity to about 2 in gemstones and diamonds.

The refractive index for infrared radiation can have greater values, and materials that are opaque to visible light may be quite transparent to infrared.

Most light detectors are based on two different technologies, photon detectors or thermal detectors (explained below). Also, the sensitivity of light detectors can be quoted two ways: (1) as the output at a given illumination in lux, or (2) as a figure of power (W/cm²) falling on the cell per square centimeter, a quantity known as irradiance (1 mW/cm² = 200 lx).

2.7.1 Photosensors

Photosensors are essentially photon detectors that are sensitive to photons with energies greater than a given intrinsic gap energy of the sensor. Some of the most common photon detectors make use of properties of cadmium and lead, such as cadmium sulfide (CdS), cadmium selenide (CdSe), lead sulfide (PbS), and lead selenide (PbSe) devices. Their response times can be less than 5 µsec. The main classes of photosenors are photodiodes/phototransistors, photovoltaic cells, and photoresistors/photoconductors.

2.7.1.1 *Photodiodes/Phototransistors*

Photodiodes and phototransistors are semiconductor devices that are sensitive to light. In a photodiode, as the incident light falls on a reverse-biased *pn*-junction, the photonic energy carried by the light creates an electron–hole pair on both sides of the junction, causing a current to flow in a closed circuit, as illustrated in Figure 2.19. The output voltage of photodiodes may be highly nonlinear, thus requiring suitable linearization and amplification circuits.

The photodiodes are constructed like any other diodes, but without the opaque coating, so that light can fall onto the junction. The current generated at the junction can be linearly related to the amount of illumination falling on the junction. The response time of photodiode sensors is very short, typically 250 nsec. The short response time of photodiodes allows these devices to be used in modulated light beam applications.

Wide ranges of photodiodes are available, such as *pn*-photodiodes, *PIN* photodiodes, Schottky photodiodes, and avalanche photodiodes.

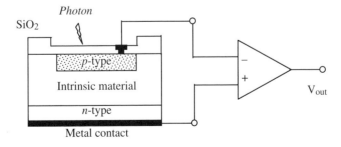

FIGURE 2.19
A typical structure of a photodiode.

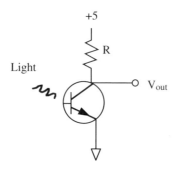

FIGURE 2.20
A phototransistor for sensing light.

Phototransistors are the most used light sensors. In addition to converting photons into charge carriers, they have current gain properties. Particularly, Darlington phototransistors possess high sensitivity and high current gain characteristics.

A phototransistor, Figure 2.20, is a form of transistor in which the base-emitter junction is sealed with a transparent package. As in the case of photodiodes, this junction is affected by incident light and acts as a diode. The light at the junction promotes electrons from valence to conduction band. The current in this junction is then amplified by the normal transistor action to provide a much larger collector current. The spectral responses of these devices are limited to red and near-infrared portions of the spectrum. The response of the phototransistor can be adjusted by applying an appropriate bias voltage to the base. The response times of phototransistors are much larger than those of photodiodes, in microseconds. Phototransistors are manufactured in IC forms, containing light sensors integrated with amplifiers and other processing circuits.

2.7.1.2   Photovoltaic Cells

Photovoltaic sensors are based on silicon cells that generate voltages in response to a beam of light. The response time of these sensors is slow, and it takes a long time to stabilize the output (up to 20 sec). The wavelength response of photovoltaic silicon cells covers the whole visible spectrum, which makes them very useful for environmental light measurements.

The first form of photovoltaic device was the selenium cell, which produced voltage across the cell in proportion to the illumination. Modern photovoltaic devices are constructed from silicon, thus acting as silicon photodiodes with large-area junctions and used without bias, as in Figure 2.21. These devices find applications in camera exposure controls and control of light levels in manufacturing processes because with some suitable filtering, the response curves can be made very similar to the response of the human eye to light.

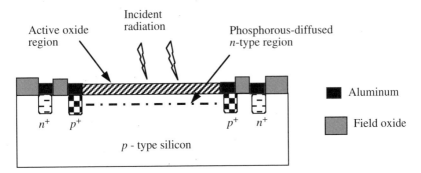

FIGURE 2.21
An Si photodiode.

2.7.1.3 Photoresistors/Photoconductors

The resistance of many materials changes when exposed to light. The materials are known as light-dependent resistors (LDRs), such as CdS and CdSe. In photoresistors, cadmium sulfide, for example, is deposited as a thread pattern on an insulator in a zigzag form. The cell is then encapsulated in a transparent resin or encased in a glass to protect the cadmium sulfide from contamination from the atmosphere. A typical sensor has a spectral response of 610 nm, corresponding to a color in the yellow–orange region; the dark resistance is 10 MΩ, and the cell resistance typically decreases from 2.4 kΩ at 50 lx to 130 Ω at 1000 lx. However, changes in the resistances against illumination are not linear, thus requiring additional linearization circuitry.

The response time of photoconductors is fairly slow, and the resistance changes in response to light might take a few seconds, making these sensors unsuitable in applications where fast responses are required.

2.7.2 Solid-State Light Emitters

Although they are not sensors, at this point it is worth introducing light-emitting diodes (LEDs). LEDs are solid-state transducers that are used extensively in electronic portable instruments. LEDs are diodes in which voltage drop for conduction is comparatively large. In a junction of compound semiconductors such as gallium arsenide, the combined action of electrons and holes releases energy, which can be radiated if the junction is transparent. The majority of LEDs in use are of gallium phosphide or gallium arsenide phosphide construction, and all are diodes with high forward voltages. The materials determine the colors of LEDs, and the predominant colors are red and green. LEDs are obtainable in a considerable variety of physical forms and sizes. The standard dot and bar types are predominant. Bar-shaped LEDs are used extensively for alphanumeric displays. However, excessive power consumption (typically 15 to 20 mA) of LEDs makes them not so attractive for some portable instruments.

2.7.3   Image Sensors

An image sensor is a device that transforms an optical image into electrical signals. A standard image sensor converts the light photons falling onto the image plane into the corresponding spatial distribution of electric charges. The accumulated charge at the point of generation is transferred suitably and converted to usable voltage signals. Each of these functions can be accomplished by a variety of approaches, such as vacuum tube technology and solid-state technology. Solid-state sensors are primarily based on photodiodes and charge-coupled devices.

The simplest image sensors are the parallel types, in which a matrix of photodiodes is used in small-scale applications such as optical character recognition. A photodiode consists of a thin surface region of *p*-type silicon formed on an *n*-type silicon substrate. A negative voltage applied to a surface electrode reverses the biases of the *pn* junction. This creates a depletion region in the *n* silicon, which contains only immobile positive charge. Light penetrating into the depletion region creates electron–hole pairs, which discharge the capacitor linearly in time. Solid-state image sensors are considerably complex and made in IC forms, and there are two basic types, the serially switched photodiode arrays, as in Figure 2.22, and the charge-coupled photodiode arrays. In these arrays, the basic principle is to use light intensity to charge a capacitor and then to read the capacitor voltage by shifting it through the register.

The most developed example of such devices is the charge-coupled device. The CCD is a form of MOS device in which a clock pulse can pass a charge from one plate to another, similar to analog shift register. In these devices, small areas of silicon are exposed to the light, thus causing a local charge, called a packet, to be generated. The synchronized switching of electrodes on the surface of the CCD, the packets are shifted in a serial fashion to an

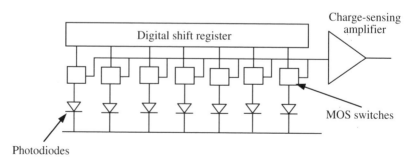

FIGURE 2.22
An IC image transducer.

output device. CCDs are fundamentally linear devices; therefore, they are suitable in array operations. CCDs are used in one-dimensional or two-dimensional arrays. One-dimensional arrays are useful for scanning applications, and two-dimensional arrays are useful in low-resolution cameras.

CCDs have a long life and high reliability. Their spectral sensitivities cover most of the visible wavelength. They are small in size and immune to damage from bright lights.

2.7.4 Spectrometers

The term *spectroscopy* has a few meanings. It applies to radiation of energy of any wavelength and sometimes to the analysis of material particles (mass spectroscopy). The electromagnetic radiation emitted by matter is measured by spectrometers, also known as spectroscopes. The selection and use of appropriate spectrometers for a particular region of a spectrum is important, as they find many diverse applications in electronic portable instruments. Some of these applications are:

*   Emission and absorption of light by matter
*   Determination of atoms and molecules present in samples of matter
*   Electronic structures of atoms and molecules
*   Characteristics of terrestrial bodies such as mass, temperature, size, and velocity

2.7.4.1 Spectroscopic Instruments

A spectroscope is an optical device that separates light into its constituents to produce a spectrum. It uses prisms, diffraction gratings, or interferometers to observe the spectrum visually, record photographically, or detect by photoelectric or radiometric means in spectral regions. The wavelength of radiation can be read directly or indirectly from a dial. If the radiation is recorded photographically, it is called a spectrograph, and the photograph produced is the spectrogram.

There are three methods of wavelength separation by spectroscopes: (1) refractive dispersion by prisms made from glass, (2) diffractive dispersion by gratings (made from engraved diamonds and mirrors), and (3) interferometric dispersion, such as the Fabry–Perot, Lummer–Gehrcke plate, and Michelson echelon interferometers. They generally consist of optical parts such as slits, dispersing devices (e.g., prisms), gratings or interferometers, and focusing lenses and mirrors that concentrate monochromatic beams. Spectroscopes range in size from tripod-mounted instruments of no more than 30 cm to devices filling rooms more than 10 m long.

Photography is often used for recording spectra at wavelengths between 1 and 1200 nm. The performance can be improved by use of dyes and phosphorescence phenomena. In this method, many spectrum lines can be recorded simultaneously, and the energy in a spectral line can be integrated over the entire period of the exposure.

Generally, photoelectric cells (e.g., barrier layer and phototube types) are used for measuring spectra in the visible and ultraviolet regions. In the phototube types transparent envelopes of quartz or special ultraviolet transmitting glasses are used. Cesium hydride or other suitable metallic materials can be selected to serve as photosensitive surfaces. Photoconductive cells such as thalofide and zinc sulfide are also used for the near-infrared region. The photocell is usually held fixed, while the spectrum emerging from an exit slit is scanned by rotation of the prism or grating.

Sources of radiation are very important in the application of spectroscopy. The sources include flames and other incandescent sources, electric arcs, electric sparks, and discharge tubes. The yellow flame of a candle emits continuous radiation because most of its light comes from incandescent solid carbon particles that have been heated to about 1370°C. The incandescent sources can be solid or liquid that emits a continuous spectrum. The energy distribution of the emission depends on the temperature of the emitter: the higher the temperature, the shorter the average wavelength of the radiation. In an electric arc, the emitted light contains sharp lines that can be related to characteristics of the atoms from which the electrodes are made. The discharge tubes emit lines or bands, characteristic of the atoms or molecules of the gases where the discharge is taking place.

The wavelength of the light can be determined by the position of a line in a spectrum or on a spectrogram. Depending on the instrument, precision of measurement can be 1 part in 1000 with a small spectrometer, to 1 part in 2 million with a large diffraction grating, to 1 part in 50 million with a Fabry–Perot interferometer.

Intensities of radiation are measured by thermoelectric, photoelectric, and photographic methods or, in some cases, by visual methods. In thermoelectric methods the sensitivity is low, but they have good linearities and uniform sensitivities at various wavelengths. Although photoelectric cells are linear in response and very sensitive, they need to be standardized to determine their response at various wavelengths. Similarly, the photographic emulsion methods are sensitive too, but must also be standardized and calibrated for a reasonable response to intensity for each desired wavelength since they are highly nonlinear.

The most common application of spectroscopy is the qualitative and quantitative analysis of materials. For example, burning in an arc or spark of chemical element produces known spectrum lines that dissociate almost any sample of material into its constituent elements. When this spectrum from the sample is studied, the constituents are easily revealed. Burning a few milligrams of a sample in an electric arc can usually reveal the presence or absence of some 70 or so of the chemical elements. Sodium, potassium, the other alkali metals, and the alkaline earths are very sensitive to spectroscopic analysis, often being detectable in concentrations of 1 part in 10 million or less. Many metallic elements too can be detected in concentrations as small as 1 part in 10 million. Nonmetallic elements are somewhat less sensitive to spectroscopic detection. However, some spectroscopic methods are available for detection of sulfur, selenium, fluorine, and other halogens in amounts of less than 1 part in 100,000.

As an example of spectroscopic methods, a typical portable spectroradiometer, equipped with an on-board notebook computer and a battery pack, weighs about 12 kg, having dimensions of $45 \times 25 \times 20$ cm^3. The device contains fiber-optic connectors allowing the use of transmission and reflectance probes. The spectrometer has a continuous spectra range from 35 to 200 nm. It has a 512-element photodiode array sensor and two thermoelectrically cooled, graded index extended InGaAs photodiodes, with a linearity of ±1%.

2.7.5 Optical Fiber Sensors

In electronic portable instruments, optical fibers are used as sensors in varying forms, such as temperature sensors and optical gyroscopes. They are based on the properties of light as well as properties of optical materials that are constructed suitably for specific applications. The operation principles of the sensors vary considerably. Some of the sensors make use of scattering properties of light, while others make use of attenuations and losses.

An optical fiber is a circular dielectric glass waveguide that can efficiently transport optical energy by means of total internal reflections. It consists of a central glass core surrounded by a concentric cladding material with a slightly lower (≈1%) refractive index. Since the core has a higher index of refraction than the cladding, light is confined to the core if the angular condition of total internal reflection is met. The attenuation of the light begins at the moment light enters the fiber. The acceptance and transmission of light depends greatly upon the angle at which the light rays enter the fiber. The angle must be less than the *critical acceptance angle* Φ_{max} of the particular fiber being used, as shown in Figure 2.23.

There are three basic types of fibers: single-mode step index, multimode graded index, and multimode step index. The characteristics of optical losses in these three types of fibers vary slightly due to differences in constructions

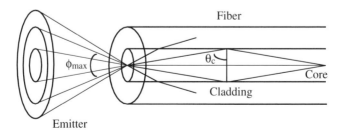

FIGURE 2.23
An optical fiber.

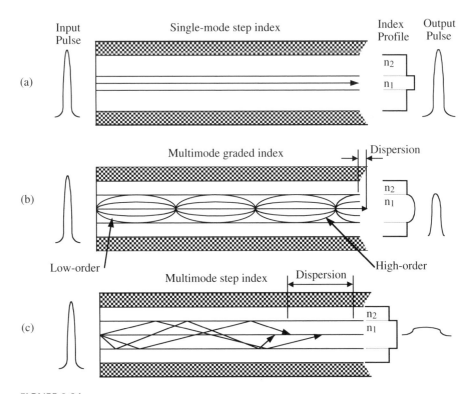

FIGURE 2.24
Types of fibers used in instruments. (a) Single-mode step index fibers. (b) Multimode graded index fibers. (c) Multimode step index fibers.

and nature of propagation of light. For example, in multimode step index fibers, rays striking the core-cladding junction at an angle higher than the angle of internal reflection pass through and are absorbed by the opaque jacket, as illustrated in Figure 2.24(c). This represents a significant source of attenuation, limiting the maximum length of such optical fibers.

Single-mode fibers (Figure 2.24(a)) are used in transmitting broadband signals over large distances. Their attenuation is generally very small, and their transmission band is very large. Owing to material properties, low attenuations can be expected for wavelengths around 1.3 to 1.6 μm. Additional attenuation arises from splices and fiber bending.

Multimode graded index fibers have medium size cores (50 to 100 μm) and a refractive index that decreases radially outward. The two optical materials with different refractive indices are mixed together in such a way that the index of refraction decreases smoothly with distance from the fiber axis. The graded index causes the light rays to gradually bend back and forth across the axis in a sinusoidal manner, as shown in Figure 2.24(b). This greatly reduces the light losses from the fiber. The smaller core size also means better bandwidth and efficiency, while the graded index allows simpler splicing techniques.

LEDs are the primary source of light in fiber-optic links for speeds up to 200 Mb/sec and distances up to 2 km. For higher speeds and longer distances, diode lasers are preferred. LEDs used in fiber-optic applications operate at three narrowly defined wavelength bands — 650, 820 to 870, and 1300 nm — as determined by optical fiber transmission characteristics.

2.7.6   Optoisolators

Other useful devices in portable instruments are the optoisolators. In this case, a light emitter and a light detector are housed in a plastic dual in-line package (DIP) case so that they can be used for electrical isolation. The incoming electric signal activates the emitter (simply a LED) to generate light with intensity proportional to the incoming electric signal. The emitted light passes through a transparent dielectric that activates a phototransistor detector. The voltage output of the phototransistor is proportional to the intensity of the incident light, hence duplicating the incoming electrical signal. The dielectric between the emitter and detector can have breakdown voltages of many kilovolts.

2.7.7   Photoemissive Sensors and Photomultipliers

Photoemissive sensors are vacuum tubes with treated metal surfaces. Light falling on the surface causes electrons to be ejected to be picked up by an anode. Photoemissive sensors are available with many different spectral sensitivities. They have low power demands but require a high voltage that makes them difficult to apply in some portable instruments. The most commonly used photoemissive device is the photomultiplier.

In photomultipliers, the emitted electrons are attracted to a nearby plate that is biased at a high voltage. The electrons are accelerated to a high energy level so that when they strike the plate, they splash a large number of electrons from it. These electrons are attracted to another plate biased at an even higher voltage than the previous one. Each stage causes amplification; therefore, the more stages the device has, the higher the resulting signal becomes. This is the reason that photomultipliers are the most useful devices for detecting extremely weak lights. But they are expensive and require voltages up to 1000 V.

2.8   Radiation Sensors

Because of the health and safety implications to humans and other living organisms, the correct sensing of radiation historically has been taken very seriously. Radiation can exist anywhere; therefore, associated instruments need to be portable. Hence, today, radiation sensors constitute an important section of electronic portable instruments.

Many radiation sensors use ionization as the basic physical principle behind the operations. Basically, there are two types of ionizing radiation: (1) particle radiation such as alpha particles, beta particles, neutrons, and cosmic rays, and (2) electromagnetic radiation such as x-rays and gamma rays.

2.8.1 Particle Radiation

The alpha particles, beta particles, neutrons, and cosmic rays are forms of particle radiation. Alpha and beta particles are emitted during the normal radioactive decay phenomenon, but the neutrons are emitted in a nuclear reaction. They are also produced when a deuteron (a proton and a neutron) and a triton (a proton and two neutrons) collide during the reaction. An intense source of neutrons results from the fission reaction in nuclear reactors. But the neutrons are highly unstable, and they change spontaneously to protons, electrons, and neutrinos.

Alpha particles also are spontaneously emitted from heavy atomic nuclei, called *alpha decay*. They are positively charged and consist of two protons and two neutrons, which is identical to the nucleus of helium-4. *Beta radiation* refers to the emission of electrons and positrons, known as beta decay. A positron is a positively charged electron. There are two kinds of beta particles: beta-minus and beta-positive. The energy of a beta particle has the same magnitude as gamma rays. Beta particles ionize atoms in a way similar to that of heavy positive particles. But they can interact with matter in excitation of electrons by displacing electrons to more energetic orbits, but not entirely removing them.

2.8.2 Electromagnetic Radiation

The x-rays and gamma rays are a form of electromagnetic radiation. The x-ray radiation is emitted when an electron drops from an outer orbit into a lower orbit. Similarly, a gamma ray is emitted when the nucleus of an atom existing in an excited state decays to a lower energy state. They both can be visualized as photons, which are a form of traveling packets of electromagnetic energy without mass or electrical charge. X-rays and gamma rays have shorter wavelengths; therefore, they carry greater energies than other forms of electromagnetic radiation.

Gamma and x-rays interact with matter by either photoelectric effect, Compton scattering process, or pair production process. In the photoelectric effect, all the photon energy is transferred to orbital electrons. In Compton scattering, the photon passes a portion of its energy to the electrons. In pair production, the gamma ray disappears and an electron and positron become emitted in its place.

Accuracy in radiation measurements is necessary since radiation is not perceivable by human senses; therefore, we are largely dependent on instrument readings. Humans are exposed continuously to naturally occurring or, from time to time, man-made radiation. The main sources of the natural radiation come from cosmic radiation, and soils and rocks on the Earth's crust, such as uranium-238, thorium-232, and potassium-40. Man-made sources of radiation are very common. Radiation is used often in medicine for diagnosis, cure, and physical examinations. Nuclear reactors generate beta, gamma, and neutron radiation originating in the fission process. Other sources of man-made radiation include detonations of nuclear weapons. In this case, some radioactive materials enter the atmosphere and are transported by the winds to be carried to other parts of the Earth. They are then brought down to the surface by rain known as fallout.

It is important to understand the units of radiation in order to make sense of the results displayed by the instruments. At first sight, a confusing number of units are used to measure radiation, but the logical arrangements of units are as follows: The radiation activity refers to the disintegration rate of a sample of radioactive material. This is measured by curie (Ci) (Marie Curie, 1867–1934), which is equal to 37 billion disintegrations per second. The Système International d'Unités for radiation activity is the becquerel (Bq) (Antoine Becquerel, 1852–1908). One becquerel is the quantity of the radioactive element in which there is one atomic disintegration per second; therefore, 1 Bq $= 2.7 \times 10^{-11}$ Ci. The term *radiation intensity* is the strength of a radiation field at a given point in space. It describes the number of radiation particles passing through a unit area per unit time. Exposure describes the strength of a gamma ray or x-ray field falling on a body at a point in space. It is measured in terms of the number of ions produced in the atmosphere adjacent to the body. Exposure is sometimes measured in units of coulombs per kilogram of air. However, a more common unit is the roentgen (R) (Wilhelm Conrad Roentgen, 1845–1923) defined as the charge of one electrostatic unit produced by photons in 1 cm^3 of air at standard temperature, humidity, and pressure conditions. One roentgen equals 0.0088 J/kg.

The units of absorbed dose are the gray (Gy), after Stephen Gray, and the rad (rad). One gray equals 1 J of radiation absorbed per kilogram of tissue and 1 rad equals 100 ergs/g of tissue; therefore, 1 rad = 0.01 Gy. The unit of the rad is applied to any type of radiation and to any material. The biological effects of radiation are not directly proportional to energy and the way that this energy is deposited in an organism. The damage depends on the energy of the radiation, the nature of the tissue, and the absorbed dose and dose rate. For instance, alpha particles cause more biological damage than gamma rays, for the same absorbed dose. That is because the alpha energy is deposited over a more localized volume.

The biological effect of different types of radiation depends on the quality factors. These factors are as follows: x-rays and gamma rays equal 1; beta particles, 1 to 1.7; neutrons, 2 to 11; alpha particles, 10; and heavy nuclei, 20. The unit of radiation used to describe biological effects is the dose equivalent, which is the product of the quality factor and the absorbed dose, sometimes called the biological dose. The units of dose equivalent of radiation are the sievert (Sv) and rem (rem). That is, dose of radiation in gray multiplied by the quality factor. Therefore, 1 Sv is the amount of radiation that is equivalent to the biological effectiveness of 1 Gy of gamma rays. Similarly, 1 rem is the amount of radiation that is equivalent to the biological effectiveness of 1 rad of gamma rays; hence, 1 Sv = 100 rem. In one year, a radiation worker should not be allowed to accumulate a dose of more than 5 rems to the whole body, 30 rems to the skin, and 75 rems to hands, forearms, feet, and ankles. It is possible for a human to recover from a dose of several hundred rems. Statistics show that for doses in the range of 600 to 1000 rems, death would occur within two months for 80 to 100% of those so exposed. For certain animals, such as insects, the maximum dose may be as high as 10,000 to 100,000 rems.

There are four basic types of radiation detectors: gas-filled types, scintillation types, semiconductor types, and thin-film types. In this book, the first three type detectors will be explained. The film detectors primarily consist of chemical thin films, which are sensitive to the incoming radiation. The intensity of the darkening of the film indicates the amount of radiation that a person who is wearing the detector has been exposed to.

2.8.3 Gas-Filled Detectors

A simple gas-filled detector consists of a cylindrical container filled with an inert gas (e.g., argon). A voltage is applied in a wire placed in the cylinder and the cylinder wall. Radiation entering the container ionizes some of the gas atoms to give negatively charged and positively charged ions. Free electrons travel to the anode (the wire) and the cathode (the cylinder wall). The resulting electronic pulse is amplified and recorded as a count. In some gas-filled detectors, electrons traveling to the anode can cause further ionization, magnifying the incoming radiation. Instruments based on these principles can be very sensitive to small amounts of incoming radiations.

A typical and widely used example for the gas-filled detectors is the Geiger–Müller detector, also known as the Geiger counter, shown in Figure 2.25. In this arrangement, the anode is a tungsten or platinum wire, while the cylindrical tube acts as the cathode for the circuit. The tube is filled with argon with a slight mixture of a hydrocarbon gas. Internal pressure is kept slightly less than atmospheric pressure. The radiation particle is transmitted through the cathode material. Interaction of the particle produces ionization of the gas molecules, thus causing a voltage pulse for each particle.

In some detectors, boron trifluoride (BF3) gas is used to detect neutrons. In this case, the boron nucleus absorbs the neutron and emits an alpha particle with a nucleus of lithium. The alpha particle and the lithium nucleus together produce ionization in the BF3 gas. Sometimes a solid coating of boron is used on the detector walls to produce the nuclear reaction as described above. Another kind of neutron detector, a fission counter, has a

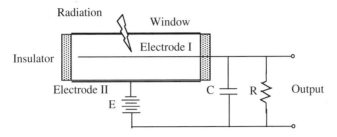

FIGURE 2.25
Geiger–Müller radiation detector.

lining of uranium-235 or similar material rather than boron. On absorbing a neutron the uranium nucleus fissions, and the fission fragments initiate the ionization of the gas.

2.8.4　Semiconductor Radiation Sensors

Relatively new detectors are largely made from semiconductor materials. When an electric field is placed across the semiconductor, a region is created such that no free electrons exist. An incoming radiation ionizes atoms in this region, producing free electrons, which can be collected and counted. The semiconductor sensors are highly sensitive and accurate and have a high resolution, compared to gas-filled counterparts. They can distinguish between radiation particles that are close together in energy contents.

The semiconductor sensors can be manufactured from germanium or silicon. Silicon sensors are capable of operating at room temperatures, and they are fabricated from slices of a single silicon crystal. The basic principle is that the energetic heavily charged particles lose their kinetic energy along a linear path in an appropriate absorbing material.

There are many different types of semiconductor radiation sensors: diffused junction, ion implanted, strip, drift, CCD sensors, and so on. As a typical semiconductor radiation sensor, a position-sensitive silicon strip type sensor is illustrated in Figure 2.26. This sensor is suitable to detect radioactive particles with better than 5 μm localized hit rates. They are often used in particle physics experiments.

2.8.5　Scintillation Detectors

These detectors work on the principle of fluorescence. In this case, the energy is absorbed by a substance and reemitted as visible or near-visible light. A photomultiplier tube senses the amount of light. An example is illustrated in Figure 2.27. In this figure, the scintillator is attached to the front end of a photomultiplier. The front end contains many plates called *dynodes* positioned inside a photomultiplier tube in an alternating pattern. Each dynode

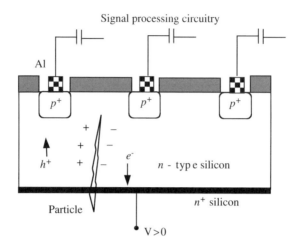

FIGURE 2.26
A semiconductor radiation detector.

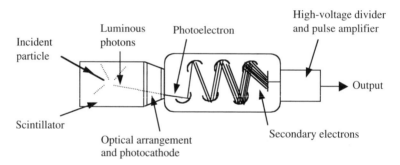

FIGURE 2.27
A scintillation detector.

is attached to an increasing voltage source having the highest positive potential on the last one.

The most widely used scintillators are inorganic alkali halide crystals (e.g., sodium iodine), some plastics, and suitable organic-based liquids. The organic scintillators are sensitive but slow. A general limitation of scintillation detectors is that they have relatively inferior energy resolution, thus, comparatively, finding limited applications in portable instruments.

2.9 Chemical Sensors

Chemical sensors and analyzers constitute large and important classes of electronic portable instruments. They are used for identifying chemical compounds and elements in industrial, environmental, food processing, and domestic applications. Some of the applications are: air and water quality measurements, pollution level determination, detection of chemical and gas leaks, determination of toxicity levels, prospecting of minerals in mining and metallurgy industries, finding explosives, detecting drugs, fire warning, and many other different types of domestic, medical and industrial uses, and health- and safety-related measurements.

Chemical sensors are classified as contact types and noncontact types. The contact types involve the physical interaction of the sensor with the chemicals under observation. The chemicals can be in gas, liquid, solid, or mixture forms. The noncontact types use remote sensing techniques, which are based on the analysis of the gaseous part of the chemical substances under investigation.

Chemical sensors have properties of *selectivity*, which is defined as the ability to primarily respond to only one chemical element or species (compound) in the presence of other species. Therefore, the performance of chemical sensors is evaluated by its selectivity or its ability to eliminate or minimize the effect of interfering species.

Many different methods of sensing chemicals include: catalytic sensors, enzyme sensors, chemosensors, chromatographic sensors, electrochemical sensors such as potentiometric and amperometric sensors, radioactive sensors, thermal sensors, optical sensors, mass sensors, concentration sensors, titration sensors, and semiconductor sensors. Out of this vast repertoire of sensors, only those most relevant to modern electronic portable instruments will be briefly discussed below.

2.9.1 Enzyme Sensors

Enzymes are a special type of catalyst — proteins that are found in living organisms. Enzymes exist in an aqueous environment in the form of immobilization matrices as gels or hydrogels. As sensors, enzymes tend to be very effective in increasing the rate of some chemical reactions and also are strongly selective to a given substrate.

Figure 2.28 illustrates a typical enzyme sensor. The sensing element can be a heated probe or an electrochemical or optical sensor. The enzyme, acting as a catalyst, immobilizes inside the layer into which the substrate diffuses; hence, it reacts with the substrate and the product is diffused out of the layer into the sample solution.

2.9.2 Catalytic Chemical Sensors

These sensors are based on the liberation of heat as a result of the catalytic reaction taking place at the surface of the sensors. For example, when the combustible gas reacts at the catalytic surface, the heat increases the temperature to be detected by suitable temperature sensors such as platinum or

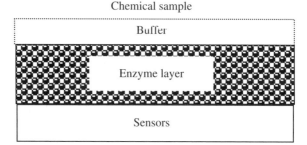

FIGURE 2.28
Arrangement of an enzyme sensor.

other RTDs. As the generated gas contacts the surface of the platinum filament, it becomes oxidized without combustion. The heat of the oxidation
raises the filament temperature, and the heat change is detected as an
increase in its electrical resistance. Some of the catalytic sensors contain
heating elements and temperature sensors employed side by side.

A different type of catalytic sensor is the Taguchi sensor, which is often
used to detect carbon monoxide or natural gases. It consists of a heating
element in a hot tin oxide pellet. When gases come in contact with the hot
pellet, their combustion causes the resistance to change. The current flow
through the pellets becomes proportional to the combustion rate. These
sensors are often used in domestic and industrial gas-sensing alarm systems.

There are catalytic sensors that can detect *flowing gases* and *fluids*. A typical
example is the thermal conductivity sensor, which is a warm filament in a
steam of flowing gas. The loss of heat from the filament depends on the flow
rate of the gas or fluid and thermal conductivity properties of the arrangement. As these sensors do not consume much power, they are extremely
suitable for portable instruments.

2.9.3   Electrochemical Sensors

Electrochemical sensors are the most used sensors. They can be classified as
voltage sensors (potentiometric), current sensors (amperimetric), and resistivity or conductivity sensors (conductometric). In all cases, special electrodes are used as part of a closed electric circuit.

Most modern electrochemical sensors are produced from semiconductors.
Although many electrochemical semiconductor sensors are suitable for fluid
applications, details on these sensors are explained in Section 2.10.3.

2.9.4   Chromatographic Sensors

In chromatographic sensors, the chemical samples are introduced in the form
of a gaseous or vaporized mixture into a carrier gas stream, which passes in

a suitable column over a solid-adsorbing surface or nonvolatile liquid applied to a solid surface. The individual components of the sample move through the column at rates corresponding to their different distribution coefficient factors between fixed and mobile phases. Detectors that are based on different operational principles, such as thermal conductivity, electron capture, etc., determine the adsorption of the gas. This technique is very sensitive, and there are many different versions available commercially.

2.9.5 Radioactive Chemical Sensors

The radioactive method of analyte analysis is most commonly used domestically in smoke detectors. In this arrangement, two metal plates exist inside the chamber, and a small radioactive source emits alpha radiation that ionizes the air in between the metal plates and allows a current to flow. If chemicals or particles in the air enter the space between the plates and bind with the ions, then the amount of normal current is altered, thus triggering an alarm. This method may be applied only to certain analytes.

2.9.6 Chemosensors

A chemosensor is a molecule that can bind selectively and reversibly to the analyte of interest. A chemosensor binds or unbinds easily with the dissolved gas molecules of interest. The constituents and physical properties of the chemosensor may be unique to each analyte, depending on the chemical composition. That is, a different chemosensor will be required for each analyte. The analyte can be made from various compounds, for example, methane, which dissolves into the chemosensor solution and is randomly bound to the chemosensor.

2.9.7 Sensors Based on Light and Properties of Optics

These sensors are based on the interaction of electromagnetic radiation with matter. As a result of this interaction, some properties of light are altered, such as intensity, polarization, and velocity. Optical chemical sensors are designed and built in variety of ways, some of which will be discussed in detail here.

Figure 2.29 illustrates a typical semiconductor-based optical chemical sensor. This sensor consists of two chambers, one acting as a reference containing a reference concentration chemical, and the other as the sensor. These chambers are illuminated by a common light-emitting diode. The surfaces of chambers are metallized to improve internal reflectivity, and the bottoms of chambers are covered by glass. One of the chambers has slots covered with a gas-permeable membrane. The slots allow the penetration

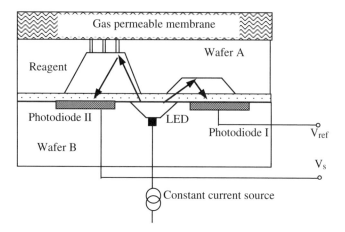

FIGURE 2.29
An optical semiconductor chemical sensor.

of gas to be measured (e.g., CO_2) into the chamber. Wafers A and B form optical waveguides. The chamber filled with reagent is used to monitor the optical absorbency for comparison with that obtained from the reference chamber.

The *luminescence* principle is one method of chemical detection. In this case, a solution is used that allows the "analyte" to be dissolved into the solution. In the bound state, certain molecules become luminescent. This means that the bound molecules will absorb and reemit light. When the solution is exposed to light, the amount of light reemitted is directly proportional to the amount of bound molecules, which is proportional to the amount of analyte in the medium. By measuring this luminescence with a photometer, the amount of analyte can be determined. In the predictions, the probability theory is used and the known concentration of the chemosensor is related to the estimated amount of analyte.

A typical example of luminescence sensors is the *flame ionization detector*. This device has an incorporated regulated fuel, air, and sample delivery system, an internal burner, and associated electronics for measuring the ion current produced by pollutants in the airstream. The increase in light emission from the flame, also known as the flame chemiluminescence, is detected by carefully filtering the suitable wavelength ranges of emission by photomultipliers. Other similar sensors are *flame photometric* sensors and *thermionic* sensors. Many of these sensors require compressed gases and mechanically stable conditions.

Photoionization is another type of optical chemical sensor that responds to many different gases. It operates on the principles of ultraviolet light generated by the gases under investigation. Although it requires high voltages and specially constructed tubes, it still finds applications in portable gas analyzers.

2.9.7.1 Absorption and Emission Sensors

Absorption and emission sensors act as a binding agent of analytes to some chemosensors. The absorption and emission spectrum gives distinctive bands of colors, which may be analyzed and compared against known spectra for the determination of the chemical composition. This method is also used in the analysis of gases, such as oxygen.

2.9.7.2 Spectrometric Sensors

Spectrometric sensors consist of a beam of light aimed through the gas or fluid. The intensity of light emerging from the other side is measured. The decrease in the intensity is related to the color or cloudiness of the sample under investigation. Some typical applications are the measurement of the amount of smoke in air or sediments in flowing waters.

2.9.7.3 Colorimetry or Photometry

Chemical sensors convert the substances into colored compounds, and the intensity of color is related to the amount of chemical under investigation. A large number of specific and very sensitive ranges of color reactions are available for comparative and accurate measurements. The measurements may be conducted with spectrophotometry of pure monochromatic light, which is obtained by high-resolution white light with the aid of suitable prism mechanisms.

2.10 Gas Sensors

Gas sensors could be included in the section of chemical sensors since gas, in chemical terms, is any matter that is in the gaseous state, characterized by high molecular kinetic energy and a tendency to expand out of an open container. However, gas sensing constitutes a very significant part of electronic portable instruments; therefore, special attention will be paid here.

In practice, the word *gas* takes distinct and separate meanings. For example, *natural gas* is a fossil fuel that is collected in the gas phase, and *gasoline*, which is also called gas, is a liquid mixture of hydrocarbon fuels that become gaseous in the combustion chamber of an engine. Here, gas will be taken in the chemical sense. For the purpose of detection of gases, there are primarily two types of gases:

1. *Inorganic gases*, such as sulfur dioxide, hydrogen sulfide, nitrogen oxides, hydrochloric acid, silicon tetrafluorade, carbon monoxide and carbon dioxide, ammonia, ozone, etc.

2. *Organic gases*, such as hydrocarbons, terpines, mercaptans, formaldehyde, dioxin, fluorocarbons, etc.

Modern gas detectors are based on semiconductor technology and microprocessors. In recent years, the functionality of gas detectors has increased
considerably, and nowadays most gas detectors can detect many different
types of gases.

The most popular methods of gas detection are infrared and fluorescent
spectrometry and light absorption methods, colorimetry or photometry, chromatographic methods, flame ionization, indicator tubes, test papers, titration,
semiconductor methods with electric conductivity and amperimetric methods, and ring oven methods. Some of these methods have already been
explained above. The other most used sensors is briefly visited below.

2.10.1   Infrared, Fluorescent, and Light Absorption Gas Sensors

Infrared spectrometry is based on the absorption of infrared energy. This
method is suitable for both automatic and nonautomatic measurements. This
technique involves the determination of the differences in the absorption of
infrared energy between gas samples. The sensor contains a gas sample path
and a reference path. The sample and reference paths are separated in time
or space. In one version, the device contains two cells, one for the sample
gas and the other for the reference gas. A nichrome radiation source, an
infrared optical system, a detector mechanism, and appropriate electronic
circuitry for signal processing are included in the same unit.

Fluorescence spectrometry, on the other hand, measures the fluorescent light
emitted by certain molecules of gas when excited by a radiation source of
appropriate energy or wavelength. Organic molecules containing conjugated
double bonds are the most commonly encountered fluorescing materials and
hence are suitable for application of this method.

The *light absorption method* is based on the absorption of light by gases,
particles, and aerosols. Depending on the samples under investigation, the
absorbed light can be in the visible spectrum or in the infrared or ultraviolet
spectral regions. Usually, two beams of light are used and the differences in
the absorptions are processed to identify the gases or particles. These devices
are extremely sensitive, from 0 parts per million (ppm) to several hundred
ppm pollutant levels.

2.10.2   Indicator Tube Gas Sensors

Indicator tubes are portable and inexpensive devices that are used to detect
gases in the atmosphere. A pollutant gas causes a change in the color of a
reagent packed in a tube. Comparison of change in color with the standard
colors indicates the amount of concentration of the pollutant gas. For quantitative results, a known volume of air must be drawn through the tube at

a fixed rate. The influence of other gases must be eliminated by the correct choice of the reagent. Potential sources of errors are the variations in tube diameters, temperature change, and inaccurate estimation of color intensity. Possible inconsistencies in the manufacturing in the reagents must also be considered. Indicators may contain tubes with several layers of impregnated compounds to remove other interfering substances. Many indicator tubes come in kit forms containing pumps and other necessary ancillaries. They can be sensitive to detect gases, e.g., carbon monoxide, down to 1 to 10 ppm.

2.10.3   Semiconductor Gas Sensors

There are many different types of chemical and gas sensors that are based on semiconductors — *conductometric, amperimetric,* and *potentiometric* sensors.

2.10.3.1   Conductometric Sensors

Conductometric sensors are based on the changes of electrical resistance or conductivity of a material due to the presence of some chemical element or compound. The conductivity changes because of traces of ions generated or carried by the chemical substance under investigation.

Pure water, for example, is a poor conductor, but a small trace of ions in water increases its conductivity considerably. Many conductivity sensors can detect the concentration of ions as well as the type of ions, since different ions affect conductivity differently. In many cases, only very small liquid volumes of samples are required for measurable changes in conductivity. However, care must be exercised for adequate temperature compensations. Figure 2.30 illustrates a conductivity-based gas sensor. In this figure, the reactions at the oxide surface cause changes in resistance between the filament and the electrode.

A standard conductivity sensor can be made from platinum–metal electrodes. In these sensors, the platinum is electroplated in suitable shapes and sizes onto the surface of the metal so that the surface of the electrode is increased. A standard sensor contains two electrodes of platinized wires each having an approximate 1-cm^2 surface area separated by 1-cm distance. The electrodes are dipped into a solution and calibrated against a standard cell to obtain a cell constant for a particular application. The cell constant is used as a multiplier of the conductivity to obtain the true value for that solution.

2.10.3.2   Amperimetric Sensors

Amperimetric chemical and gas sensors are based on the generation of electric current or potential by free elements or compounds. For example, sulfur dioxide and other reducing gases in the air sample react with electrically generated free iodine or bromine in a detection cell. The cell has an anode and a cathode electrode and contains a constant titrant concentration. Any sulfur compound introduced into the cell reacts with the titrant and changes

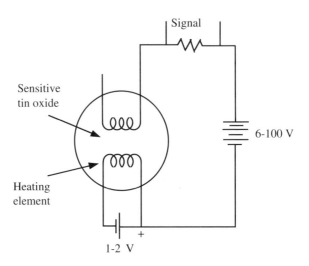

FIGURE 2.30
A tin oxide catalytic gas sensor.

the concentration to produce a potential change in the solution to be picked up by the electrodes. The range of the sensor is 0.0001 to 0.2 ppm.

In amperimetric sensors, the chemical reaction occurs at the active electrodes by means of exchange of electrodes leading to flow of current. A typical amperimetric cell contains three electrodes inserted in the solution to be measured. The desired reaction takes place in the working electrodes (WEs). The working electrodes are usually made from materials such as mercury, platinum, carbon, or gold showing different responses to different chemicals. The potential of the WE relative to the solution is measured by a reference electrode (RE). The WE carries current, but the RE does not. In some cases, the chemical reaction is encouraged by setting a potential difference between the electrodes by counterelectrodes; this is called the potentiostat. The basic components of amperimetric sensors are illustrated in Figure 2.31. The cathode is made from platinum. The reference electrode, called the Clark oxygen sensor, is made from AgCl that results in the generation of electrons. The cathode is maintained at a fixed reference condition by establishing constant chloride ion concentration. The cathode is separated from the sensing fluid by an oxygen-permeable membrane made from silicone polymer, which allows oxygen transfer.

The most critical parameter of an amperimetric cell is the WE vs. RE potential. Once the parameter is set correctly, the cell responds to some chemicals but not to others, thus giving selective responses to identify contents of chemical mixtures. The amperimetric cells are often used for batch analysis, where samples are placed in cells and analyzed separately.

Amperimetric cells are also used to measure chemical vapors in air. In such arrangements, a porous plastic membrane is in contact with air, but the surface tension of the membrane prevents the electrolyte from flowing out.

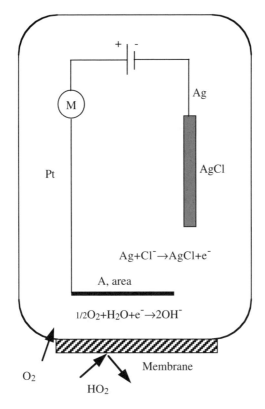

FIGURE 2.31
An amperimetric oxygen sensor.

The working electrode is usually made from gold or platinum deposits located on the wet side of the membrane. The reference and counterelectrodes are placed in the bulk solution. The gas (such as carbon monoxide) under investigation diffuses through the membrane and comes in contact with the electrolyte and WE material, thus causing responses. These types of amperimetric sensors are very sensitive devices. They can be used for analysis of a diverse range of gases such as carbon monoxides and other compounds of carbon, hydrogen sulfides, and so on.

2.10.3.3   *Potentiometric Sensors*

Potentiometric chemical or gas sensors act as an electric battery. A typical example of potentiometric sensors based on semiconductors is given in Figure 2.32. These sensors are based on the use of ion-selective field-effect transistors (ISFETs). They are derived from MOSFETs with hydrogen ion-sensitive elements. These silicon chips contain a pH-responsive membrane much like the glass electrodes. The signals are amplified and processed on the same chip.

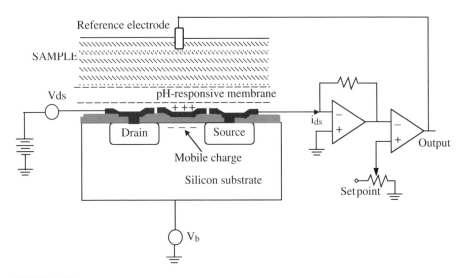

FIGURE 2.32
A pH sensor based on ion-selective FET.

In these sensors, the potential or current signals generated by the electro-chemical process between the solution and the sensors are processed. Potentiometric sensors have two important characteristics: (1) very small current needs to be drawn during measurements, thus implying alterations in the characteristics, and (2) the response is not linear but usually logarithmic. The potentiometric relationship is described by the Nernst equation as

$$E = E_0 \frac{RT}{nF} \ln \frac{C_o}{C_R} \tag{2.35}$$

where E is the electrode potential, E_0 is the potential at the reference state, n is the number of electrons transferred, F is Faraday's constant, R is the gas constant, T is the absolute temperature, and C_o and C_R are the concentrations of the oxidant and reduced product, respectively.

The measurement of the potential E of the sensors is made under zero current or quasi-equilibrium conditions. This requires a high-input imped-ance amplifier.

The most common type of potentiometric sensor is the pH sensor measur-ing the concentration of the hydrogen ions in a solution. In these sensors, the interior of the electrode is separated from the exterior by a thin glass membrane. But the glass membrane is manufactured in such a way that hydrogen ions can penetrate through. The glass membrane also acts as an electric insulator for having a high resistance in the region of 1 GΩ. These sensors are expensive and physically fragile.

Many other potentiometric electrodes are available to measure dissolved substances such as fluoride, cyanide, potassium ions, and so on.

2.10.4 Ring Oven and Titration Methods

Inorganic ions are identified by microchemical methods. High-volume paper tape samplers, electrostatic precipitators, and other methods are used to collect samples such as the airborne particles and aerosol pollutants. One drop of sample solution is placed in the center of a piece of filter paper; i.e., passing hydrogen sulfide through the paper will precipitate sulfides. The solution, evaporated at appropriate temperatures, leaves a soluble substance on the paper. This method is suitable for the determination of iron, aluminum, copper, nickel, zinc, lead, vanadium, cobalt, antimony, beryllium, selenium, phosphate, sulfate, etc. These ions are detected in nanogram to microgram quantities.

2.10.4.1 Titration

In this method, the sample to be measured is dissolved in a liquid or gas. The amount of the sample is determined by adding a known volume or weight of a reagent solution that just neutralizes or completely reacts with the sample under test. Titration is a method often used in air pollution analysis.

2.10.5 Application of Gas Detectors

An important application of the gas detector is in the measurement of air pollution. An accurate prediction of pollution is also necessary to observe its effect on humans, plants, vegetation, animals, the environment, and properties. Nevertheless, precise estimation of substances responsible for air pollution is difficult due to geographical, physical, and seasonal variations. For this purpose, automatic instruments based on continuous sampling or batch type sampling are developed mainly based on the following techniques: electrolytic conductivity; electrolytic titrimetry; electrolytic current or potential; colorimetry; turbidimetry; photometry; fluorimetry; and infrared or ultraviolet absorption, spectrometry, and gas chromatography. Nonautomatic methods are suitable to determine important air pollutants such as acrolein, aldehydes, ammonia, arsenic, beryllium, cadmium, carbon disulfide, carbon monoxide, chlorine, chromium trioxide and salts of chromic acids, fluorides, and so on.

A typical portable handheld gas detector is a microprocessor-based device that runs on a disposable 9-V battery with a digital display. It has multiple electrodes and printed circuit board (PCB)-mounting sensors with sensitivities ranging from 0 to 1000 ppm. Full-scale accuracy is in the region of 95 to 98% with a zero drift of ±2 ppm. It detects a diverse range of gases simultaneously, such as ammonia, ethylene oxide, carbon monoxide, chlorine, hydrogen and its compounds, oxygen, ozone, etc. Some options include a sample draw pump and alarm. For the detector, other sensors are available to extend the detection range to less common gases such as hydrogen cyanide and various acid vapors.

2.11 Biomedical and Biological Sensors

Sensors and portable instruments in biomedical and biological applications are well developed. Some of the application examples are:

- Biopotential and electrophysiology determination
- Blood pressure measurements
- Blood flow measurements
- Body temperature measurements
- Body weight and composition measurements, etc.

Traditionally, biological detectors require human intervention in a laboratory environment. Some of these laboratories are portable. However, in recent years, automatic devices and robots are involved in biological applications, such as in the detection of microorganisms and their concentration levels. For example, for the detection of microorganisms in air, three different methods are used:

1. *Biochemical,* which detects a DNA sequence and protein that are unique to a bioagent through its interaction with test modules
2. *Chemical,* e.g., mass spectrometry, which works by breaking down a sample into its components, such as amino acids, and then comparing their weights with those of known bioagents and other molecules
3. *Biological tissue-based systems,* in which a bioagent or biotoxin affects live mammalian cells, causing them to undergo some measurable response

Some classes of detectors rely on comparing the DNA taken from a microorganism with the DNA of a known agent. These detectors are built on the prior knowledge of the bioagent. In automatic systems, a detector collects samples and then multiple copies of the DNA are made by using a polymerase chain reaction (PCR). In this method the samples are heated and cooled periodically to produce copies of the DNA. As the DNA is copied, the resulting strands are mixed with fluorescent DNA probes. The probe bound with the bioagent glows under ultraviolet light. By using different probe markers, several PCRs can run at the same time, thus enabling detection of more than one agent at a time.

There are microfluidic devices, which contain tiny channels, valves, and chambers in a single chip. Once the microorganism is in the chip, their cells are cracked open ultrasonically. The PCR is applied by means of small thin-film heaters.

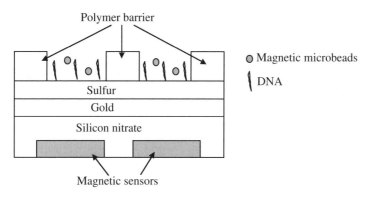

FIGURE 2.33
A magnetic biosensor.

Apart from optical and fluorescent methods, magnetic methods, illustrated in Figure 2.33, are also used to detect DNA. These devices comprise an array of wire-like magnetic field microsensors. These sensors are coated with single-stranded DNA probes specific for a gene from a bioagent. Once a strand of bioagent DNA in a sample binds with a probe, the resulting double strand binds a single magnetic microbead. When a magnetic bead is present above a sensor, the resistance of the sensor decreases in proportion to the number of microbeads.

Some sensors use fluorescent antibodies to bind to bacterial cells. In these devices, the sample passes through a portable flow cytometer, which counts cells by measuring their fluorescent or other properties as they move in a liquid.

Another commonly used method in biodetectors is the use of live tissues. Many toxins trigger measurable or differentiable reactions in living cells. Mammalian cells, such as heart cells, are cultured in a lab and then seeded into a cartridge containing a microelectrode array. When biotoxin is introduced, the cells create voltages detectable in millivolts at the electrodes.

2.12 Environmental Sensors

2.12.1 Time

An important environmental concept is *time*. Traditionally, time measurement relied on mechanical methods. Nowadays, mechanical methods have severe limitations in precision and accuracy, particularly in subsecond time measurements such as microseconds and below. Therefore, recently electronic methods are well developed. The majority of electronic time measurements make use of crystal-controlled oscillators, exemplified in Figure 2.34.

Figure 2.34 illustrates a crystal oscillator with quartz and a coupling mechanism. Generally, the piezoelectric properties of the quartz crystals make a

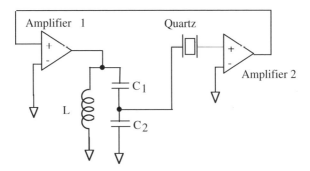

FIGURE 2.34
Block diagram of a crystal oscillator.

good vibration transducer, and the crystal is cut such that it can vibrate mechanically. This arrangement acts as an electrical resonator with practically no losses. In these resonators, the frequency is very stable and the oscillations can be maintained with very little energy input. Crystal oscillators are used in domestic electronic clocks and watches where precision may not be all that important. For more precise timing, the crystals should be kept at a constant temperature, and for utmost stability and precision, maser oscillators are available.

2.12.2 Electromagnetic Field Sensors

Magnetic field measurements are frequently used in geomagnetic studies, navigation, manufacturing processes, medicine, etc. There are two classes of magnetic detectors: nondirectional or scalar sensors, and directional or vector magnetometers.

Many different types of electronic portable instruments based on different sensors are available to measure magnetic field strengths. Also different techniques can be employed to measure the magnetic fields. Each technique has unique properties that make it suitable for a particular application. These applications can range from sensing the presence of the magnetic field in a closed area to the determination of scalar and vector properties of the magnetic fields in open spaces. The vector type magnetic field sensors can be divided into two parts: those that measure low fields, less than 1 mT in range, called magnetometers, and those that measure high fields that are greater than 1 mT, called gaussmeters.

In the vector measuring instruments, the fluxgate and induction coil sensors are most commonly used. Other sensors are the fiber-optic magnetometers, Hall-effect devices, magnetoresistive sensors, anisotropic magnetoresistors (AMRs), giant magnetoresistors (GMRs), and proton precision magnetometers.

Induction coil, fluxgate, and fiber-optic magnetometers are used in low-field vector measuring instruments. Hall-effect sensors are used in gaussmeters, and they are most common in high-field vector applications. Magnetoresistive sensors can be applied in both high- and low-field sensing. The GMRs are replacing the traditional fluxgate magnetometers for their improved sensitivities. Another type of magnetometer is the proton (nuclear) precision magnetometer, which is used to measure scalar magnetic field strength, particularly in geological exploration and aerial mapping of geomagnetic fields.

Most of these magnetic sensors have been explained in Section 2.2. The most important magnetic sensors — fluxgate sensors and search coil sensors — are explained in this section.

2.12.2.1 *Fluxgate Magnetometer*

The fluxgate magnetometer is one of the most commonly used magnetometers. It is made from two coils wound on a ferromagnetic material. One of the coils is excited with a sinusoidal current, which drives the core into a saturation state. At this stage, the reluctance of the core to an external magnetic field increases, thus repelling the external flux and reducing the effect of the external field on the second coil. As the core becomes unsaturated, the effect of the external field increases. The increase and reduction of the effect of the external field on the second coil are sensed as harmonics of the induced voltage. These harmonics are then directly related to the strength and variations in the external field. A typical magnetometer based on fluxgate principles is illustrated in Figure 2.35. The sensitivity of this type of device is dependent on the magnetic properties and shape of the saturation curve of the core material. The measurement range can vary from 10^{-10} T to 10 mT, operating up to the 0- to 50-kHz frequency range.

In electronic portable instruments, the fluxgate magnetometers are constructed as microprocessor-based instruments with digital displays. They operate on 9-V alkaline batteries with a typical power consumption of 150 mW. The bandwidth is from DC to 400 Hz with an accuracy of ±0.5%. Its resolution is 1 nT to 0.2 T. An AC magnetometer operating on a 9-V battery has a range from 2 to 200 μT. The frequency bandwidth is from 0 Hz to 50 kHz.

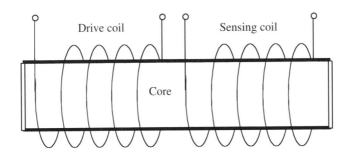

FIGURE 2.35
A fluxgate magnetometer.

2.12.2.2 Search Coil Magnetometers

Search coil magnetometers operate on the principle of Faraday's law of induction. The flux through the coil changes if the magnetic field varies in time or the coil is moved through the field. The sensitivity depends on the properties of the core material, dimensions of the coil, number of turns, and rate of change of flux through the coil. These magnetometers are manufactured from 4 cm in dimension to 100 cm. They can sense weak fields as low as 10^{-10} T within the frequency range of 1 Hz to 1 MHz. The upper limit is dependent on the relative magnitudes of resistance and inductance of the coil. During signal processing, they can be used as part of a bridge or a suitable resonant circuit.

2.12.3 Humidity and Moisture Sensors

Humidity and moisture measurements are very important in agricultural applications and international food trading. Moisture is the amount of water, in some cases amount of liquids, in materials. The presence of moisture in a gas is the humidity. The absolute humidity of a gas is the mass of water per unit mass of gas. The maximum humidity that can be attained is called the saturation humidity. The saturation humidity heavily depends on temperature. For many purposes, relative humidity is important. The relative humidity is the ratio of absolute humidity to saturation humidity at a particular temperature.

Another measure of relative humidity is the dew point. When a surface is cooled and in contact with gas, a temperature will be reached such that gas-containing water deposits the water onto the surface by condensation. This surface temperature is called the dew-point temperature.

There are many different methods available for moisture and humidity sensing. These methods range from changes of electrical properties to changes in optical properties or even changes in mechanical characteristics of the sensors. Different techniques may be designed and developed for specific application in a particular process depending on the characteristics they exhibit.

The oldest method of measuring humidity is the *hygrometer*. A human hair, free from oil or grease, is strongly affected by humidity, shrinking in dry conditions and expanding in moist conditions. Relative humidity sensors are constructed by slightly tensioned hair along with some method of detecting changes in lengths of materials. Some of these methods may be the use of a linear potentiometer, LVDTs, capacitive techniques, or some forms of oscillatory circuits.

Use of *lithium chloride* for moisture measurement is very common. Lithium chloride has a high resistance in its dry state, but the resistance drops considerably in the presence of water with a reasonable linearity in proportion to the amount of water. The lithium chloride cell is made partly of a measuring bridge, and the output of the bridge is processed suitably.

Another method for measuring humidity and moisture content is by using *microwave energy*. Microwave energy at various frequencies, e.g., 2.45 GHz, is absorbed strongly by water. The attenuations and phase shifts of the microwave across the moist media become a good way of measuring the water content of the media. Microwave techniques are extensively used in agricultural products such as grain moisture determinations.

Moisture of solids is usually sensed in terms of conductivity of the materials, changes in permittivity, or microwave absorption. For moisture contents in materials of fixed composition, resistance measurements between the conductors set at fixed distances are sufficient, but lengthy calibration procedures are needed.

Most electronic methods of measuring humidity are based on either the properties of dew points or the characteristics of moist materials. Two basic electronic humidity sensors are discussed in detail:

1. Conductive sensors
2. Capacitive sensors

2.12.3.1   Conductive Humidity and Moisture Sensors

The resistance of some of the materials used in semiconductor structures changes significantly under varying humidity conditions. The sensing elements are made from many different materials, such as the polystyrene films (Pope cells) treated with sulfuric acid. In some cases, solid polyelectrolides are used. Solid-state humidity sensors are fabricated from silicon substrates, as illustrated in Figure 2.36.

In these sensors, the silicon substrate has a high conductance. The gold porous material on the top of the substrate allows the penetration of water molecules to provide electrical contact with the sensor. An oxide layer deposited on top of the aluminum layer becomes anodized and affected from the porous oxide surface. In this way an electrical conduction mechanism is obtained from the porous gold to the silicon substrate. This yields the resistance of the system to be directly proportional to the number of water molecules in the porous material.

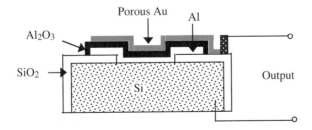

FIGURE 2.36
An aluminum oxide thin-film conductive moisture sensor.

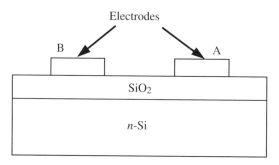

FIGURE 2.37
A capacitive thin-film humidity sensor.

2.12.3.2   Capacitive Humidity and Moisture Sensors

Capacitive humidity and moisture sensors are extensively used in electronic portable instruments because of their ease of use and cost-effectiveness. Therefore, a detailed treatment of the subject is given here.

In capacitive methods, the changes in the permittivity of atmospheric air as well as the changes in the permittivity of many solid materials are utilized. The permittivities of most of the dielectric materials are functions of moisture contents and temperature. The capacitive humidity sensor is based on the changes in the permittivity of the dielectric material between plates of capacitors.

Thin-film capacitive sensors can be fabricated on silicon substrates as shown in Figure 2.37. In this construction, an SiO_2 layer is deposited on the n-type silicon substrate. The two metal electrodes, shaped in interdigitized patterns, are located on the SiO_2. These electrodes may be made from aluminum, tantalum, chromium, or phosphorous-doped polysilicon. Since there are many types of capacitive humidity sensors operating on similar principles, only the aluminum and tantalum types will be introduced as typical examples.

Aluminum type capacitive humidity sensors constitute the majority of capacitive humidity sensors. In these types of sensors, high-purity aluminum is chemically oxidized to produce a prefilled insulating layer of partially hydrated aluminum oxide, which acts as the dielectric. A water-permeable but conductive gold film is deposited onto the oxide layer, usually by vacuum deposition, which forms the second electrode of the capacitor.

In another type of aluminum, aluminum oxide, the sensor has a pore structure. The oxide with its pore structure forms the active sensing element. Moisture in the air, reaching the pores, reduces the resistance and increases the capacitance. The decreased resistance can be thought of as being due to an increase in the conduction through the oxide. An increase in capacitance can be viewed as due to the changes in the dielectric constant. The quantity measured can be resistance, capacitance, or impedance. The high humidity end is best measured by variations in capacitance since resistance changes may be very small in this region.

Tantalum capacitive humidity sensors are made from plates, one of which consists of a layer of tantalum deposited on a glass substrate. A layer of polymer dielectric is then added, followed by a second plate, which is made from a thin layer of chromium. The chromium layer is under a high tensile stress so that it cracks into a fine mosaic structure, which allows water molecules to pass into the dielectric. A sensor of this type has an input range of 0 to 100% relative humidity (RH). The capacitance is 375 pF at 0% RH and a linear sensitivity of 1.7 pF/%RH. The error usually is less than 2% due to nonlinearity and 1% due to hysteresis.

Capacitive humidity sensors enjoy wide dynamic ranges, from 0.1 ppm to saturation points. They can function in saturated environments for longer lengths of time, which would adversely affect many other humidity sensors. Their ability to function accurately and reliably extends over a wide range of temperatures and pressures. Capacitive humidity sensors also exhibit low hysteresis and high stability with little maintenance requirements. These features make capacitive humidity sensors viable for many specific operating conditions and ideally suitable for a system where uncertainty of unaccounted conditions exists during the operations. However, in some applications, contaminants can block the flow of water vapor into the sensing element, thus affecting the accuracy of the instrument. Many sensors come with some form of special casing to provide protection.

Many capacitive moisture sensors, based on the changes in the permittivity of granular or powder type dielectric materials such as wheat and other grains containing water, are readily available. Usually the sensor consists of a cylindrical chamber. The chamber is filled with samples under test. The variations in capacitance values with respect to water content are processed. The capacitor is incorporated as a part of an oscillatory circuit operating at a suitable frequency.

Capacitive moisture sensors need to be calibrated if they are applied on samples that are made from different materials, as the materials themselves demonstrate different permittivities. Accurate temperature is necessary as the dielectric constant may be highly dependent on temperature. Most of these devices are built to operate at temperature ranges of 0 to 50°C, supported by tight temperature compensation circuits.

2.12.4 Soil Acidity and Alkalinity

Acidity and alkalinity of water is an important factor for water supplies, industrial and domestic water users, generation stations, and agriculture and horticulture applications. The acidity and alkalinity of water is measured on the pH scale, which is based on free hydrogen ions in the water. Natural water has a pH value of 7, fairly strong acid solutions have a pH of 2, and fairly strong alkaline solutions have a pH of 12.

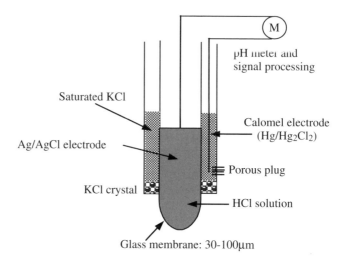

FIGURE 2.38
A pH sensor.

The sensing of pH makes use of changes in ionization. Natural water with 7.0 pH levels has a very high resistivity, but ionization causes the resistivity to drop very sharply. The standard electrical system of sensing acidity depends on the glass electrode system, as shown in Figure 2.38; in this arrangement, a thin glass bulb contains a mildly acidic solution with good conducting properties. A suitable reference electrode, calomel, and a sensing element inside the bulb provide electrical signals for pH levels.

A relatively recent development in pH measurement is the introduction of semiconductor sensors as illustrated in Figure 2.32.

2.12.5   Fire and Smoke Sensors

Smoke detection is important in domestic and industrial applications. These detectors operate on two main principles: ionization and optical detection. The ionization detector uses a radioactive source, usually americium-241, with a low activity level, typically 0.8 μCi. A source plate and an electrode, separated by a uniform gap, are in a permanent conducting state. In the presence of smoke, molecules entering the ionization chamber (uniform gap) are struck by the alpha particles that cause ion currents. Although the ion current is very small, about 10 pA, it causes a reduction in the permanent current that activates the alarm.

The optical type detector operates on the Tyndall effect of scatter. When a beam of light passes through clean air, the beam is invisible; therefore, looking along an axis at right angles to the beam results in no detectable light. In the presence of smoke, light is scattered; hence, there is an appreciable amount of light visible on the transverse axis. The scattered light in the transverse direction is detected, and resulting electrical signals are processed appropriately.

2.13 Pollution Sensors

Discussions on pollution sensors can be included in a section on environmental sensors, but pollution sensors have specific characteristics. Historically, pollution sensors constituted an important part of electronic portable instruments, and they still do; therefore, more discussions will be presented here, stressing specific characteristics that separate them from other sensors. Also, this section gives a good insight into how portable instruments can be applied in specific situations. It gives information on other factors affecting the measurements and clarifies the importance of such measurements.

2.13.1 Air Pollution

Air pollution is defined as unwanted change in the quality of the Earth's atmosphere caused by the emission of gas, solid, and liquid particulates. It is considered to be one of the major causes of climatic change (greenhouse effect) and ozone depletion, which may have serious consequences for all living organisms in the world. Polluted air is carried everywhere by winds and air currents and is not confined by national boundaries. Therefore, air pollution is a concern for everybody irrespective of what and where the sources are; hence, instrumentation for air pollution measurements is a very well-developed area.

The seriousness of air pollution was realized when 4000 people died in London in 1952 due to smog. In Britain, the Clean Air Act of 1956 marked the beginning of the environmental era, which spread to the U.S. and Europe soon after. The Global Environmental Monitoring System (GEMS), established in 1974, has various monitoring networks around the globe for observing pollution, climate, ecology, and oceans. Concentrations of atmospheric pollutants are monitored routinely in many parts of the world at remote background sites and regional stations, as well as urban centers. Since the establishment of GEMS, some interesting findings have been reported. Some examples are as follows: Overall, only 20% of people live in cities where air quality is acceptable. More than 1.2 billion people are exposed to excessive levels of sulfur dioxide, and 1.4 billion people to excessive particulate emission and smoke. In 1996 there were more than 64,000 deaths in the U.S. for which the cause could be traced to air pollution.

Air pollutants can be classified according to their physical and chemical composition as:

- *Inorganic gases* — sulfur dioxide, hydrogen sulfide, nitrogen oxides, hydrochloric acid, carbon monoxide, carbon dioxide, ammonia, ozone, etc.
- *Organic gases* — hydrocarbons, terpenes, mercaptans, formaldehyde, dioxin, fluorocarbons, etc.
- *Inorganic particulates* — asbestos, lime, metal oxides, silica, antimony, zinc radioactive isotopes, etc.
- *Organic particulates* — pollen, smuts, fly ash, etc.

Accurate measurement of air pollution is necessary to establish acceptable levels and to establish control mechanisms against offending sources. Accurate prediction of pollution helps in setting policies and regulations, as well as in observing the effects on humans, plants, vegetation, animals, the environment, and properties. Nevertheless, precise estimation of substances responsible for air pollution is difficult due to geographical, physical, and seasonal variations. Currently, many studies are taking place to understand the processes involving the formation, accumulation, diffusion, dispersion, and decay of air pollution and the individual pollutants causing it. Effective national and international control programs very much depend on this understanding.

A fundamental requirement for an air pollution survey is the collection of representative samples of homogeneous air mixtures. The data must include the content of particulate and gaseous contaminants and their fluctuations in space and time. Geographical factors — horizontal and vertical distribution of pollutants, location of sources of contaminants, airflow directions and velocities, intensity of sunlight, time of day — and the half-lives of contaminants must be considered in order to determine the level of pollution in a given location. The sampling must be done by proven and effective methods and supported by appropriate mathematical and statistical analysis.

There are two basic types of methods used in determining air pollution: the spot sampling method and continuous sampling. These techniques can be implemented by a variety of instruments. *Automatic* instruments are based on one or more methods, such as electrolytic conductivity, electrolytic titrimetry, electrolytic current or potential, colorimetry, turbidimetry, photometry, fluorimetry, infrared or ultraviolet absorption, and gas chromatography. Some of these methods have been explained above. *Nonautomatic* instruments are mainly based on absorption, adsorption, and condensation.

2.13.2   Water Pollution

Water is a resource fundamental to all life, and it is important in both quantity and quality, particularly for humans. Fresh water is essential for life, and clean, unpolluted water is necessary to human health and the preservation of nature. Water is used for many purposes besides human consumption, and in arid and semiarid countries, large quantities are used for irrigation.

The determination of the quality of water is important to decide the suitability of it for consumption. Different standards of water quality are acceptable for different uses. Water for human consumption should be free of disease-causing microorganisms, harmful chemicals, objectionable taste and odor, unacceptable levels of color, and suspended materials. Most of the sensors for water quality are based on chemical methods, as explained in Section 2.9, allocated on chemical sensors.

2.13.3 Soil Pollution

Soil pollution refers to addition of soil constituents, due to domestic, industrial, and agricultural activities, that were originally absent in the system. Soil contamination is of two different kinds. One is the slow but steady degradation of soil quality (e.g., organic matter, nutrients, water-holding capacity, porosity, purity) due to contaminants such as domestic and industrial wastes or chemical inputs from agriculture. The other is the concentrated pollution of smaller areas, mainly through dumping or leakage of wastes. The sources of contamination include the weathering of geological parent materials, where element concentrations exceed the natural abundances in wet or dry deposition forms. Soil pollution is determined by chemical methods as explained previously.

2.14 Distance and Rotation Sensors

The position of an object implies the location of the object in reference to a point. The reference point may be a fixed point on Earth, the starting point of the object, or any other convenient stationary or moving point that can be related to the object. Determining the position of the object makes use of distance and direction information so that the position can be specified by using either rectangular or polar coordinates in two or three dimensions and time.

There are many methods of position determination of objects depending on their physical properties and other requirements. Broadly, two different methods of position determination are used: large scale and small scale. Also, there are many different types of sensing position and distances in relation to stationary or moving objects or to stationary or moving reference points. If the distances are stationary relative to a fixed point, this requires different methods than objects in motion. Rotation is also a form of distance measurement and requires different techniques of measurement.

In position determinations, the *direction*, which is vector quantity, is an important property. In large-scale position sensing, the sensing of direction on the Earth's surface can be achieved by using the Earth's magnetic field, gyroscopes, radio signals, and satellite systems. The most ancient method is the traditional compass, which uses the effect of the Earth's magnetic field on a small, magnetized needle that is freely suspended. In this arrangement, the needle points along the line of the field, in the direction of magnetic north and south. The use of compasses is enhanced by electronic means. The needle position within the device can be sensed by various methods such as the Hall-effect sensors to convert the position of the needle to electrical signals. By using A/D converters and digital techniques, the direction can be read out numerically in degrees, minutes, and seconds.

Gyroscopes are alternative devices that are used for direction determination; detailed information on gyroscopes will be given in Section 2.15.2. A gyroscope does not have an inherent electric output. The electric output can be obtained by placing slight mechanical loading on the gyro wheel that is transformed into electrical signals, or optically.

Radio waves have been used for position determination and navigational purposes for a few decades, in the form of radio beacons that are used in much the same way as the light beams. In Section 2.15, more information on this topic will be provided. Also, satellite direction finding will be explained in detail.

Many sensors that are useful for measuring distances are also useful for measuring orientation. Orientation with vertical lines can be determined by devices such as mercury switches. Also, angles between two objects can be determined by inductive coupling methods. In this case, the coupling is minimum when the inductors are perpendicular and maximum when they are parallel.

2.14.1   Large-Scale Distance Sensors

As far as portable instruments are concerned, the predominant method for measuring distances to a target point in a large scale is based on the reflections of optical, radio, or sonar waves from objects. The basic principle is that a pulse of a few waves is sent out from the transmitter, reflected at some distant object, and the returning signals are detected as shown in Figure 2.39.

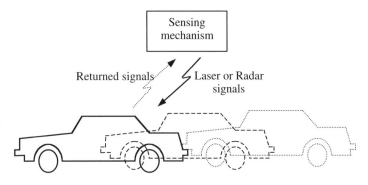

FIGURE 2.39
A large-scale distance sensing system.

Since the speed of the wave in the propagation media is known, the distance of the reflector can be calculated from the time that elapses between sensing and receiving. This time can be very short (a few microseconds) so that the duration of the wave pulse must be measured very carefully. Large-scale distance measurement leads to the sensing of distance traveled or the velocity of objects by measuring the variations in the distance against time. These types of devices are predominantly used in aviation for target detection such as radar, land surveying, satellite-based detection systems, military applications, and road traffic control. The speeds of moving objects can easily be determined from continuous measurements of distances, as in the case of detection of speed of vehicles used by the police.

2.14.2 Small-Scale Distance Sensors

There are many different methods for small-distance sensing. Some of these sensors are based on resistive, capacitive, or inductive principles, while others make use of optical interferometers and millimeter-wave radar techniques.

A simple way of sensing distance is the use of a linear potentiometer. The moving object is connected to the slider of the potentiometer, so that each position along the axis corresponds to movement of the object. These position sensors can give a precision of 0.1%. Alternative methods are capacitive, induction, or other magnetic sensors. For very small distances, the strain gauges can be used. Laser interferometers are applicable when very precise changes must be sensed. One particular advantage of the laser interferometer is that its output can readily be converted to digital form, since it is based on counting of the wave peaks.

Linear digital encoders give an output, which is a binary number that is proportional to the distance of the encoder relative to a fixed point. They provide a set of binary digital signals directly, but they require a slide, which is very precisely printed, and a good light source and a light detector.

2.14.3 Rotation Sensors

Sensing and measurement of rotational movement is necessary in many industrial applications since many industrial machines have rotating shafts. The simplest form of sensing angular velocity is by using AC or DC generators commonly known as tachogenerators.

In some applications, signals are generated for each revolution of the wheel or shaft optically or by means of piezoelectric or magnetic pulsing arrangements. These pulses enable the determination of the angular velocities as well as positions of shafts or wheels.

Angular methods for displacement measurements can be grouped to be small or large angular measurements. Small displacements are measured by the use of capacitive, strain gauge, or piezo sensors. For larger displacements, potentiometric or inductive techniques are more appropriate. The rotary digital encoders also find a wide range of applications.

2.15 Navigational Sensors

Electronic portable instruments constitute a significant portion of navigational instruments since navigation naturally requires portability. The original meaning of the word *navigation* is "ship driving." In ancient times when sailing boats were used, navigation was a process of steering the ship in accordance with some means of directional information, and adjusting the sails to control the speed of the boat. The objective was to bring the vessel from location A to location B safely. In present days, navigation is a combination of science and technology. The term is no longer limited to the control of a ship on the sea surface, but applied to land, air, underwater, and space navigational systems.

2.15.1 Inertial Navigation Sensors

Inertial navigation is the technique of using a self-contained system to measure a vehicle's movement and to determine how far it has moved from its starting point. Acceleration, a vector quantity involving magnitude and direction, is often used for this purpose. A single accelerometer measures magnitudes but not direction. Typically, it measures the component of acceleration along a predetermined line or direction. Gyroscopes that provide a reference frame for the accelerometer are used to supply the direction information. Unlike other position determination methods that rely on external references, an inertial navigation system (INS) does not require communication with any other stations or reference points. This property enables crafts to navigate in unknown territories.

Inertial navigation can be described as a process of directing the movement of a vehicle, rocket, ship, aircraft, robot, etc., from one point to another with respect to a reference axis. The vehicle's current position can be determined from *dead reckoning* with respect to a known initial starting reference position. On Earth's surface, the conventional reference will be north, east, and down or up. This is referred as the *Earth's fixed axes*. A vehicle such as an aircraft or a marine vessel will have its own *local axes* as roll, pitch, and yaw shown in Figure 2.40.

The controlling action is based on the sensing components of acceleration of a vehicle in known spatial directions, by instruments that mechanize Newtonian laws of motion. The first and second integrations of the sensed acceleration determine velocity and position, respectively. A typical inertial navigation system includes a set of gyros, a set of accelerometers, and appro-

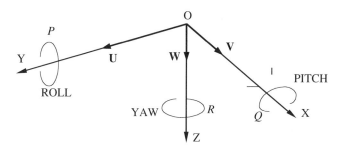

FIGURE 2.40
Degrees of freedom of a moving object such as an aircraft.

priate signal processing units. Although the principle of the systems may be simple, the fabrication of a practical system demands a sophisticated technological base. System accuracy is not dependent on altitude, terrain, and other physical variables, but is limited almost purely by the accuracy of its own components. Traditional inertial navigation systems relied on mainly mechanical gyros and accelerometers, but today there are many different types available, such as optical gyroscopes, piezoelectric vibrating gyroscopes, active and passive resonating gyroscopes, etc. Also, micromachined gyroscopes and accelerometers are making important impacts on modern inertia navigation systems. Brief descriptions and operational principles of gyroscopes that are suitable for inertial navigation are given below.

2.15.2 Gyroscopes

There are two broad categories of gyroscopes: (1) mechanical and (2) optical. Within these categories, there are many different types. Only a few basic types are discussed here.

An example of a gyroscope is shown in Figure 2.41. This gyro is called the double-axis flywheel gyro. In these types, an electrically driven rotor is suspended in a pair of precision low-friction bearings at either end of the rotor axle. The rotor bearings are supported by a circular ring known as an *inner gimbal ring*, which in turn pivots on a second set of bearings that is attached to the *outer gimbal ring*. The pivoting action of the inner gimbal defines the horizontal axis of the gyro, which is perpendicular to the spin axis of the rotor. The outer gimbal ring is attached to the instrument frame by a third set of bearings that defines the vertical axis of the gyro, which is perpendicular to both the horizontal axis and the spin axis. This type of suspension has the property that it always preserves the predetermined spin-axis direction in inertial space.

Optical gyroscopes are based on the inertial properties of light instead of Newton's law of motion. They operate on the Sagnac effect, which produces interferometer fringe shift against rotation rate. In this case, two light waves circulate in opposite directions around a path of radius R, beginning at source

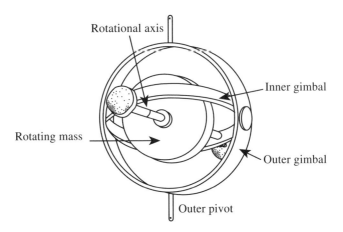

FIGURE 2.41
A double-axis flywheel gyro.

S. A typical arrangement for the illustration of operation principles is shown in Figure 2.42.

In Figure 2.42, when the gyro is stationary, the two beams arrive at the detector at the same time and no phase difference will be recorded. Assume that the source is rotating with a velocity ω so that light traveling in the opposite direction to rotation returns to the source sooner than that which travels in the same direction. Thus, any rotation of the system about the spin axis causes the distance covered by the beam traveling in the direction of rotation to lengthen, and the distance traveled by the beam in the opposite

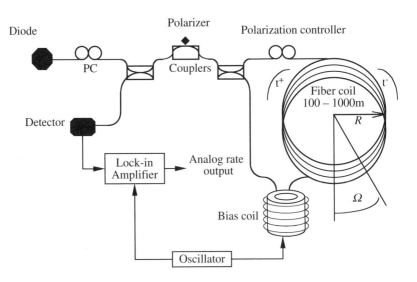

FIGURE 2.42
An optical gyro.

direction will be shortened. The two beams interfere to form a fringe pattern, and the fringe position may be recorded, or the phase differences of the two beams may be sensed. This phase difference is directional proportional to the angular velocity. Usually photodetectors are used to measure phase shift.

There are two different types of optical gyros: active or passive, sometimes referred to as resonant or nonresonant. In passive gyrosensors, the Sagnac phase is measured by some external means, whereas in active gyros, the Sagnac phase causes a frequency change internal to the gyro, which is directly proportional to the rotation rate.

2.15.3   Satellite Navigation and Radiolocation

The use of electronic portable instruments in satellite-based navigation systems is gaining momentum. They find extensive applications in position determination and movement detection of objects, animals, or humans in military, commercial, and recreational purposes. Modern electronic navigation systems can be classified by range, scope, error, and cost. The range classifications can be short, medium, and long ranges, within which exact limits are rather indefinite. The scope classifications can be either self-contained or externally supported, or active (transmitting) or passive (not transmitting) mode of operation.

Today, there are two types of navigation systems: (1) systems that use ground stations such as radiobeacons, radar, Decca, Loran-C, Consol, and Omega, and (2) advanced navigation systems based on satellites such as the global positioning system (GPS) and Glonass. A brief description of both methods is given below.

Short-range systems include radar and radiobeacon-based systems such as Decca. Medium-range systems include Decca and certain types of extended-range radars. Long-range systems include Loran-C, Consol, and Omega. All of these systems depend on active RF transmissions, and all are externally supported with respect to the object being navigated, with the exception of radar. In addition to these, there is another category of systems, which is called advanced navigation systems; the transit satellite navigation systems, Glonass, and GPS are typical examples. GPS navigation systems are replacing the radiolocation systems rapidly; therefore, GPS will be discussed in detail.

Irrespective of the method selected, in electronic navigation systems, three types of accuracy are important: (1) *predictable* or *absolute accuracy*, the accuracy of a position with respect to the geographic coordinates of the Earth; (2) *repeatable accuracy*, the accuracy with which the user can return to a position whose coordinates have been determined at a previous time with the same navigation system; and (3) *relative accuracy*, the accuracy with which a user can measure position relative to that of another user of the same system at the same time. These accuracies are highlighted next.

2.15.4   Global Positioning Systems

The Global Satellite Navigation Systems are second-generation satellites evolved primarily from the naval global positioning system. They provide a continuous three-dimensional position-finding capability (i.e., latitude, longitude, and altitude), in contrast to the periodic two-dimensional information of the transit system. Twenty-four operational satellites constitute the system. Each satellite orbit is circular, about 22,200 km high above the Earth's surface. They are inclined at angles of 55° with respect to Earth's axis.

   Position determination using GPS is based on the ability of receivers to accurately determine the distance to the GPS satellites above the user's horizon at the time of fix. If accurate distances of two such satellites and the heights are known, then the position can be determined. In order to do this, the receiver would need to know the exact time at which the signal was broadcasted and the exact time that it was received. If the propagation speeds through the atmosphere are known, the resulting range can be calculated. The measured ranges are called *pseudoranges.* Nowadays, normally, information is received from at least four satellites leading to accurate calculations of the fix. The time errors plus propagation speed errors result in range errors, common to all GPS receivers. Time is the fourth parameter that is evaluated by the receiver if at least four satellites can be received at a given time. If a fifth satellite is received, an error matrix can be evaluated additionally.

   Each GPS satellite broadcasts simultaneously on two frequencies for the determination and elimination of ionosphere and other atmospheric effects. The frequencies of transmission are at 1575.42 and 1227.6 MHz designated as L1 and L2 in L-band of the ultrahigh frequency (UHF) range. Both these signals are modulated by 30-sec navigation messages transmitted at 50 bits/ sec. The first 18 sec of each 30-sec frame contains *ephemeris* data for that particular satellite, which define the position of the satellite as a function of time. The remaining 12 sec is the *almanac* data, which define orbits and operational status of all satellites in the system. GPS receivers store and use the ephemeris data to determine the pseudorange, and the almanac data help determine the four best satellites to use for positional data at any given time. However, the "best four" philosophy has been overtaken slowly by an all-in-view philosophy.

   The L1 and L2 satellite navigation signals are also modulated by two additional binary sequences: *C/A code,* for acquisition of coarse navigation, and *P-code,* for precision ranging. The L1 signal is modulated by both the C/A and P-codes, and the L2 signal only by the P-code. Positional accuracy of about 20 m rms is usual in using C/A codes alone. The P-code, however, is not available for civilian users. The P-code is redesignated to be a Y-code, decipherable only by high-precision receivers having access to encrypted information in the satellite message. Nevertheless, it is fair to comment that civilians have figured out how to benefit from the P/Y signals without actually knowing the codes, but at a lower signal-to-noise ratio (SNR). Further, C/A codes are degraded by insertion of random errors such that positional accuracy is limited to 50 m rms for horizontal values and 70 m for vertical values. These intended errors are supposed to be lifted by the year 2006. Civilian users have availability, called *Standard Positioning Service* (SPS), accurate to 50 m rms, while U.S. and NATO military users will use *Precise Positioning Service* (PPS).

In enhancing SPS accuracy, differential techniques may be applied to the encrypted GPS signals. Since the reference receiver is at a known location, it can calculate the correct ranges of pseudoranges at any time. The differences in the measured and calculated pseudoranges give the correction factors. Accuracy less than 1 m can be obtained in the stationary and moving measurements. The differential techniques are employed in airplanes and surveying applications.

Differential navigation is also applied where one user set is navigating relative to another user set via a data link. In some cases, one user has been at a destination at some prior time and is navigating relative to coordinates measured at that point. The true values of this receiver's navigation fix are compared against the measured values and the differences become the differential corrections. These corrections are transmitted to area user sets in real time, or they may be recorded for postmission use so that position fixes are free of GPS-related biases.

Differential navigation and GPS find applications in en route navigations for commercial and civil aviations; military applications; navigation of ships, especially in shallow waters; station keeping of aircraft; seismic geophysical explorations; land surveying; and vehicle and traffic controls.

2.15.4.1 GPS Receivers

There are currently three basic types of GPS receivers designed and built to address various user communities: *slow-sequencing*, *fast-sequencing*, and *continuous-tracking* receivers. The least complicated and lowest cost receiver for most applications is the slow-sequencing type, wherein only one measurement channel is used to receive sequentially L1 C/A code from each satellite every 1.2 sec, with occasional interrupts to collect ephemeris and almanac data. Once the data are received, computation is carried out within 5 sec, making this system suitable for stationary or near stationary fixes.

Fast-sequencing receivers have two channels — one for making continuous pseudorange measurements, and the other collection for ephemeris and almanac data. These receivers are used in medium dynamic applications such as position sensing of ground vehicles.

Continuous-tracking receivers employ multiple channels (at least five) to track, compute, and process the pseudoranges to the various satellites being utilized simultaneously, thus obtaining the highest possible degree of accuracy, making it suitable for high dynamic applications such as aircrafts and missiles. The parallel channel receivers are so cost-effective nowadays that other types will disappear.

There are a number of companies that produce highly sophisticated GPS receivers. For example, a typical GPS receiver operates on two and five channels for military applications. It provides features like precise time, interfacing with digital flight instruments, RS-422 interface, altimeter input, and self-initialization.

Software implementation satellite management functions are offered by many manufacturers and have different features. In the majority of GPS receivers, three functional elements are implemented: (1) database management of satellite almanac, ephemeris, and deterministic correction data; (2) computation of precise satellite position and velocity for use by navigation software; and (3) the use of satellite and receiver position data to periodically calculate the constellation of four satellites with optimum geometry for navigation.

2.15.5   Satellite Relay Systems

Satellite communication systems are very important for portable instruments since they allow the communication of remote instruments with a base station. The use of satellites is a highly developed technology utilized extensively throughout the world. In the past two decades, it has progressed from a quasi-experimental nature to one with routine provisions of new services. Satellites take advantage of unique characteristics of *geostationary satellite orbits* (GSOs). The design of satellite systems is well understood, but the technology is still dynamic. Satellites are useful for long-distance communication services, for services across oceans or difficult terrain, and point-to-multipoint services such as television distribution.

There are other satellite systems operated by different organizations and different countries. Every year some new ones are added, while others become obsolete and taken out of service. Some of the operational satellite systems are Iridium of Motorola, Globalstar of Loral Corporation, Intelsat, CS-series of Japan, Turksat of Turkey, Aussat of Australia, Galaxy and Satcom of the U.S., Anik of Canada, and TDF of France. Many of these satellite systems are developed mainly for communication purposes and data transmissions, which have important implications in the development of some electronic portable instruments. Some of these communication satellites are suitable for navigation purposes. However, there are only a handful of systems that specifically designed for navigation. The most established and readily accessible by civilian and commercial users are the GPS of the U.S., Galileo of Europe, and the GLONASS of Russia.

The International Telecommunication Union (ITU) controls frequency allocation for satellites. In the U.S., the Federal Communications Commission (FCC) makes the frequency allocations and assignments for nongovernment satellite usage. The FCC imposes a number of conditions regarding construction and maintenance of in-orbit satellites.

2.16   Mechanical Variables Sensors

At first sight, it appears that electronic portable instruments may not have much to do with mechanical variables. But this perception is not correct historically; many portable instruments were developed to measure mechanical variables, such as weights and densities of the objects. In fact, many mechanical variable measuring devices built in ancient times were portable.

Mechanical variable measurements can be divided into two broad categories: solid variable measurements and fluid variable measurements. There are many portable instruments in both categories. Primarily, solid variable measurements consist of stress and strain measurements; mass, weight, and density measurements; and acceleration, velocity, force, torque, and power measurements. Mechanical variable measurements for fluids can be categorized as pressure and sound measurements; velocity, flow, and level measurements; and viscosity and surface tension measurements. In this section, some sensors associated with mechanical variables are discussed.

2.16.1   Position, Displacement, and Proximity Sensors

The measurement of distances or displacements is an important aspect in many industrial, scientific, and engineering applications. Sensors for position and displacement sensing may be based on many different principles such as capacitive, inductive, magnetic, optical, ultrasound, and those based on variations in resistances.

There are many different types of position and displacement sensors that are used in electronic portable instruments. Designing and selecting position and displacement sensors depends on the type of displacement (linear, nonlinear, rotary), the required resolution, type of material being applied for, environmental conditions, availability of power, and so on. For example, microswitches are small devices and they can be easily mounted on surfaces of objects; hence, they are used in many position and displacement sensing applications. Alternatively, interruption of light beams can be used instead of mechanical switches. In some cases, the location of a reflected beam can be used to measure the distances of objects. Other methods may include potentiometers, ferrite core inductors with movable cores, and variable capacitors to track the movements of an attached object.

In this section, a brief visit will be paid to capacitive, magnetic, optical, ultrasonic, and microwave displacement sensors since they are used most frequently. In many applications, these sensors satisfy the requirements of high linearity, good sensitivity, and wide range, which can vary from a few centimeters to a few nanometers in distance.

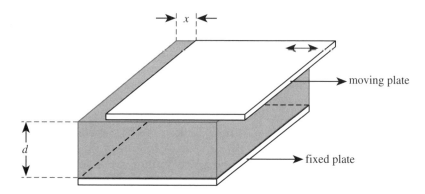

FIGURE 2.43
A variable-area displacement sensor.

2.16.1.1   *Capacitive Position, Displacement, and Proximity Sensors*

The basic sensing element of a typical displacement sensor consists of two simple electrodes with capacitance, C, as shown in Figure 2.9. The capacitance is a function of the distance, d (cm), between the electrodes of a structure; the surface area, A (cm²), of the electrodes; and the permittivity, ε (= 8.85 pF/m for air) of the dielectric between the electrodes. Therefore,

$$C = f(d, A, \varepsilon) \tag{2.36}$$

There are three basic methods for realizing a capacitive displacement sensor, i.e., by varying d, A, or ε. As an example, the variable-area and variable-distance displacements sensor are discussed below.

2.16.1.1.1   *Variable-Area Capacitive Displacement Sensor*

The displacements may be sensed by varying the surface area of the electrodes of a flat-plate capacitor, as illustrated in Figure 2.43. In this case, the capacitance would be

$$C = \varepsilon_r\varepsilon_0 \, (A - wx)/d \tag{2.37}$$

Then the transducer output is linear with displacement x. This type of sensor is normally implemented as a rotating capacitor for measuring angular displacement. The rotating capacitor structures are also used as an output transducer for measuring electric voltages as capacitive voltmeters.

2.16.1.1.2   *Variable-Distance Capacitive Displacement Sensor*

If a capacitor is made from two flat coplanar plates separated by a variable distance x, then, ignoring fringe effects, the capacitance can be expressed by

$$C(x) = \varepsilon \, A/x = \varepsilon_r\varepsilon_0 \, A/x \tag{2.38}$$

where ε is the dielectric constant or permittivity, ε_r is the relative dielectric constant (in air and vacuum, $\varepsilon_r \approx 1$), $\varepsilon_0 = 8.854188 \times 10^{-12}$ F/m ($10^{-9}/36\pi$ F/m) is the dielectric constant of vacuum, x is the distance of the plates, and A is the effective area of the plates.

The capacitance value is a nonlinear function of distance x, having hyperbolic transfer function characteristics. The sensitivity of capacitance to changes in plate separation may be found as

$$dC/dx = -\varepsilon_r\varepsilon_0 \, A/x^2 \qquad (2.39)$$

Equation 2.39 indicates that the sensitivity increases as x decreases. Nevertheless, from Equation 2.38 and Equation 2.39, it can be proven that the percent changes in C are proportional to the percent changes in x, which can be expressed as

$$dC/C = -dx/x \qquad (2.40)$$

These types of sensors are often used for measuring small incremental displacements without making contact with the object.

2.16.1.1.3 Capacitive Proximity Sensors

Capacitive proximity sensors are based on stray capacitance that exists between a metal plate and Earth, a quantity that can be altered by the presence of materials, which could be in solid, liquid, or powder form. Often, a bridge or resonant circuit detection methods are used.

2.16.1.2 Magnetic Position, Displacement, and Proximity Sensors

Various forms of magnetic sensors are used for position and displacement sensing. Many of these sensors have been explained in Section 2.2. In addition to those, other popularly used magnetic position and displacement sensors used in portable instruments can be listed as follows:

- Inductive sensors
- Hall-effect sensors
- Magnetoresistive sensors
- Eddy current proximity sensors

2.16.1.2.1 Inductive Proximity Detectors

These sensors operate on the principle that inductance of a coil changes in the presence of metal in the core or close by it. The coil can be part of a bridge circuit or the inductor of a tuned circuit. When used as part of a bridge, the presence of metal close to the coil forces the bridge output to be off balance. In the case of tuned circuits, change in the tuned frequency can easily be processed.

2.16.1.2.2 *Hall-Effect Sensors*

Hall-effect sensors are usually fabricated as monolithic silicon chips and encapsulated into epoxy or ceramic packages. There are two types that are used in portable instruments: linear Hall-effect sensors and threshold Hall-effect sensors. Linear sensors incorporate an amplifier and other electronic circuits for interfacing with the peripheral. Compared to basic Hall-effect sensors, they tend to be more stable and operate over wider voltage ranges. Threshold sensors have a Schmitt trigger detector with a built-in hysteresis to eliminate spurious oscillations.

2.16.1.2.3 *Magnetoresistive Sensors*

Magnetoresistive sensors are based on the changes in resistance of some materials, such as permalloys, in the presence of magnetic fields. In these materials, current passing through the material magnetizes the material in a particular magnetic orientation. The resistance is highest when the magnetization is parallel to the current and lowest when it is perpendicular to the current. Hence, depending on the intensity of the external magnetic field, the resistance of the permalloy changes in proportion. The magnetoresistive sensors are manufactured as thin films and are usually integrated as a part of an appropriate bridge circuit. They have good linearity and low temperature coefficients. These devices have a sensitivity ranging from 10^{-6} T to 50 mT. By choice of good electronic components and suitable feedback circuits, the sensitivity can be as low as 10^{-10} T. They can operate from DC to frequencies in several gigahertz.

2.16.1.2.4 *Eddy Current Sensors*

Eddy current sensors are inductive type sensors. Most of them have two coils in the shape of probes. One of the coils, known as the active coil, is influenced by the presence of the conducting target. The second coil, known as the balance coil, serves as a bridge that balances the circuit. It also provides temperature compensation. The magnetic flux from the active coil passes into the conductive target from the probe. When the probe is brought close to the target, the flux from the probe links with the target, producing eddy currents within the target. The eddy currents in the target alter the magnetic properties of the sensors, thus yielding unbalanced conditions of the bridge.

2.16.1.3 **Other Position, Displacement, and Proximity Sensors**

There are many other sensors that are used for position, displacement, and proximity sensing. Most of these sensors can also be used as motion detectors by suitable signal processing and display techniques. Some of these sensors are:

- Optical sensors
  - Proximity detectors with polarized light

- Photobeam sensors
- Fiber-optic sensors
- Grating sensors such as encoded discs
- Position-sensitive detectors
- Ultrasonic sensors
- Microwave and Doppler radar sensors

2.16.1.3.1 Optical Sensors

Some optical sensors are already explained in Section 2.14, dedicated to large-scale position and distance sensors. Here, only the position-sensitive detectors (PSDs) and photobeam sensors will be explained.

2.16.1.3.1.1   Position-Sensitive Detectors — Optical systems operating near an infrared region can be very effective in sensing short- and long-range displacements and positions. As in the case of cameras, PSDs are used in conjunction with light-emitting diodes. A PSD operates on the principle of the photoelectric effect. The surface resistance of a silicon photodiode changes depending on the light that falls onto it. It can be a one-dimensional or two-dimensional sensor. As illustrated in Figure 2.44, a one-dimensional sensor is fabricated with high-resistance silicon with p-and n^+-type layers. It has two electrodes to form contact with the p-type. The photoelectric effect occurs in the upper pn-junction.

2.16.1.3.1.2   Photobeam Sensors — Optical position, displacement, and proximity sensors use mostly reflection techniques; the transmitted beam reflects from the stationary or moving object. The light source is usually a LED working in the visible red or invisible infrared region. Visible red makes the setting up easier, but the infrared type is less affected by interference from other light sources. The sensors make use of the pulse-modulated signals and polarizing filters to avoid stray light and multiple reflection effects.

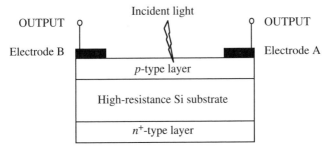

FIGURE 2.44
A one-dimensional silicon position-sensitive detector.

2.16.1.3.2   Ultrasonic Sensors

Ultrasonic type sensors can make use of diffuse-scan mode or through-scan mode. The diffuse-scan mode uses a single transducer as a transmitter and receiver, whereas the through-scan mode uses two transducers. For instance, in proximity detection, applications' using an ultrasonic frequency of 215 kHz is an industrial norm. The transmitted signal is modulated, using frequencies in the range of 30 to 360 Hz to minimize interference.

2.16.1.3.3   Microwave and Doppler Radars

In microwave and Doppler radars, the sensors contain Gunn diode oscillators (about 10-mW output) located in launching horns, and the receiving horns contain mixer diodes. The microwave beam, typically 10.7 GHz, is generated by the Gunn diode, and a low-frequency signal is generated when any object in the field of view moves. This signal is extracted suitably to work as a proximity detector of the moving object.

2.16.2   Pressure Sensors

Pressure is defined as the normal force exerted on the unit surface area of an object. The SI unit of pressure is the Pascal (Pa), after Blaise Pascal (1623–1662), which is defined as newtons per square meter (Nm^{-2}). There are three types of pressure measurements that can be performed:

- Absolute pressure — the pressure difference between the point of measurement and perfect vacuum where pressure is zero
- Gauge pressure — the pressure difference between the point of measurements and the ambient pressure
- Differential pressure — the pressure difference between two points

Some pressure sensors involve large mechanical parts and permanent linkages that make them unsuitable for portable instruments. However, many electronic portable instruments are available, particularly for gas pressure measurements. The type of sensors suitable for a particular application depends on characteristics of the object under investigation, such as gas, air, or liquid. Also, there are different types of sensors for low-pressure, high-pressure, and differential pressure measurements.

Most pressure sensors operate on the principle of converting pressure to some form of physical displacement, such as compression of a spring or deformation of an object. If displacement is used, there are many options for converting displacement into electrical signals. However, because of the heavy reliance on the physical displacements, many pressure sensors are sensitive to vibration and shock, which introduces a drawback to their use in portable instruments.

Pressure can be sensed directly or indirectly. Indirect methods rely on the action of pressure to cause displacement of a diaphragm, a piston, or other

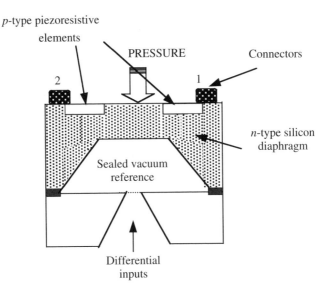

FIGURE 2.45
A semiconductor's piezoresistive pressure sensor.

device, so that the electronic sensing of displacement can be related to the pressure. The most commonly employed method is indirect pressure sensing. However, there are both direct and indirect pressure sensors for atmospheric pressure sensing (about 101.3 kPa).

Indirect pressure sensors are commonly known as aneroid barometers. There are many different versions of aneroid barometers depending on the sensing elements of the electrical components. Figure 2.45 illustrates a semiconductor piezoresistive indirect pressure sensor. In this sensor, *p*-type doping material is introduced into the *n*-type silicon diaphragm using ion implementation technology. Another method of sensing pressure is the use of piezoelectric crystals such as PZT. These sensors are useful to measure pressures that vary in time.

A common method of measuring pressure is the *aneroid barometer*, which is a sealed bellows that expands and contracts as the external pressure changes. Some suitable displacement sensors detect the movement of the bellows.

The pressure sensors can also be divided into two main groups: low-pressure sensors and high-pressure sensors (discussed below).

2.16.2.1   Low-Pressure Sensors

Low-pressure sensing involves much more specialization tailored for the characteristics of a particular application. The most common sensors are the Pirani gauges and the ion gauges.

A low-pressure Pirani gauge uses the principles of thermal conductivity of gases that decrease in proportion to pressure for a wide range of low pressures. It uses a hot wire element. The principle of the device is that less heat is conducted through the gas as the gas pressure around the wire is reduced. The resistance of the wire, which is dependent on the heat transferred by the gas, is measured by a suitable bridge network.

Ionization gauges use a stream of electrons to ionize the gas of which the pressure is measured. The positive gas ions are then attracted to a negative-charge electrode, and the amount of current carried by these ions is measured. Since the number of ions per unit volume depends on the number of atoms per unit volume, the readings of currents can be related to the pressure. Ionization gauges are suitable for very low pressure measurements, down to 10^{-7} Pa. There are many different types of these gauges, but the principles of operation are the same.

2.16.2.2   High-Pressure Sensors

A common device used in electronic portable instruments is the *capacitance manometer*. The capacitance manometer contains a large diaphragm serving as one plate of the capacitor. The movement of the diaphragm causes the capacitance to change, and the change is detected and processed by an on-chip electronic circuit. Capacitive manometers are available for measuring pressure differences as well as absolute pressures.

Many semiconductor capacitive pressure sensors are made from two parallel metal plates. One of the metal plates is fixed, and the other plate forms a flexible diaphragm, as shown in Figure 2.46. The flat diaphragm is clamped around its circumference and is bent into a curve by applied pressure, P. The deflection, y, of this system at any radius r is given by

$$y = 3 \, (1 - v^2) \, (a^2 - r^2) \, P/16 \, Et^3 \tag{2.41}$$

where a is the radius of the diaphragm, t is the thickness of the diaphragm, E is Young's modulus, and v is Poisson's ratio.

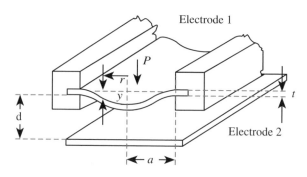

FIGURE 2.46
A two-plate capacitive pressure sensor.

Deformation of the diaphragm means that the average separation of the plates is reduced. Hence, the resulting increase in the capacitance, ΔC, can be calculated by

$$\Delta C/C = (1 - v^2)\, a^4\, P/16\, Et^3 \qquad (2.42)$$

where d is the initial separation of the plates and C is the capacitance at zero pressure.

There are many semiconductor pressure sensors that operate entirely on silicon chips. These integrated circuits contain the sensors and the associated signal processing electronics in them. They can measure pressure in the range of 1 to 100 psi.

2.16.3 Force Sensors

Force sensors make use of physical behavior of the body under external force. Force can be measured by many different methods, for example:

1. Balancing the force against a standard mass through a system of levers
2. Measuring the acceleration of a known mass
3. Equalizing the force to magnetic force by suitable coil and magnet arrangements
4. Measuring the pressure caused by the force on a surface
5. Converting the force into the deformation of an elastic element

Consequently, there is a vast range of sensors suitable for force measurements:

- Strain gauge load cells
- Piezoelectric sensors
- Resistive methods
  - Force sensing resistors (conductive polymers)
  - Magnetoresistive force sensors
- Inductive force sensors
- Magnetoelastic force sensors
- Piezotransistors
- Capacitive force sensors

The basic principles of most of these sensors have been discussed in various related places of this book.

2.16.4   Acceleration Sensors

Acceleration is measured by accelerometers as an important parameter for general-purpose absolute motion measurements and vibration and shock sensing. Modern accelerometers are well advanced and are manufactured as intelligent sensors representing an important family of advanced IC technology. Accelerometers find a diverse range of applications and are commercially available in a wide variety of ranges and types to meet specific requirements. They are manufactured to be small in size, light in weight, and rugged and robust to operate in harsh environments. They can be configured as active or passive sensors. An active accelerometer (e.g., piezoelectric) gives an output without the need for an external power supply, while a passive accelerometer only changes its electric properties (e.g., capacitance) and requires an external electrical power. In applications, the choice of active or passive type accelerometers is important, since active sensors cannot measure static or DC mode operations. For true static measurements, passive sensors must be selected.

Accelerometers can be classified in a number of ways, such as deflection or null-balance types, mechanical or electrical types, and dynamic or kinematic types. The majority of industrial accelerometers can be classified as either deflection types or null-balance types. Those used in vibration and shock measurements are usually the deflection types, whereas those used for measurements of motions of vehicles, aircraft, etc., for navigation purposes may be either type. In general, null-balance types are used when extreme accuracy is needed.

A large number of practical accelerometers are the deflection type; the general configuration is shown in Figure 2.47. There are many different deflection type accelerometers. Although the principles of operation are similar, they differ in minor details, such as the spring elements used, types of damping provided, and types of relative motion transducers employed.

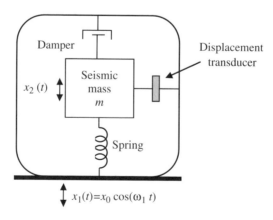

FIGURE 2.47
A typical seismic accelerometer.

Dynamic accelerometers are based on the principles of measuring the force required to constrain a seismic mass to track the motion of the accelerated base. Although applicable to all, the mathematical treatment of the dynamic response of an accelerometer is a second-order system, which can lead to oscillations. A good example of such accelerometers is the spring-constrained slug type. The kinematic accelerometer is based on the timing of the passage of an unconstrained proof mass from the spaced points marked on the accelerated base. They find highly specific applications such as in interspace spacecrafts and gravimetry type measurements.

For practical purposes, accelerometers can also be classified as mechanical or electrical type. This classification depends on whether the restoring forces and the measuring mechanisms are based on the mechanical properties, for example, law of motion, distortion of a spring, or fluid dynamics, or on electrical or magnetic properties.

2.16.4.1 Seismic Accelerometers

Seismic accelerometers make use of a seismic mass that is suspended by a spring or lever inside a rigid frame. The schematic diagram of a typical seismic accelerometer is shown in Figure 2.47. The frame carrying the seismic mass is connected firmly to the vibrating source whose characteristics are to be measured. As the system vibrates, the mass tends to remain fixed in its position so that the motion can be registered as a relative displacement between the mass and the frame. An appropriate transducer senses this displacement, and the output signals are processed further. Nevertheless, the seismic mass does not remain absolutely steady, but for selected frequencies it can satisfactorily act as a reference position.

2.16.4.2 Piezoresistive Accelerometers

Piezoresistive accelerometers are essentially semiconductor strain gauges with large-gauge factors. High-gauge factors are obtained since the material resistivity is dependent primarily on the stress, and not only on the dimensions of the device. This effect can be greatly enhanced by appropriate doping of semiconductors such as silicon. The increased sensitivity is critical for vibration measurements since it allows miniaturization of the accelerometer. Most piezoresistive accelerometers use two or four active gauges arranged in a Wheatstone bridge form. Extra precision resistors are used as part of the circuit in series with the input for controlling the sensitivity, balancing, offsetting and temperature compensation.

2.16.4.3 Piezoelectric Accelerometers

Piezoelectric accelerometers are used widely for general-purpose acceleration, shock, and vibration measurements. They are basically motion transducers with large output signals and comparatively small size. They are available with very high natural frequencies and are therefore suitable for high-frequency applications and shock measurements.

2.16.4.4   Microaccelerometers

By the end of the 1970s, it became apparent that planar processing and IC technology could be modified to fabricate three-dimensional electromechanical structures, called micromachining. Accelerometers and pressure sensors were among the first IC sensors. The first accelerometer was developed in 1979. Since then the technology has been progressing steadily, and now an extremely diverse range of accelerometers are readily available. Most sensors use bulk micromachining rather than surface micromachining techniques. In bulk micromachining, the flexures, resonant beams, and all other critical components of the accelerometer are made from bulk silicon in order to exploit the full mechanical properties of silicon crystals. With proper design and film process, bulk micromachining yields extremely stable and robust accelerometers.

An example of microaccelerometers is the vibrating beam accelerometer, also termed resonant-beam force transducer; it is made in such a way that acceleration along a positive-input axis places the vibrating beam in tension. Thus, the resonant frequency of the vibrating beam increases or decreases with the applied acceleration. A mechanically coupled beam structure also known as a double-ended tuning fork (DETF) is shown in Figure 2.48.

In DETF, an electronic oscillator capacitively couples energy into two vibrating beams to keep them oscillating at their resonant frequency. The beams vibrate 180° out of phase to cancel reaction forces at the ends. The dynamic cancellation effect of the DETF design prevents energy from being lost through the ends of the beam. Hence, the dynamically balanced DETF resonator has a high Q factor, which leads to a stable oscillator circuit. The acceleration signal appears as an output from the oscillator in the form of a frequency-modulated square wave, which can be used for digital interface.

There are other accelerometers with operational principles very similar to those of capacitive force–balance or vibrating beam accelerometers, discussed earlier. Manufacturing techniques may change from one manufacturer to another. Nevertheless, in general, vibrating beam accelerometers are preferred because of better air gap properties and improved bias performance characteristics.

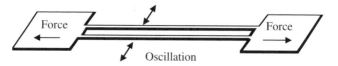

FIGURE 2.48
A double-ended tuning fork accelerometer.

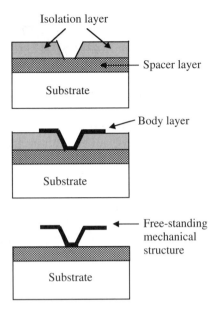

FIGURE 2.49
Steps of surface machining.

In general, the selective etching of multiple layers of deposited thin films, or surface micromachining, allows all kinds of movable microstructures to be fabricated on silicon wafers. With surface micromachining, layers of structure material are disposed and patterned as shown in Figure 2.49. The structure is formed by polysilicons and sacrificial materials such as silicon dioxides. The sacrificial material acts as an intermediate spacer layer and is etched away to produce a freestanding structure. Surface machining technology also allows smaller and more complex structures to be built in multiple layers on a single substrate.

2.16.4.5 *Vibrating Beam Accelerometer*

The vibrating beam accelerometer is similar in philosophy to the vibrating string accelerometer with an advantage that the frequency output provides easy interface with digital systems. Also, measurement of the output voltage phase leads to velocity determinations by a simple step of integration. Static stiffness eliminates the tension and makes the device much smaller. The recent trend is that most vibrating beam accelerometers are manufactured as micromachined devices. With differential frequency arrangements, many common mode errors can be eliminated, including the clock errors within the chip.

The frequency of resonance of the system must be much higher than any input acceleration, and this limits the measurable range. In a micromachined accelerometer, used in military applications, the following typical characteristics may be observed: range of ±1200g, sensitivity of 1.11 Hz/g, bandwidth of 2500 Hz, and unloaded DETF frequency of 9952 Hz. The frequency at +1200g is 11,221 Hz, the frequency at –1200g is 8544 Hz, and the temperature sensitivity is 5 mg/°C. The accelerometer size is 6 mm in diameter by 4.3 mm in length, with a mass of about 9 g. It has a turn-on time of less than 60 sec, the accelerometer is powered with +9 to +16 V DC, and the nominal output is 9000-Hz square waves.

Surface micromachining has also been used to manufacture application-specific accelerometers, such as those suitable for air bag applications in the automotive industry. In one type of such an accelerometer, a three-layer differential capacitor is created by alternate layers of polysilicon and phosphosilicate glass on a 0.38-mm-thick, 100-mm-long wafer. A silicon wafer serves as the substrate for the mechanical structure. The trampoline-shaped middle layer is suspended by four supporting arms. This movable structure is the seismic mass for the accelerometer. The upper and lower polysilicon layers are fixed plates for the differential capacitors. The glass is sacrificially etched by hydrofluoric acid.

2.16.4.6   Differential-Capacitance Accelerometers

Differential-capacitance accelerometers are based on the principle of the change of capacitance in proportion to applied acceleration. They come in different shapes and sizes. In one type, the seismic mass of the accelerometer is made as the movable element of an electrical oscillator, as shown in Figure 2.50. The seismic mass is supported by a resilient parallel-motion beam arrangement from the base. The system is set to have a certain defined nominal frequency when undisturbed. If the instrument is accelerated, the frequency varies above and below the nominal value depending on the direction of acceleration.

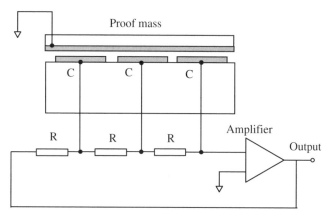

FIGURE 2.50
A typical differential capacitive accelerometer.

2.16.5 Stress and Strain Sensors

Strain is a fractional change in the dimensions of an object. An example of the strain is the change in the length of an object divided by its original length; therefore, the strain does not have any dimensions. The stress, on the other hand, is the force applied on an object divided by the area. Consequently, strain takes place as a result of stress. There are different types of stresses, such as tensile stress and bulk stress. Tensile stress is the applied force divided by the area over which it is applied. Bulk stress is the force per unit area. The most common strain sensors are produced for tensile strain measurements. The measurement of strain also allows calculation of the amount of stress by using the knowledge of the elasticity of the objects. The information of elastic modules such as Young's modulus, shear (twisting) modulus, and bulk (pressure) modulus of the objects is useful in conducting such calculations.

Sensing tensile strain involves the measurement of small changes in the length of a sample. The most common form of strain measurement uses resistive strain gauges, semiconductor strain gauges, piezoelectric strain gauges, and optical interferometry.

The principles behind the resistive, semiconductor, and piezoelectric types are explained elsewhere in this book. An advanced method measuring the strain is achieved by interferometry and fiber-optic sensors. In these methods, generally high sensitivities are achieved when lasers are used. Laser constitutes a coherent light beam resulting in a good interference effect. Other methods involve various types of fiber-optic sensors.

2.16.6 Mass and Weight Sensors

Mass is a quantitative measure of inertia of a body at rest. It is a product of density and volume. Volume is the amount of space that an object occupies. Weight, on the other hand, is the force with which a body is attracted toward the Earth, and it is determined by the product of mass and acceleration of gravity. The mass of an object is constant, but its weight on the surface of the Earth is not quite constant, because gravity varies slightly from one place to another depending on the height. In space, the weight of an object is almost zero.

There are many different types of devices to measure the mass of bodies such as equal arm balances, spring scales, beam balances, and electronic balances. The oldest method is the equal arm balance, which basically determines the force of gravity to find the difference in the weights of two objects. The modified version of the equal arm balance is the beam balance. Another type of mass measurement is the spring scale, which again uses the force of gravity as a reference scale. An adaptation of the spring scale is the torsion balance, which is suitable to measure very small objects, down in the few micrograms range. Nowadays, most of the mechanical-based balances are equipped with electronic components for display, accuracy, and precision purposes.

Electronic mass and weight sensors are based largely on strain gauges. As explained above, the strain gauge is an electrically resistive wire element that changes resistance when a force is applied to it. In the weight measurements, the strain gauge is bonded to a steel cylinder (or a cubic structure) that shortens when a body is placed on it. Since bonded to the cylinder, the gauge is shortened with the cylinder, varying the electrical resistance in proportion to the compression. The change in the resistance of the gauge can electronically be made to read out the mass in required units.

2.16.7   Density Sensors

The density of a substance is defined as the mass per unit volume ($\rho = m/V$) under fixed conditions. The term is applicable to solids, liquids, and gases. Density depends on temperature and pressure. This dependence is much greater in gases. Although, there are many different units in use, usually the values of densities are given in terms of grams per cubic centimeter.

Specific gravity (SG) is an abstract number expressing the ratio of the density of one substance to the density of another reference substance, both obtained at the same temperature and pressure. For solids and liquids, water is taken as the reference substance, whereas air is the reference for gases. The specific gravities of solids, liquids, and gases under reference conditions may be expressed by

Liquid (or solid) SG = density of liquid (or solid)/density of water

Gas SG = density of gas/density of air

Care must be taken to define standard conditions under which the densities or specific gravities are measured, so as not to introduce errors due to variations in measurement conditions. Commonly accepted sets of conditions are *normal temperature and pressure* (NTP) and *standard temperature and pressure* (STP). NTP is usually used for solids at the temperature of 0°C with a pressure of 760-mm mercury. STP is used for solids and fluids at a temperature of 15.6°C with a pressure of 1 atm (or 101.325 kPa).

Density measurements are a significant part of electronic portable instruments. In many processes, the density is taken as the controlling parameter for the rest of the process; therefore, accurate measurements are necessary. Density measurements are made for at least two important reasons: (1) to determine the mass and volume of products, and (2) to assess the quality of products. In many industrial applications, density measurement ascertains the value of the product.

In many modern applications, densities are obtained by sampling techniques. However, there are two basic approaches: *static density measurements* and *dynamic (on-line) density measurements*. Within each concept, there are many different methods available, depending on physical principles and process characteristics. In many cases, application itself and process conditions determine the best suitable method to be employed. Generally, static methods are well developed, lower in cost, and more accurate. Dynamic samplers are expensive and highly automated. Nevertheless, nowadays many static methods are also computerized, offering easy-to-use, flexible, and self-calibrating features.

There is no single universally applicable density measurement technique available. Different methods must be employed for solids, liquids, and gases. The measurements of densities of fluids are much more complex than those of solids; therefore, there are many different techniques developed. Hydrometers, pycnometers, hydrostatic weighing, flotation methods, drop methods, radioactive methods, optical methods, etc., are typical examples of measuring liquid densities. Flask methods, gas balance methods, optical methods, and x-ray methods are typical techniques employed for gas density measurements. Some of these sensors that are relevant to electronic portable instruments, particularly in portable laboratory instruments, will briefly be explained next.

2.16.7.1 *Magnetic Methods*

Magnetic methods are used for both liquids and gases. These methods allow the determination of effects of pressures and temperatures down to the cryoscopic range. Basically, these devices contain a small ferromagnetic cylinder, encased in a glass jacket. The jacket and ferromagnetic material combination constitutes a buoy or float. The cylinder is held at a precise height within the medium by means of solenoid, which is controlled by a servo system integrated with a height sensor. The total magnetic force on the buoy is the product of the induced magnetic moment by the solenoid and the field gradient in the vertical direction. The total magnetic force at a particular distance in the vertical direction in the solution compensates the difference in the opposing forces of gravity (downward) and buoyancy (upward) exerted by the medium, through Archimedes' principle. The magnetic force is directly proportional to the square of the current in the solenoid. If the buoyant force is sufficient to make the ferromagnetic assembly float on the liquids of interest, the force generated by the solenoid must be downward to add to the force of gravity for equilibrium.

2.16.7.2 *Vibrational Methods*

Vibrational methods make use of the changes in the natural frequency of the vibration of a body containing fluid in it or surrounded by it. The natural frequency of the vibrating body is directly proportional to the stiffness and inversely proportional to the combined masses of the body and the fluid. It also depends on the shape, size, and elasticity of materials, induced stresses, and the total mass and mass distribution of the body. There are different types, such as vibrating tube densitometers, vibrating cylinder densitometers, and tuning fork densitometers. These densitometers make use of the natural frequency of a low-mass tuning fork. In some cases, the liquid or gas is taken into a small chamber in which the electromechanically driven forks are situated. In other cases, the fork is inserted directly into the liquid. Calibration is necessary in each application.

2.16.7.3   Pycnometric Densitometers

Pycnometers are static devices used for measuring densities of liquids and gases. They are manufactured as fixed-volume vessels that can be filled with sample fluids. The density of the fluid is measured by weighing the sample with the vessel. The simplest version consists of a glass vessel in the shape of a bottle with a long stopper containing a capillary hole. The volume and variation of volume with temperature have accurately been determined. The capillary is used to determine the exact volume of the liquid, thus giving high resolution when filling the pycnometer.

The pycnometers have to be lightweight and strong enough to contain samples. They need to be nonmagnetic for accurate weighing to eliminate possible ambient magnetic effects. Very high resolution balances have to be used to detect small differences in weights of gases and liquids. Although many pycnometers are made of glasses, they are also made of metals to give enough strength for the density measurements of gases and liquids at extremely high pressures. In many cases, metal pycnometers are necessary to take samples from the line of some rugged processes.

2.16.7.4   Hydrostatic Weighing Densitometers

Hydrostatic weighing densitometers are suitable for solid and liquid density measurements. The density of solid is often measured by weighing it first in air, and then again after having submersed it in a suitable liquid of known density. The latter weighing is done by suspending the solid under the pan of a precision balance by means of a very thin wire.

The hydrostatic weighing methods of liquids give continuous readings for two phase liquids, such as slurries, sugar solutions, powders, etc. They are rugged, give accurate results, and are used for the calibration of the other liquid density transducers. However, they must be installed horizontally on a solid base; hence, they are not flexible enough to adapt for any process, and the process must be designed around them.

2.16.7.5   Hydrometers

Hydrometers are direct reading instruments, most commonly used for measurement of the density of liquids. They are used so much that their specifications and procedure of use are described by national and international standards such as ISO 387. The buoyancy principle is used as the main technique of operation. Almost all hydrometers are made from high-grade glass tubing. The volume of fixed mass is converted to a linear distance by a sealed bulb-shaped glass tube containing a long-stem measurement scale.

The bulb is ballasted with a lead shot and pitch, the mass of which is dependent on the density range of the liquid to be measured. The bulb is simply placed into the liquid, and the density is read from the scale. The scale may be graduated in density units such as kg/m^3. Hydrometers can be calibrated for different ranges for surface tensions and temperatures. Temperature corrections can be made for set temperature such as 15, 20, and 25°C. ISO 387 covers a density range of 600 to 2000 kg/m^3.

2.16.7.6   Column Type Densitometers

Column type densitometers are used for liquid density measurements. There are a number of different versions of column methods. A known head of sample liquid and water from their respective bubbler pipes is used. A differential pressure measuring device compares the pressure differences, proportional to relative densities of the liquid and the water. By varying the depth of immersion of the pipes, a wide range of measurements may be obtained. Both columns must be maintained at the same temperature in order to avoid the necessity for corrections of temperature effects.

A simpler and the most widely used method of density measurement is achieved by the installation of two bubbler tubes. The tubes are located in the sample fluid such that the end of one tube is higher than that of the other. The pressure required to bubble air into the fluid from both tubes is equal to the pressure of the fluid at the end of the bubbler tubes. The openings of the tubes are fixed; hence, the difference in the pressure is the same as the weight of a column of liquid between the ends. Therefore, the differential pressure measurement is equivalent to the weight of the constant volume of the liquid, and calibrations can be made in direct relationship to the density of the liquid. This method is accurate to within 0.1 to 1% specific gravity. It must be used with liquids that do not crystallize or settle in the measuring chamber.

2.16.7.7   Refractometric Method

The refractometric method is suitable for density measurement of gases and clear liquids. It essentially uses optical instruments operating on the principles of refraction of light traveling in liquid or gas media. Depending on the characteristics of the samples, measurement of refractive index can be made in a variety of ways; critical angle, collimation, and displacement techniques are just a few. Usually, an in-line sensing head is employed whereby a sensing window, commonly known as a prism, is wetted by the product to be measured. In some versions, the sensing probes must be installed inside the pipelines or in tanks and vessels. They are most effective in reaction type process applications where blending and mixing of liquids takes place. For example, the refractometers can measures dissolved soluble solids accurately.

Infrared diodes, lasers, and other lights may be used as sources. However, this measurement technique is not recommended in applications in processes containing suspended solids, high turbidity, entrained air, heavy colors, poor transparency and opacity, or extremely high flow rates. The readings are automatically corrected for variations in process temperature. The processing circuitry may include signal outputs adjustable in both frequency and duration.

2.16.7.8   Absorption Type Densitometer

Absorption techniques are also used for density measurements in specific applications. X-rays, visible light, ultraviolet (UV), and sonic absorptions are typical examples of this method. Essentially, attenuation and phase shift of a generated beam going through the sample are sensed and related to the density of the sample. Most absorption type densitometers are custom designed for applications having particular characteristics. Two typical examples are (1) UV absorption or x-ray absorption, used for determining the local densities of mercury deposits in arc discharge lamps, and (2) ultrasonic density sensors, used in connection with difficult density measurements such as the density measurement of slurries. The lime slurry, for example, is a very difficult material to handle. It has a strong tendency to settle out and to coat all equipment it comes in contact with. An ultrasonic density control sensor can fully be emerged into agitated slurry, thus avoiding the problems of coating and clogging. Since the attenuation of the ultrasonic beam is proportional to the suspended solids, the resultant signal is proportional to the specific gravity of the slurry. Such devices can give accuracy up to 0.01%. The ultrasonic device measures the percentage of the suspended solids in the slurry by providing a close approximation of the specific gravity.

2.16.8   Viscosity Sensors

The viscosity of a liquid or gas is the quantity that corresponds to friction between solids. Electronic sensors of viscosity make use of damping of mechanical oscillations by viscous liquids or gases. A typical example involves the vibration of a plunger by a piezoelectric transducer, and amplitude of vibration is sensed by another transducer. This arrangement constitutes part of an oscillating circuit; the damping effect of liquid on the amplitude of oscillation is dependent on the viscosity of the liquid.

3

Digital Aspects: Hardware, Software, and Electronic Portable Instruments

Introduction

Today's modern electronic portable instruments are almost exclusively digital. There are only a few that are nondigital. This is mainly due to the advantages in the market position and to keep the cost of the instrument down. This section is dedicated entirely to digital aspects of portable instruments. Much of the information is common to all digital devices, computers, and microcontroller and microprocessor systems, as well as to electronic portable instruments. In this chapter, basic architectural aspects of digital systems relevant to modern electronic portable instruments will be highlighted and discussed in detail.

Most sensors, transducers, and actuators are basically analog devices. That is, they generate or operate on analog signals. In all types of instruments, including the portables, signals generated by sensors can be processed in three different ways: (1) by analog techniques that directly deal with analog signals, (2) by converting analog signals into digital forms and implementing the systems as digital instruments, and (3) by dealing with the signals purely in digital forms as digital-to-digital inputs and outputs. These aspects, together with signal conversion methods, will be elaborated in greater detail in the sections to follow.

Digital systems have many advantages over their analog counterparts, particularly if the system is complex. This is due to the fact that these systems are based on digital processors, which provide powerful signal processing and data handling capabilities, which can be achieved simply by software and firmware implementations. Another important advantage of digital systems is that once the signals are converted to digital forms, they can be managed by all types of computers and other digital systems. This provides a wide range of possibilities for data processing, communications, storage, and visual displays.

Nowadays, due to easy and cost-effective availability of advanced microprocessors and their supporting components, portable instruments are based

FIGURE 3.1
Block diagram of a digital instrument.

on powerful digital systems integrated at various stages of the device starting right from sensing elements to complex communication capabilities. Being a stand-alone device is no longer a disadvantage for electric portable instruments. The integration of fast central processing units (CPUs), vast memories, and advanced remote or wireless communication techniques allows the portable instruments to function just like any other sophisticated nonportable device. Therefore, advanced modern digital hardware, software, and firmware are just as applicable to portable instruments as they are to any other digital system. For this reason, portable instruments are going through a revolutionary phase and replacing many classical nonportable counterparts. In doing so, at the same time, they are offering greater flexibility in their applications. In this chapter, general concepts of digital aspects will be discussed and information specifically applicable to portable instruments will be emphasized. The general aspects of digital systems will be introduced next.

3.1   Microprocessors, Microcontrollers, and Computers

Digitals systems are based on digital components such as microprocessors, microcontrollers, memory devices, digital inputs/outputs, and digital communication components. However, the operation of all digital instruments is governed by common mathematical theories, such as the sampling and the digital signal conditioning theories. The whole process of carrying out sampling, conversion, and signal conditioning is referred to as the data acquisition. In this chapter, the basic functions of data acquisition systems and the theoretical foundations will be discussed.

Microprocessors and microcontrollers constitute the heart of almost all types of modern digital instruments and instrumentation systems. They play an important role in data acquisitions, data processing, and control. A typical digital instrument consists of a number of subsystems, illustrated in Figure 1.5. The simplified form is shown in Figure 3.1.

Applications of microprocessors and microcontrollers in electronic portable instruments and instrumentation systems can be categorized according to the following roles:

1. *Data handling functions*: data acquisition, data processing, information extraction, data compression, interpretation, recording, storage, and communication

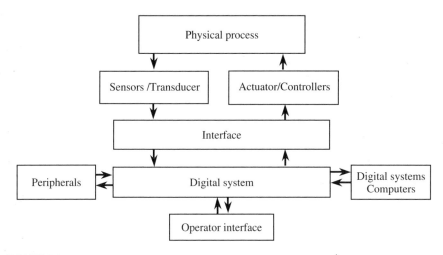

FIGURE 3.2
A typical digital process control system.

2. *Instrumentation control*: sensors, actuators, system resources, and process controls
3. *Human–machine interface*: one of the significant roles of digital systems to provide a meaningful and flexible user interface to the operators for ergonomic information and control display
4. *Experimentation and procedural development*: commissioning, testing, and general prototyping of the targeted system under investigation
5. *Reporting and documentation*: an important part for formulating the operational procedures and data recording

A branch of digital systems is the digital control system in which many instruments may be integrated. Mainly due to recent advances in communications technology, electronic portable instruments are gaining wider acceptance in digital control systems. Therefore, some background information will be provided.

In the case of digital control systems, we talk about instrumentation systems operating as one complete unit. Digital control systems are largely based on microprocessors, microcontrollers, and computers, and they enable the interface and control of external events in the digital environment. One of the main features of digital control systems is the inclusion of interface units that convert analog signals to digital, and digital information to analog forms, as illustrated in Figure 3.2. This figure also illustrates basic components of an electronic portable instrument.

3.1.1 Microprocessors and Microcontroller Architecture

In order to be able to understand the functionality of modern digital portable instruments, it is important to appreciate the basic architecture of micropro-

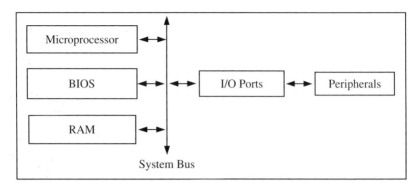

FIGURE 3.3
Block diagram of a computer.

cessors and microcontrollers. Microprocessors are integrated circuits that handle and process data in binary format. A general-purpose microprocessor is used as the CPU in computers. The most common manufacturers of microprocessor are Intel, Motorola, AMD, Philips, Zilog, Atmel, Hitachi, and many others. Microprocessors are the basic building blocks of modern computers, as illustrated in Figure 3.3.

Microcontrollers, on the other hand, are special microprocessors that have built-in memory and interface circuits within the same IC. Due to smaller sizes, simplicity and low costs, and power efficiency, microcontrollers are commonly used in electronic portable instruments and the associated instrumentation systems. Therefore, in this book, detailed information on microcontrollers will be provided. Typical families of such microcontrollers are the Motorola MC68HC11 and the PIC microcontroller. The basic architectures of a microcontroller are given in Figure 3.4.

The most commonly used microcontrollers, microprocessors, and computers in portable instruments and instrumentation systems can be categorized

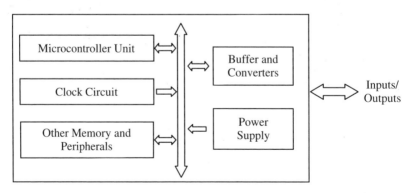

FIGURE 3.4
Block diagram of a microcontroller.

into four basic types: microcontrollers, embedded controllers, dedicated computers, and digital signal processors.

3.1.2 Microcontrollers

Prior to the design of any portable instrument, one needs to understand the features and limitations of the target microcontroller. During the designing stages, one might need to consider the number of input/output (I/O) ports needed for performing the required tasks, the amount of memory that is available to the user, and the number of data bits that the controller can handle, which in turn determines the accuracy and the speed of the system.

A microcontroller can be defined as a single-chip device, which has memory for storing information and is able to control read/write functions and manipulate data. The performance of a microcontroller is classified by its size, i.e., the number of bits that the CPU can handle at a time. There are many types or families of microcontrollers that are offered by a diverse range of manufacturers. Therefore, it is vital to understand the differences in the characteristics of different families in order to select an appropriate microcontroller for a specific task. Some of the microcontrollers available in the marketplace are listed in Table 3.1. This is by no means an exhaustive list; every day many microcontrollers are introduced in the market with new or improved features. Therefore, it is advised that designers of electronic portable instruments should conduct a comprehensive search on the availability of microcontrollers in the marketplace before they commit themselves to a particular one.

With reference to Figure 3.4, the architecture of a microcontroller can further be expanded as shown in Figure 3.5. The main components of a microcontroller perform the following functions:

1. *CPU* — Can be considered the main "brain" of the system. It executes instructions that are specially written to perform certain tasks. There are a number of functional units inside a CPU, such as the program counter; the registers are used to get the instructions and data bytes. Other registers in the CPU are used to store specialized data or address information.

2. *ROM* — The read only memory stores permanent program information that cannot be deleted. The ROM normally holds the application

TABLE 3.1

Types of Microcontrollers

Number of Bits	Family
8	Atmel's AVR 8-bit RISC, Cypress's CY8C25xxx/26xxx
16	Hitachi's H8/300H, Motorola MC9S12DP256
32	AT91 ARM Thumb, Motorola MC68F333

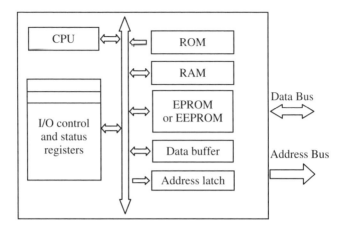

FIGURE 3.5
Block diagram of a microcontroller in expanded mode.

program, which can be fabricated through the manufacturing process.

3. *RAM* — The random access memory stores data, which can be subject to continual changes during the execution of the program. The data remain in the memory whenever the power supply is turned on. Each location of this type of memory can be accessed in the same amount of time.

4. *EPROM* — The erasable programmable ROM, which means that it has the same functionality as ROM; the only difference is that it can be erased and reprogrammed. This type of memory is nonvolatile as the name suggests. Exposing the EPROM to ultraviolet light for a period of time erases the program residing inside. Reprogramming requires special devices such as an EPROM programmer.

5. *EEPROM* — This type of memory can be erased electrically; therefore, it is called electrically erasable programmable ROM. It has the same functionality as EPROM, but the data can be programmed while the device is in circuit.

EPROM and EEPROM are used for two purposes. The first is for storing the customized data needed when different parameters are required. A typical use in instrument is the configuration and setup of the device for a particular application. The second purpose of this type of memory is that it has the ability to be reprogrammed, which provides a degree of flexibility for the designer to modify the program or use it as a backup system of the essential data in the event of power failure.

Programs for microcontrollers to perform specific tasks are written in either assembly codes or high-level languages such as Basic or C. The program written on other digital devices can be downloaded into the system memory through the serial communication port of the microcontroller. All high-level programs are compiled into machine language for execution. There are significant numbers of compilers available in the market for the different families of microcontrollers. In addition, microcontroller emulators are also available. The emulator of a specific microcontroller essentially is software that can be programmed as the microcontroller itself on a computer.

The emulator is capable of simulating the execution of programs being developed. Generally, the emulator software simulates all the I/O ports, registers, memory addresses, and features of the target microcontroller. During simulation, the designer is able to debug and test out the program. This helps the programmer to easily rectify any errors by checking the performance of the program against the status of the I/O ports and the registers. The emulation process avoids waste, as there may be no need for hardware devices until the programs are proven to be operational and correct.

The general features of microcontrollers consist of the following:

- Fabrication techniques
- Architecture features
- Advanced memory
- Power management
- I/O
- Interrupts

3.1.2.1   Fabrication Techniques

There are two common ways of fabricating microcontrollers: complementary metal oxide semiconductor (CMOS) and post metal programming (PMP) of National Semiconductor. CMOS is the most common technique and requires less power; hence, it permits longer battery operation. This technique also provides the advantage of possessing a sleep mode. It is resistant to noise such as power fluctuations or spikes. On the other hand, PMP is a process, which permits the ROM to be programmed after metalization. Usually the ROM pattern is specified in the early production stages.

3.1.2.2   Architecture Features

There are four types of architectures that microcontrollers are built upon:

1. Von Neumann architecture — Microcontrollers built upon this architecture have a single data bus, which is used to fetch instructions and data from a common main memory. When the microcontroller addresses the memory, two separate fetches occur; i.e., first instructions are fetched and then data are fetched. This type of architecture slows down the microcontroller's operation.

2. Harvard architecture — There are two separate buses in a microcontroller: data and instructions buses. Therefore, the fetching occurs in parallel, which is faster than the Von Neumann, but the complexity of the architecture increases.

3. Complex instruction set computer (CISC) — In a typical CISC microcontroller, there are 80 instructions within the structure to perform specific control tasks. These instructions include addressing registers, recognizing addressing modes, etc. The advantage of this architecture is that many of the instructions are macro-like where the programmer can use internal instructions in place of many simpler ones.

4. Reduced instruction set computer (RISC) — This type of architecture simply reduces the instructions within the structure. As a result, the chip is smaller and has lower power consumption and more space for enhancement features. Some of the features of a typical RISC processor are separate instruction and data buses, instruction pipelining, and orthogonal (symmetrical) instruction set for programming simplicity.

Common to most microcontrollers, the following are available:

1. *Advanced memory* — In addition to common memory such as ROM, EEPROM, and EPROM, advanced memory includes flash memory, battery backup-upstatic RAM, filed programming/reprogramming, and one-time programming.

2. *Flash memory* — A better solution than regular EEPROM when there is a large amount of nonvolatile program memory. It is faster and permits more erase/write cycles than EEPROM. The advantages are in-application programming, improved write/erase and data retention performance for flash (allowing users to define their own preferred programs), faster programming and erase times of the flash memory, flexible block protection and security, it can be used to emulate EEPROM, reduced code obsolescence/scrapped product, end-of-line customization for regional variations in consumer demands, and standardized platforms (reduces product variability).

3. *Battery backup-upstatic RAM* — Battery backup-upstatic RAM is useful when a large nonvolatile program and data space is required. A major advantage of static RAM is that it is much faster than other types of nonvolatile memories, so it is well suited for high-performance applications. Also, there are no limits as to the number of times that different programs may be written to; therefore, it is perfect for applications that keep and manipulate large amounts of data locally.

4. *Field programming/reprogramming* — This nonvolatile memory allows the device to be reprogrammed in the field without removing the

microcontroller from the system. Reprogrammable nonvolatile program memory on a portable instrument allows the instrument program to be modified during routine service to incorporate the latest features or to compensate for such factors as aging or even fixing bugs.

5. *One-time programmable (OTP)* — It is a programmable read only memory (PROM) device. The program cannot be erased or modified once it is written into the device with a standard EPROM programmer. This is usually used for limited production runs before a ROM mask is done in order to test codes. An OTP part uses standard EPROM. However, one problem with mask ROM is that it is economical only when the system manufacturer purchases large quantities of identically programmed micros.

6. *Power management* — Nowadays, there are many microcontrollers that run on low voltages, e.g., 3 V, which are most suitable for electronic portable instruments. Apart from electronic portable instruments, the use of low-voltage microcontrollers and supporting ICs is becoming popular in many other application areas too. As the trend continues, the voltage may even be lower in the near future.

Most microcontrollers can be placed into IDLE/HALT mode by the software control. In both HALT and IDLE modes, the state of the microcontroller remains unchanged, RAM is not cleared, and outputs remain unaltered. The terms *idle* and *halt* often have different definitions, depending on the manufacturer. Some might call idle, others may call halt. In the IDLE mode, all activities are stopped except:

- Associated on-board oscillator circuitry
- Watchdog timers
- Clock monitors
- Free-running timers

The power supply requirement on the microcontroller in the IDLE mode is typically around 30% of normal operational power. IDLE mode is exited by a reset, or some other stimulus (such as timer interrupt, serial port, etc.). The idle timer causes the chip to wake up at a regular interval to check on the system. The chip then goes back to sleep. IDLE mode is extremely useful in electronic portable instruments and in remote and unattended data-logging systems (such as remote weather station or power meter) where the microcontroller wakes up at regular intervals to take measurements, logs the data, and then goes back to sleep.

In HALT mode, all activities are stopped, which includes timers and counters. The only way to wake up the system is by a reset or device interrupt (such as an I/O port). The power requirements of the device are minimal, and the applied voltage can sometimes be decreased below operating voltage without altering the state (RAM/outputs) of the device. The current consumption is typically less than 1 μA. A common application of HALT mode is in laptop keyboards. In order to have maximum power saving, the controller is halted until it detects a keystroke via a device interrupt. It then wakes up, decodes and sends the keystroke to the host, and then goes back into HALT mode when it is not in use, waiting either for another keystroke, or information from the host.

3.1.2.3 Input/Output

The data communication between a microcontroller and the outside world is through I/O pins. The format of data transmission of a microcontroller can be universal asynchronous receiver transmitter (UART), a port adapter for asynchronous serial communications. There is a diverse range of I/O hardware and software. Some of the inputs and outputs suitable for electronic portable instruments are explained below.

1. *Universal synchronous/asynchronous receiver transmitter (USART)* — A serial port adapter for either asynchronous or synchronous serial communications. Communications using a USART are typically much faster (as much as 16 times) than those with a UART.

2. *Synchronous serial port* — Synchronous serial ports do not require start/stop bits, and they can operate at much higher clock rates than asynchronous serial ports. They are used to communicate with high-speed devices such as memory servers, display drivers, external analog-to-digital (A/D) ports, etc. In addition, they are suitable to implement simple microcontroller networks.

3. *Serial peripheral interface (SPI)* (Motorola) — A synchronous serial port. Another version is the serial communications interface (SCI), an enhanced UART (asynchronous serial port).

4. *I²C bus (interintegrated circuit bus)* (Philips) — A simple two-wire serial interface developed by Philips. It was developed for 8-bit applications and is widely used in consumer electronics and automotive and industrial applications. The I²C bus is a two-line multi-master, multislave network interface with collision detection. Up to 128 devices can be connected on the network, and they can be spread out over 10 m. Each node (microcontroller or peripheral) may initiate a message, and then transmit or receive data. The two lines of the network consist of the serial data line and the serial clock line. Each node on the network has a unique address, which accompanies any message passed between nodes. Since only two wires are needed, it is easy to interconnect a number of devices.

5. *Controller area network (CAN)* — A multiplexed wiring scheme that was developed jointly by Bosch and Intel for wiring in automobiles.

The CAN specification is being used in industrial control both in North America and Europe.

The I/O has other functions apart from simply inputting and outputting data, such as A/D and digital-to-analog (D/A) conversions.

3.1.2.4 Interrupts

Most microcontrollers have at least one external interrupt, where interrupts for rising or falling edges can be selected or externally triggered. The advantages of interrupts are that the speed of response to external events is faster and the software overhead of continually asking peripherals is reduced if they have any data ready, compared to the polling approach.

3.1.3 Examples of Microcontrollers

Some examples of 8-bit, 16-bit, and 32-bit microcontrollers that are suitable to be used in electronic portable instruments discussed next. It is important that these examples are given to highlight the features of different micro-controllers. It has to be remembered that the microcontroller market is dynamic, and many different types may be available at the time of design of your electronic portable instrument.

3.1.3.1 Cypress CY8C25xxx/26xxx

Cypress's microcontrollers are built upon the programmable system-on-chip technology, namely PSoC blocks. These analog and digital PSoC blocks implement the general-purpose microcontroller unit (MCU) peripherals as user modules. A user module is a specific peripheral function (i.e., timing, counting, pulse width modulation, etc.). It is created by setting the personality of one or several analog or digital PSoC blocks and then adjusting the parameters for the desired function. Using the PSoC, one is able to define the desired peripherals by selecting the corresponding user modules. This is done by using the PSoC designer software tool. The advantage of this is to enable designers to generate custom devices. These PSoC blocks can be connected in parallel or serial form to provide different functionalities. As a typical example of illustration of microcontroller architecture, the block diagram of an 8-bit microcontroller is shown in Figure 3.6. The functionalities of the microcontroller include:

- A/D and D/A converters
- Multipole filters and various gains
- Timer/counter
- Serial receiver/transmitter
- Cyclic redundancy code (CRC) generator

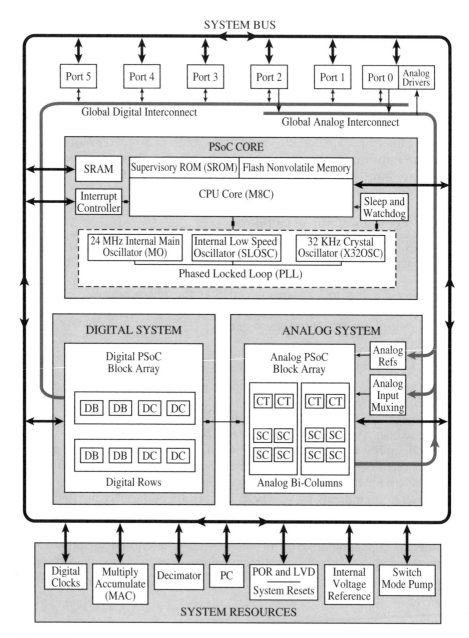

FIGURE 3.6
Block diagram of a microcontroller. (Courtesy of Cypress Corporation.)

TABLE 3.2

Key Feature of the Family of Cypress Microcontrollers

	8C25122A	8C26233A	8C26443A
Operating frequency	93.7 kHz–24 MHz	93.7 kHz–24 MHz	93.7 kHz–24 MHz
Operating voltage	3.0–5.5V	3.0–5.5 V	3.0–5.5 V
Program memory (Kbytes)	4	8	16
Data memory (bytes)	128	256	256
Digital PsoC blocks	8	8	8
Analog PsoC blocks	12	12	12
I/O pins	6	16	24
External switch	No	Yes	Yes
Available packages	8 PDIP	20 PDIP	28 PDIP
		20 SOIC	28 SOIC

Note: PDIP = Plastic dual-in-line packaging. SOIC = Small outline integrated circuit.

- Pseudo-random number generator
- Serial peripheral interfaces

The assemblers/compilers are M8C language assembler and PSoC designer C compiler.

The key feature of the family of Cypress microcontrollers is tabulated in Table 3.2.

3.1.3.2 Atmel's AVR 8-Bit RISC

This 8-bit microcontroller utilizes AVR reduced instruction set computer architecture, and it has the following features:

- 32 × 8 general-purpose working registers
- 1 Kbyte of in-system programmable flash and 84 bytes of in-system EEPROM
- One 8-bit timer/counter
- On-chip analog comparator
- Programmable watchdog timer
- 15 pins of programmable I/O lines, i.e., Port B (PB0 to PB7) and Port D (PD0 to PD6); Port D pins are tristate

The assemblers/compilers are ImageCraft C for AVR, IAR ICCA90 compiler, and CodeVisionAVR C compiler.

3.1.3.3 Hitachi's H8/300H

This 16-bit microcontroller has sixteen 16-bit general registers and 62 basic instructions. It has 4 timers:

- Timer A — an output pin for divided clocks

- Timer V (8 bits) — output pin for waveforms generated by output compare
- Timer W (16 bits) — functions include output compare output, input capture input, and pulse width modulation (PWM) output
- Watchdog timer

The assembler/compiler is Hitachi Embedded Workshop.

Other characteristics of H8/300H include: 29 general I/O pins, 8 input pins for A/D converter, EEPROM interface, I²C bus interface, serial communication interface and various power-down states.

3.1.3.4   *Motorola MC9S12DP256*

The Motorola MC9S12DP256 microcontroller hash 16-bit central processing unit (HSC12 CPU) consists of:

- 256 Kbytes of flash EEPROM
- 12 Kbytes of RAM
- 4 Kbytes of EEPROM
- Two SCIs
- Three SPIs
- Eight channels of input capture and output capture enhanced timer
- 10-bit A/D converters
- Eight channels of PWM
- Digital byte data link controller (BDLC)
- 29 discrete digital I/O channels (ports A, B, K, and E), where 20 of them have the interrupt and wake-up capability
- 5 CAN 2.0 — a software-compatible module (MSCAN12)
- I²C bus

The assemblers/compilers are Embedded Workbench and ImageCraft ICC12 V6.

This microcontroller is able to operate in two different modes: normal and emulation operating mode, and special operating mode.

3.1.3.5   *AT91 ARM Thumb*

The AT91 ARM Thumb microcontroller family is a 32-bit system. Its architecture is based on the RISC processor. The key features of this microcontroller are:

- 8 Kbytes of on-chip standby RAM (SRAM), which is directly connected to 32-bit data bus and is single-cycle accessible

- Fully programmable external bus interface (EBI) enables connection of external memories and application-specific peripherals; eight levels of priority, individually maskable, vectored interrupt controller to reduce the latency time
- 32 programmable I/O lines where user can define the lines as inputs or outputs; these I/O lines can be programmed to detect an interrupt on a signal on each line
- USART, which permits the user to select communication mode, i.e., asynchronous or synchronous at high baud rates
- Three channels of 16-bit timer counter (TC) that are programmable and able to capture waveforms
- Tristate mode that is for the debug process; this enables users to connect an emulator probe to an application board without having to desolder the device from the target board
- Five peripheral registers: control register, mode register, data register, status register, and enable/disable/status register

The assemblers/compilers are Multi 2000 Development Environment, ARM SDT and ARM ADS Development Environment, and IAR Embedded Workbench.

3.1.3.6   *Motorola MC68F333*

A Motorola 32-bit family microcontroller has the following features:

- CPU32, which is a 32-bit architecture microprocessor with virtual memory implementation, table lookup and interpolate instruction, and improved exception handling for controller applications
- Single-chip integration module (SCIM), which consists of programmable chip select outputs, external bus support, system protection logic, watchdog timer, clock and bus monitors, phase locked loop (PLL) clock system, and six general I/O ports
- Intelligent 16-bit time processor unit (TPU)
- 8/10 bits A/D converter with eight channels, eight result registers, eight conversion modes, three result alignment modes, one 8-bit digital input port, and one 8-bit digital output port
- Two flash EEPROM modules of sizes 16 and 48 Kbytes, respectively
- 512 bytes SRAM

3.1.4   Embedded Controllers

Embedded systems are essentially microprocessors or microcontrollers with added components to perform functions within a purposely built device

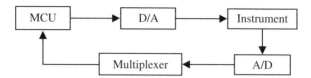

FIGURE 3.7
An instrument using an embedded controller.

such as electronic portable instruments. In most electronic portable instruments, microcontrollers are used due to their size, cost, and power efficiency. As mentioned earlier, microcontrollers with built-in memory and interface circuits have reduced the size of the final instrument. A program that is written in both low-level and high-level languages is compiled and downloaded onto the memory of the controllers. A typical embedded controller used in instruments consists of components as illustrated in Figure 3.7.

Generally, embedded controllers in instrumentation systems are stand-alone units. They perform data acquisition processing and activate actuator tasks in response to the input data.

In reference to the closed-loop system given in Figure 3.7, the following features can be observed:

- *MCU* — An embedded controller that typically consists of memory for storing programs, I/O ports for input/output, and communication ports for data transfer.

- *D/A converter* — Used to convert digital signal from the MCU. Actions or outputs are determined by the control algorithm in response to input data.

- *Instrument* — Can be any measurement devices, servomotors, or drive systems. It monitors the external environment such as temperature, humidity, speed, etc. It also performs some dedicated tasks such as display and man–machine interface.

- *A/D converter* — Converts analog signals from the outside world to digital signals. Theses signals will then be applied to the MCU.

- *Multiplexer* — This electronics device selects a particular signal from a sensor in multisensory-based systems.

Once embedded controllers are implemented, instruments may have limited flexibility. When programs are written and burnt into the ROM or EEPROM, the system will be inflexible for further alterations of the programs. This is due to the fact that programs written for a dedicated system may not be compatible for other families of microcontrollers or other configuration settings. Furthermore, the programs are constrained by the I/O ports of that particular embedded controller and I/O characteristics of external devices.

To illustrate basic features of an embedded controller-based instrumentation system, let us consider a remote weather monitoring station. The station will consist of an array of sensors for measurements and a microcontroller system for the execution of the control algorithms. Environmental data such as temperature, rainfall, and relative humidity sensors are mostly likely to generate analog signals. Therefore, these analog signals have to be converted into digital formats. Some microcontrollers have built-in A/D converters. Therefore, the analog signal can directly be applied to the MCU, and conversion of the signal can be incorporated within the control software easily. Accuracy of the data conversion through software depends on the bit size of the microcontroller in use. If an external A/D converter converts signals, the accuracy will be dependent on the structure of the converter.

3.1.5   Dedicated Computers and Electronic Portable Instruments

Dedicated computers are often used in automatic electronic portable instruments where there are many sensors and transducers involved, as in the case of some mobile air pollution monitoring systems. Such instruments consist of a laptop-like or Palm computer with a liquid crystal display (LCD) or a cathode ray tube (CRT) display. These instruments may have additional computer hardware and highly specialized software in order to perform specific measurement and control tasks. The basic features of dedicated computers in portable instruments are not much different than those of ordinary PCs. Dedicated computers are designed to perform a set of specific tasks rather than for general-purpose computing. In order to understand the functionality of the dedicated systems, the basic features of computers will briefly be described next, with some emphasis on communication capabilities.

1. *CPU* — Heart of the computer, which performs the core functions of fetching instructions from memory, decoding them, and then executing them.
2. *Memory* — Information gathered from the physical variable(s) is stored in memory either on a short-term basis or for longer periods. A computer memory stores three distinct entities: program instructions, partial (intermediate) operation results from the measurements system, and the final and overall results.
3. *Input/output* — I/O ports are essential in establishing communication between the computer and external devices. The external hardware is data transfer media between different systems. For the transfer of information efficiently, protocols or some form of communication standards is necessary. Protocols enable the flow of information smoothly without errors.

4. *System bus* — To be able to establish communication channels between the CPU, memory, and I/O, the computer must have a common connection for the system bus. Bus systems contain information on the addresses of the data sources and destinations for data to be transferred and for transmission of control signals to achieve orderly flow of information.

In addition, a typical portable instrument based on a dedicated computer consists of:

5. *Input or output interface* — For monitors, keyboards, mouses, and the measurement or control of the instrument.
6. *Processors* — To perform monitoring tasks, run the programs, and control the entire operation.
7. *Communications devices* — To establish two-way communication between different units or systems separated by some distances. The data are transferred from each remote terminal unit (RTU) to the master terminal unit (MTU). This can be done by radio frequency communication systems; off-line communication by telephone lines, as in the case of data acquisition units; modems; or specially established LANs.

Portable instruments with dedicated computer systems establish communication channels between the RTU and MTU. There are a number of ways of establishing communications. For example, in master–slave communication mode, MTU will initiate the communication and is capable of calling one of the RTUs whenever necessary. The corresponding RTU will then respond to the MTU. The RTU is not able to initiate any message transfer, but instead, it only sends a message when requested. All the data between the MTU and RTU are transferred in serial fashion. This means that single strings of binary characters are sent one after another. In order to read the signals or inputs from an analog measurement device, the inputs need to be converted to digital signals through A/D converters. In the meantime, in order to fulfill the functions, a communication protocol of sending the data in serial format is required. The protocol of the system can be defined as a set of rules that defines the meaning of a pattern of binary words. The protocol determines which word will be sent first, second, and so forth. A typical communication medium using modems between the RTU and MTU is presented in block diagrams, as in Figure 3.8.

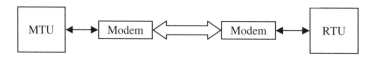

FIGURE 3.8
Block diagram of data communication and data terminal equipment.

SYNCH	Remote address	Function	Internal address	Modifier	Protocol 8-bits	DATA 0-92-bits	CRC 16-bit

FIGURE 3.9
Block diagram of IEEE-C37.1 protocol.

RTU can be considered a "black box" that takes field data in analog forms, such as voltages or currents. RTU then encodes and transmits the information to the MTU. The RTU also responds to the MTU instructions, such as turn off the power; switch on the valves, etc. Therefore, it is able to accomplish the remote control and monitoring tasks. The communication or data-transferring format of the RTU is in serial RS-232 format. For longer distances, other formats can be adopted. The MTU uses the same data transfer medium as RTU. Therefore, the only difference between the MTU and RTU is that the RTU cannot initiate conversion; it only acts as a slave.

Modems are often used as a data transfer medium between the MTUs and RTUs. For the data transfer, the system needs protocols that are established for modems. For example, the IEEE-C37.1 protocol has the layout of messages, as shown as in Figure 3.9.

The length of data of each block shown above is sent in 8-bit-long format. Referring to Figure 3.9, the definition of each block is:

- *SYNCH* — first 8 bits to be sent to the MTU
- *Remote address* — the station to which the message is sent
- *Function* — the type of message sent; there are 256 different types of messages
- *Internal address* — the address of the message directed to the sets of registers within the receiving station
- *Modifier* — modifies the internal address and defines the length of the data words in the message
- *Protocol* — contains the shorthand messages about MTU and RTU conditions
- *DATA* — a field of variable length, from 0 to 192 bits
- *CRC* — a 16-bit cyclic redundancy code based on a Bose–Chaudhuri–Hocquenghem formula for the detection of transmission errors

To establish a communication channel, i.e., when MTU or RTU needs to send messages to each other, it needs to activate the protocol driver to encode the message into the form shown in Figure 3.9. The protocol will take operational information and arrange it by the strict rules of the protocol. This will ensure that the communication link is clear before it passes the request to send a command to the modem.

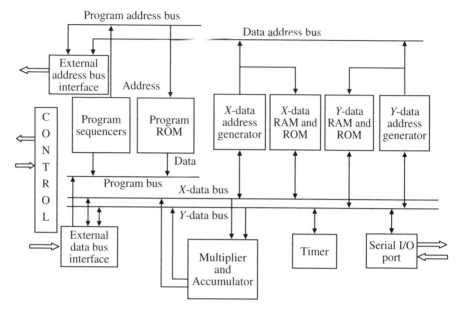

FIGURE 3.10
Block diagram of a digital signal processor.

3.1.6   Digital Signal Processors (DSPs)

The computational demands of many digital signal processing applications, including portable instruments, almost always require fast computer hardware. There are specialized types of microprocessors that involve partial computer architecture to fulfill fast operational needs; they are known as digital signal processors (DSPs). A simplified block diagram of a typical DSP is given in Figure 3.10.

DSPs incorporate special hardware features that are capable of speeding up calculation of digital filtering, fast Fourier transforms (FFTs), and other frequently based algorithms. Needless to say, modern general-purpose microcomputers can also address the needs of digital signal processing by adding some of the necessary hardware and special instructions. As a result, distinction between the DSP and microprocessors, in general, is in the degree of specialization. This particular DSP has two data memories, denoted as X and Y. For the implementation of signal processing, say finite impulse response (FIR), X-memory can be used to store the samples of input signals, and Y-memory to store the impulse response.

As compared to many other microprocessors and computer applications, the typical computational load carried by a digital signal processor can be quite demanding, but at the same time algorithmically simple. Typically, most of the processor's time is spent in processing loops with comparatively few branches and performing repetitive operations. Other important frequently used features in DSPs are the storage capability and the flexibility to move large amounts of data. Again, although required bandwidth can be very large, such movements are usually straightforward, such as manipulation of information coming from the A/D converters.

One of the main characteristics of DSPs, for which they are designed, is the capability of handling integers. Most of the existing DSPs is either 16- or 32-bit devices. For example, the Motorola DSP56300 processors are 24-bit integer processors that offer a compromise between inexpensive 16-bit and powerful 32-bit devices.

In the ensuing sections, the types of DSPs, their data handling capabilities, and implementation are dealt with in detail.

3.1.7 Types of DSPs and Examples

DSPs can be divided into three main groups: general-purpose DSPs, floating-point DSPs, and fixed-point DSPs.

A *general-purpose DSP* is fundamentally deterministic in terms of code execution timing. Given adequate information, it is possible to predict the exact number of clock cycles required to execute a specific segment of object code. These processors have straightforward architectures and are supported by tools that help the programmer determine a code fragment's execution time. Some general-purpose processors, in contrast, have extremely complex architectures (for example, superscalar architectures that dynamically select instructions for parallel execution) and lack tool support to aid programmers in predicting execution times. These factors mean that it can be extremely difficult to predict execution timing of general-purpose processor code.

In many cases, programmers writing real-time DSP applications for general-purpose processors can execute their code on the target processor and measure the run-time performance. Unfortunately, the execution timing of a specific segment of code is likely to change depending on the code that preceded it; it also depends on the locations of the code and its associated data in memory. In some cases, execution timing may be data-dependent.

Where performance is not critical, developers use high-level language compilers to quickly generate application codes. But the complexity of DSP algorithms coupled with high data rates means that in many DSP applications, most programmers directly write assembly code in order to obtain maximum performance from the processor. In such cases, it is the application programmer rather than the compiler writer who must understand the intricacies of the processor's architecture, including execution timing, in order to effectively select a processor, predict performance, and optimize the codes.

Typical examples of general-purpose DSPs are Hitachi SH-DSP, Integrated Device Technology R4650, Intel MMX Pentium, and Motorola/IBM PowerPC 604/604e. Examples of high-end general-purpose processors are the Intel Pentium and Motorola/IBM PowerPC 604.

A *floating-point DSP* can support multiple data types, such as 40-bit Institute of Electrical and Electronics Engineers (IEEE) floating-point, 32-bit IEEE floating-point, and 32-bit fixed-point data. They also provide a 32-bit address space and use unusually large 48-bit instruction words. Some of these processors are based on modified Harvard memory architecture with two memory spaces: program memory (PM) and data memory (DM), each with its own address and data buses.

Typical examples of floating-point DSPs are Analog Devices ADSP-21020, ADSP-2106x, ADSP-2116x, and ADSP-TS0xx; Lucent Technologies DSP32C and DSP32xx; Motorola DSP96002; and Texas Instruments TMS320C3x, TMS320C4x, and TMS320C67xx.

Fixed-point DSPs provide multiple distinct arithmetic units: a multiplier-accumulator, a shifter, and an arithmetic logic unit (ALU). These units can access a common register file containing 16 registers of 40 bits each. Inputs to all arithmetic operations come from the register file, and results of all arithmetic operations are delivered to the register file. The typical speed of a fixed-point DSP is about 50 mega instructions per second (MIPS) at 5.0 V. Fixed-point DSPs implement a load–store architecture in which the source operands for arithmetic operations are loaded into registers before computation. Similarly, the results of arithmetic operations are stored to a register. They contain multiple arithmetic execution units: an ALU, a multiply-accumulate (MAC) unit, and a barrel shifter. All of these units are capable of single-cycle execution, but only one unit can be active at a time.

In a typical DSP, the ALU operates on 16-bit data. It includes four 16-bit input registers, a 16-bit feedback register (to return results to the input of the ALU for further processing), and a 16-bit result register. In addition to the usual ALU operations, the ALU provides increment/decrement, absolute value, bit set/clear/toggle, add-with-carry, and subtract-with-borrow functions. ALU results are saturated upon overflow if the programmer sets the appropriate configuration bit.

Typical examples of fixed-point DSPs are Analog Devices ADSP-21cspxx, ADSP-21xx, ADSP-219x, ADSP-21535 (AID/Intel MSA), and ADSP-TS0xx; IBM MDSP2780; Lucent Technologies DSP16xx and DSP16xxx; Motorola DSP560xx, DSP561xx, DSP563xx, and DSP566xx; NEC uPD7701x; Texas Instruments TMS320C1x, TMS320C2x, TMS320C2xx, and TMS320C55xx; Zilog 893xx and Z894xx; and Zoran ZR38xxx.

3.1.7.1   Data Handling of DSPs

Historically, digital signal processing was either done off-line, after the signal had been digitized and stored, or had to be performed on custom special-purpose hardware. However, measurement frequently depends on computations that must be accomplished in a fixed period of time. The term *real-time computing* is used to describe a computing environment that sets an upper limit on the execution time. Digital signal processing functions, in particular, often must be done in real time, so that processing tasks should be completed before the next data sample arrives. Typically, an instrument's processor has to perform multiple tasks, such as signal processing, display, communications, power management, and so on, so that total available computing resources, including computer time, must be allocated to different tasks. Also, the priority of task execution may be different. This too must be taken into account when computer time is allocated. As a result, scheduling of the computer time may become quite complex.

For example, let us estimate the amount of processing required to perform a 1024-point decimation-in-time FFT algorithm on a stereo audio signal, sampling at a 48-kHz rate. The FFT requires $N/2 \times \log_2 N$ complex multiplications and $N \times \log_2 N$ additions (5120 complex multiplications and 10,240 complex additions, or 20,480 real multiplications and 20,480 real additions). In order to produce FFT data at the same rate as the incoming data stream, the FFT must be executed within $1/48,000 \times 1024$ of a second, or about 21 msec (this does not include time for windowing operation). Therefore, each multiplication and addition should take no more than about 1 μsec. Because multiplication is frequently implemented as a shift–add operation, and takes many times longer than addition, modern signal processing microcomputers incorporate a special hardware multiplier unit that replaces the shift–add approach with fast, single instruction multipliers.

From the review of most common digital signal processing algorithms, we can see that multiplication and addition (or subtraction) are the dominant operations in digital signal processing. Hence, efficient signal processors should optimize execution of these two operations. Also important is the calculation of the scalar product where one array can be constant (digital Fourier transform (DFT), FIR filter, and convolution) or variable (correlation). Efficient execution of the sum of products should also be a priority.

Taking the example of the DSP56300, the main hardware of the processor are the program execution unit, data arithmetic and logic unit, data and program memory, and peripherals, all connected by the program and data buses. The main components of the ALU are the multiplier, adder/accumulator, and two storage registers (A and B), where results of the arithmetic operations reside. The accumulator is wider than the processor buses to allow accurate accumulation of the sum-of-the-product operation without rounding off each intermediate sum. To achieve high computational throughput, data are processed in stages and in parallel. For example, if the result of the previous multiplication can be stored, then multiplication of current multiplicands and addition of a previous product to the running sum can be done simultaneously. Internal program and data memories, although limited, allow fast access to the data. Data memory may also be divided in two blocks to allow simultaneous access to the two numbers to add or multiply.

Efficient input and output of data are very important in signal processing. Typically, the source of data is an A/D converter or a standard digital data stream, as in telecommunications. Similarly, output can be directed to the D/A converters or can be formatted in a digital data stream. The simplest way to implement data I/O is to interrupt the processor when data become available for transfer, so that the processor can execute the data to move to the I/O device. For that purpose, DSPs have several interrupt lines that normally are connected to the I/O devices. Although simple, this method is inefficient because it requires interruption of the main program flow, saving the current processor state, and when I/O transfer is done, restoring the processor state and resuming execution. Saving and the restoring processor state are especially time-consuming. To address this problem, many DSPs include a direct memory access (DMA) peripheral. Whenever an I/O move is requested, the DMA peripheral suspends execution of the main program by the processor, takes over the main data bus, and transfers the data. After the transfer is finished, the DMA peripheral releases the processor, which continues program execution in the state in which it was suspended. Because there can be many sources of data, DMA peripherals usually have multiple data transfer channels.

3.1.7.2 Implementation of DSPs

Implementation of digital signal processing devices should take into account several considerations. The most important considerations are required computational resources, memory capacity, dynamic range of the signals, and the required accuracy of computation. In portable instruments, power consumption and size of the signal processor are also important considerations. Because of the limitations of the earlier computers, the first real-time digital signal processors were custom special-purpose devices optimized for specific computational tasks such as digital filters. Digital filters are explained in detail in Section 3.3.

Modern portable instruments almost exclusively rely on fast microcontrollers and microprocessors for their digital signal processing needs. The combination of high performance, very small size, and low-power consumption offered by the existing microprocessors is very attractive in the small battery-powered devices and, generally, can satisfy computational needs in a wide variety of portable measurement applications.

Basic numeric formats for representing data within the computer are the integer and floating-point formats. An integer is a signed or unsigned binary number with a bit width that is usually a multiple of eight (8-, 16-, 24-, 32-, and 64-bit wide). The floating-point number, N, consists of two parts, the signed fractional mantissa, M, and the signed exponent, E, so that

$$N = \pm(M - 1)2^{\pm(E-1)} \tag{3.1}$$

The choice of the numeric format influences two important characteristics of a processing system. The dynamic range, DR, of a system describes the system's ability to faithfully process both large and small signals, and is usually expressed as a ratio of the largest to the smallest signal that a system can correctly process:

$$DR = \frac{N_{max}}{N_{min}} \tag{3.2}$$

The dynamic range can be different at different points within the system. For example, compressor systems reduce a large dynamic range of an input signal to produce a smaller dynamic range of the output signal. Basic numeric formats provide a range of digital values available for computation, but for a given bit length, floating point provides a larger dynamic range:

$$DR_f = 2^M 2^{2^E} \quad \text{and} \quad DR_i = 2^N \tag{3.3}$$

where $N = M + E$, the bit length, and DR_f and DR_i are dynamic ranges of the floating-point and integer numbers. Resolution, R, of a processing system describes its ability to discriminate small changes in the signal values:

$$R = \frac{N_{i+1} - N_i}{N_{max}} \tag{3.4}$$

where N_{i+1} and N_i are two adjacent discernable values and N_{max} is the largest permissible value. Numeric formats also have their resolution:

$$R_f = 2^{-M} \quad \text{and} \quad R_i = 2^{-N} \tag{3.5}$$

Importantly, for the same bit length, the integer format provides better resolution.

Each format has its advantages and disadvantages. Integer processing is simpler, but must be used carefully to avoid conditions known as overflow or underflow. These conditions occur when an operation is outside the dynamic range of the chosen integer, so that false results are generated.

There are several ways of dealing with errors introduced by overflow and underflow. Both of these errors can cause severe distortions and malfunctioning of the signal processing functions. A commonly used approach is to detect overflow and replace the result with the largest or smallest valid number. Multiplication also can produce truncation errors. For example, multiplication of two 16-bit numbers can produce a 32-bit product, which, again, must fit into a 16-bit number. Aside from the overflow and underflow, this truncation can cause significant loss of precision and degradation in the performance of the instrument. Typically, because multiplication is performed as a part of the sum-of-product operation, signal processors implement a wide temporary accumulator register to hold a wide-bit sum of products.

Implementation of a digital signal processing algorithm in an integer format requires close attention to the dynamic range of signals at all possible points of overflow. Careful scaling of input and output signals must be performed to avoid the overflow and, if it occurs, to detect and handle it to minimize its consequences. Recursive filters and other feedback systems can be especially sensitive, because a single overflow may linger on in the recursive sections of a system. A commonly used strategy to overcome problems is to zero out recursive memory elements.

Selection of an optimal numeric format for the signal processing application is critical, as it places limits on the dynamic range, resolution, and accuracy of the instrument, as well as its complexity, cost, and power consumption. Most portable instruments use 8-, 16-, 24-, or 32-bit integers. Although larger integers provide for better dynamic range and resolution of computations, complexity and cost of the required hardware make the use of longer integers impractical. In this case, floating-point processing offers a better alternative. An example of a common floating-point format is the IEEE format. In this format, a 32-bit number has 24-bit mantissa and 8-bit exponent that provide a dynamic range and resolution that will satisfy most of the demanding system requirements that can be encountered in a portable instrument. Floating-point processing is powerful and convenient, but also more difficult to implement than integer processing, because it requires multiple integer operations on the mantissa and exponent. If emulated in software, floating-point processing is considerably slower than integer processing, and if performed by special floating-point hardware, it may significantly impact the cost and power consumption of the processor. Typically, the majority of portable instrumentation use 16- or 32-bit integer processing with floating-point emulation when required.

3.2 Signal Conversion

In the majority of electronic portable instruments, analog signals generated by sensors and transducers have to be conditioned first by means of analog circuits before they are converted into digital forms. Analog signal conditioning has two main goals: (1) the amplification of small signals or attenuation of large ones, and (2) the removal of signals with unwanted frequencies by suitable filters.

Amplification or attenuation is used to adjust the range of amplitude of the signal to the range required by the A/D converter. This is because the voltages generated by sensors are generally much smaller than those managed by A/D converters, which are normally from 0 to 5 V or from 0 to 10 V. In addition, the inputs of most A/D converters are unipolar; that is, one of the terminals of the converter is connected to a reference terminal. However, if the signal provided by the sensor is differential, the amplifier must convert this signal into unipolar form or a differential amplifier with a single-ended output must be utilized.

Filtering is used to modify the frequency spectrum of signals generated by sensors. There are two basic reasons for using filters in data acquisition systems: (1) to reduce interference, for example, the one produced by electric networks (50 or 60 Hz); and (2) to narrow the bandwidth of the signals, thus reducing the signals (noise) outside the band of interest and avoiding possible aliasing problems.

Sometimes analog signal conditioning includes sensor interface circuits, such as bridges, impedance converters, sensor excitation devices, etc. Additionally, each externally connected channel may have to have protection against accidental electrostatic discharges, voltage surges, and unintentional application of large signals to the input channels. Safety-related issues in the industrial and biomedical instruments might also require galvanic isolation of the sensors from the rest of the instrument, thus preventing the current flow between the instrument and external devices. More information on analog signal processing and filters relevant to electronic portable instruments will be provided in Chapter 4.

Once the analog signal conditioning is completed, signals can be converted into digital forms by using signal converters, as will be explained next.

3.2.1 Sample-and-Hold Circuits

A/D converters require a finite time, called conversion time, t_c, to successfully complete the conversion operations. If the input voltages change during the conversion process, undesired uncertainties could be introduced into the equivalent digital output. Full conversion accuracy can only be obtained if this uncertainty is kept below the conversion time, also known as converter resolution. Mathematically, this means that

$$\left(\frac{dV}{dt}\right)_{max} \leq \frac{FS}{2^n t_c} \tag{3.6}$$

where n is the number of bits of the converter and FS is the full-scale value.

This formula indicates that full conversion accuracy can never be obtained, even in the lowest cases. To overcome this difficulty, sample-and-hold (S/H) circuits are utilized. The S/H circuit acquires the value of the input analog signal (sample) and holds it throughout the conversion process until the conversion is complete. Thus, a fixed level of voltage is provided to the converter during the conversion process, and the uncertainty provoked by the variation of the input voltage is eliminated.

Figure 3.11 shows the basic structure of an S/H circuit. The circuit contains a capacitor, a switch with its corresponding control circuitry, and two amplifiers to match the input and output impedances of the external circuits. The control circuit sends the commands SAMPLE and HOLD alternately to Switch S_1. This switch closes when it receives the SAMPLE command. Then Capacitor C_h charges up (or down) until it reaches the level

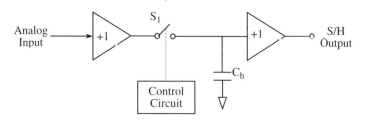

FIGURE 3.11
Basic structure of a sample-and-hold circuit.

of the input signal, and thus a sample is taken. Immediately after, the control circuit generates a HOLD command, which causes Switch S_1 to open. Then the sampled voltage is held across Capacitor C_h until a new sampling operation begins.

3.2.2 Analog-to-Digital Converters

A/D and D/A converters provide the interface between the analog environment (the real world) and the digital environment for computation and data processing purposes. Many of the applications for A/D interfaces are quite evident, as we use them often in our daily lives. Analog signals converted to digital form arise mainly from sensors of electronic portable instruments (or transducers) measuring a myriad of different real-world sources, including temperature, motion, sound, and electrical signals. When these signals are converted to digital forms, they can be processed using many different algorithms such as digital filters. When necessary, the processed signals can then be converted back to analog form in order to drive a device requiring an analog input, such as mechanical loads, heaters, or speakers. A/D and D/A converters therefore form a vital link between the real world and the world of digital signal processing.

Digitization involves two sequential steps — the sampling of a signal in time, followed by the converting of the sampled analog signals into their digital equivalents. Most modern A/D converters combine sampling and conversion in one package, usually in a solid-state integrated circuit or a hybrid device.

All modern A/D converters produce digital output values in integer format. This can be straight binary code that varies from 00.0 to 11.1 (unipolar conversion) or the 2's-complimentary code from 100.0 for negative numbers to 011.1 for positive ones (bipolar conversion). The length of the output word defines theoretical resolution of the A/D converter and the range of digital values that an A/D converter can produce. The smallest change in the input voltage that A/D can detect is

$$V_{LSB} = \frac{V_{max} - V_{min}}{2^n} \tag{3.7}$$

where n is the number of bits produced by the A/D converter and V_{max} and V_{min} are the maximum and minimum input voltages that an A/D converter can process correctly.

From the signal processing point of view, an A/D converter can be described in terms of the noise it generates. Quantization, like sampling, introduces some degree of uncertainty in the value of the analog signal before quantization itself takes place. For A/D converters with large numbers of bits (8 bits or more), this effect can be modeled as an additive noise. The amount of quantization noise relative to the signal, or signal-to-noise ratio (SNR), depends on the nature and magnitude of the input signal.

Maximum SNR is achieved when the signal amplitude fits exactly in the dynamic range of the A/D converter ($V_{max} - V_{min}$). However, this mode of operation frequently leads to high harmonic distortion of the signal. Optimal selection of the signal magnitude should also take into account statistical properties of the signal. For example, a high peak-to-average signal can be chosen to exceed the dynamic range of the A/D converter (clipped) in order to provide better noise performance and maximize the SNR. The rule of thumb is that A/D converters reduce their level of amplitude noise in half (6 dB) for each additional bit. For a sine wave signal with a peak-to-peak amplitude equal to the dynamic range of the A/D converter,

$$SNR = 6n + 1.8 \text{ dB} \qquad (3.8)$$

where n is the number of bits and SNR is expressed in decibels.

It is important to realize that quantization noise power is spread over the entire spectrum of the digital signal from –FS/2 to FS/2.

In addition to quantization, A/D converters have other sources of noise, such as internal voltage reference noise, thermal amplifier noise, etc. In practice, most converters rarely achieve their theoretical noise floor. Therefore, when selecting an A/D converter, one uses the number of bits of converter as a rough estimate of its resolution. Nevertheless, close examination of the comprehensive converter specifications should reveal its true performance, and this may be necessary in some critical applications.

Many different types of A/D converters are commercially available; the reason for this is that there are many diverse ranges of applications of the A/D converters. The most popular A/D converters are counter ramp, successive approximation (serial), flash (parallel), and delta-sigma (Δ/Σ) converters. These converters will be discussed in the following subsections, and some examples of the commercial ones will be provided.

3.2.2.1   Counterramp A/D Converters

Counterramp converters are one of the simplest forms of implementation of what are called direct A/D converters. However, the trade-off for the simplicity is the compromise of the speed. Figure 3.12 shows the block diagram

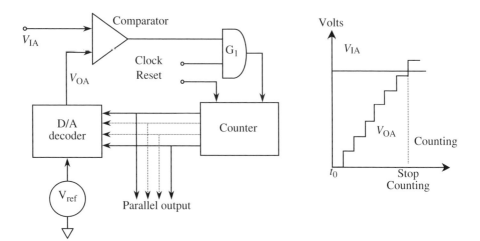

FIGURE 3.12
Counterramp A/D converter.

for a counterramp A/D converter. These converters are the slowest among all other types, requiring $(2^n - 1)$ processing steps for conversion of an analog signal into an n-bit digital equivalent.

The conversion process of the counterramp A/D converter is as follows: The reset pulse resets the counter to zero, thus driving the D/A decoder output V_{OA} to 0 V. The clock starts and the counter begins counting, driving the D/A decoder such that the output V_{OA} is proportional to the counter. V_{OA} is compared against the analog reference voltage V_{IA}, and when $V_{OA} > V_{IA}$, the comparator output changes state such that the clock is switched off and the counter stops. The parallel digital word in the counter is the quantized version of the analog input voltage.

It is clear that the input voltage V_{IA} must be constant for the whole process if successful conversion is to take place. This is achieved by a sample-and-hold operation, where the analog signal is sampled and held until the A/D converter has completed the conversion process, as explained above. But a major problem with this is that for high sampling speeds, the counter and clock must operate extremely fast, especially if the analog signal is varying rapidly. Therefore, the counting speeds limit this A/D converter from being used at high speeds.

An improvement to the counterramp A/D converter, as discussed above, is the tracking converter, which essentially has the same structure, but uses an up–down counter. The counter counts up or down depending on the change in direction of the magnitude of the analog input. For example, if the analog input voltage decreases, the output of the comparator senses the change and sends a message to the counter to count down. Similarly, if the input voltage increases, the counter starts counting up.

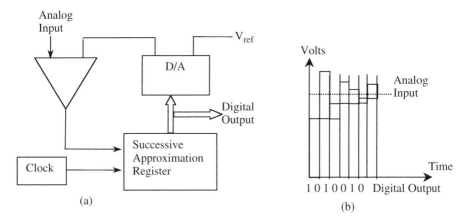

FIGURE 3.13
(a) Scheme of a successive approximation A/D converter. (b) Output bit generation in the successive approximation process.

3.2.2.2   Successive Approximation A/D Converters (Serial)

Figure 3.13(a) shows the basic design of a successive approximation A/D converter. The analog input to this system is successively compared with the voltage generated by a D/A converter. The digital input to the D/A converter, which is stored in the successive approximation register (SAR), is adjusted according to the results of each comparison. If the converter is of n bits, the conversion process requires n comparisons, and the result of the conversion is stored in the SAR as the final value.

The conversion process for this A/D converter is as follows: The programmer begins the conversion process by trying logic ONE into the most significant bit. The output of the decoder is compared against the input analog signal. If the analog signal is smaller than the decoder output V_{OA}, then logic ONE is deleted and replaced with a ZERO. If the analog signal is larger than V_{OA}, then logic ONE remains. The successful approximation register then goes to the next bit and tries logic ONE for that bit, and the process of comparison is repeated until the signal has been converted.

Figure 3.13(b) shows the conversion process of a value of an analog input. Only the calculation of the first five most significant bits of the digital output is shown. In this figure, the first comparison is carried out with the binary number 10 ... 00, which is stored in the SAR. Then, as the analog input is greater than the voltage generated by the D/A converter with this number, the most significant bit of the SAR is held in 1. Once the first step is finished, a new comparison is carried out. Then the following bit (the second) of the SAR is set to 1. At that moment the binary number in the SAR is 1100 ... 00. Consequently, the value of the analog input is greater than the voltage generated by the D/A converter, so the second bit of the output is set to 0.

This process continues in the same way until the n bits of the output are determined.

In successive approximation A/D converters, the analog inputs are sampled and held in the comparators. This means that the conversion sample rates are limited by the fact that each bit must be tested before the entire conversion is completed as a single digital word. Nevertheless, the conversion speeds of the successive approximation converters are much greater than counterramp types. But there is a trade-off between the speed and the complexity of the device. These converters require extensive programmer logic, which takes the D/A decoder through the different steps of the successive approximation processes.

A typical successive approximation A/D converter (e.g., AD7677) is a differential converter with a 16-bit resolution with no missing codes and a maximum error of 1 least significant bit (LSB). It can operate at three different speeds: a 1-Msps (mega samples per second) warp mode for asynchronous sampling applications, an 800-ksps (kilo samples per second) normal mode, and an impulse mode. The signal to noise and distortion (SINAD) is typically 94 dB, with a minimum of 92 dB. The device operates from a single 5-V power supply. The power consumption varies with the throughput rate typically 115 mW. It consumes 7 µW when in the power-down mode.

3.2.2.3   *Flash A/D Converters (Parallel)*

Flash or parallel A/D converters are the fastest, but most complicated, converters available. A flash converter, like the successive approximation converter, works by comparing the analog input signal with a reference signal, but unlike the previous converter, the flash converter has as many comparators as there are digital representation levels or digital word lengths, except for one. Therefore, $(2^n - 1)$ comparators are needed for an n-bit converter. For example, an 8-bit converter would have 256 representation levels and thus needs 255 comparators. This is shown in the block diagram representation of the flash converter in Figure 3.14.

The input analog voltage is connected to the input of every comparator so that a comparison can be made with each of the reference voltages representing the quantization levels. The outputs of the comparators then drive the encoding logic to generate the equivalent digital output. The conversion rate for flash A/D converters is very fast because for complete conversion only one step is required. Depending on the propagation time of the encoding logic, a digital output can be generated almost in real time instantaneously.

The main disadvantage of the flash converter is of course the number of comparators required to implement large-scale converters. For every additional binary bit of resolution added to the A/D converter, the number of comparators required is essentially doubled. For instance, 255 comparators are required for an 8-bit converter. If four more bits are added, for a 12-bit converter, 4095 comparators are required. Along with the number of com-

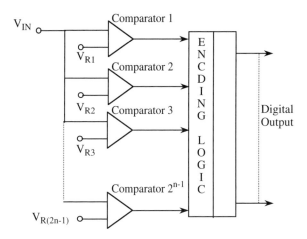

FIGURE 3.14
Flash A/D converter.

parators required, another disadvantage is the number of reference voltages needed. For a 12-bit converter, 4095 reference voltages are necessary. It must also be noted that the input capacitance increases linearly with the number of comparators implemented; thus, input impedance becomes larger, and hence it becomes impractical to incorporate too many input signal buffers.

A typical 8-bit flash A/D converter (e.g., MAX106) allows the digitization of analog signals with bandwidths up to 2.2 GHz. It integrates a high-performance track/hold (T/H) amplifier and a quantizer on board. A differential comparator and decoding circuitry are combined to reduce out-of-sequence code errors (thermometer bubbles or sparkle codes) that provide a metastable performance of one error per 1027 clock cycles. The analog input can use either differential or single-ended inputs with a ±250-mV voltage range. The supply voltage can be between +3 and +5 V for compatibility with +3.3- or +5-V referenced systems. It has a 600-Msps conversion rate with a ±0.25 LSB error.

3.2.2.4   Delta-Sigma A/D Converters

Both the successive approximation and flash converters are essentially analog devices; therefore, they are subject to the formidable difficulties of designing precise and stable analog circuits. On the other hand, the Delta-Sigma (Δ/Σ) A/D converters are based on a very different method that limits analog processing to a few simple essential steps and uses digital signal processing techniques to achieve the required performance. This approach results in very high accuracy (up to 24 bits) at the low-frequency range (1 to 10 kHz) and moderate- to high-resolution converters (12 to 16 bits) in mid-frequency range (10 kHz to 5 MHz).

A block diagram of the typical Δ/Σ converter is given in Figure 3.15. This figure represents a 1-bit Δ/Σ A/D converter. The comparator is the con-

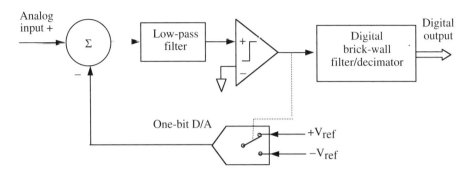

FIGURE 3.15
Block diagram of a delta-sigma A/D converter.

verter, and its output is processed digitally, thus preventing the accumulation of analog errors. The comparator is in a feedback loop with a low-pass filter and a 1-bit D/A converter. The 1-bit D/A converter can assume only one of the two values, +FS or –FS. The low-pass filter causes the loop gain to be high at low frequencies and low at high frequencies. This results in 1-bit output whose duty cycle is proportional to the input signal. An important improvement of the 1-bit Δ/Σ converters is the multibit converters in which the comparator is replaced by a flash converter with as much as a 4-bit resolution.

In Δ/Σ converters, the feed-forward path has a low resolution equipped with a high-speed A/D converter and a simple digital low-pass filter. The feedback path includes a low-resolution feedback D/A converter. The analog input is converted into a low-resolution, high-sampling-rate digital signal. Because this sampling rate is much higher than the minimum required Nyquist rate, this type of digital signal is called the oversampled signal. Typically, the modulator can sample at 10 to 100 times, or even higher, the final output rates. Oversampling produces a digital signal with a frequency band that is much wider than the input signal band. A frequency-selective digital filter in the feedback path shapes the signal frequency spectrum so that quantization noise is shifted toward the high end of the spectrum with an intention of removing this noise with the follow-on digital filtering and decimation. Most existing Δ/Σ converters use a 1-bit A/D converter, which is simply a comparator. This results in a very elegant design that combines the strengths of analog and digital design techniques in an optimal fashion. The output of the converter is fed into the digital decimator so that it reduces high sampling rates and, at the same time, increases bit length and resolution of the digital sample. Build up of the converter resolution is possible because total quantization noise power in the band of interest constitutes only a small part of the full noise power of the oversampled signal.

Perhaps the greatest advantage of the Δ/Σ converters is that they do not require complex analog anti-aliasing filters. Suppression of the potential aliasing spectral components is much easier in the oversampled signal because these components are located far from the band of interest, and therefore can be removed with very simple passive-component networks. As a result, data acquisition systems that use Δ/Σ converters tend to achieve higher degrees of integration and are smaller in size. Another attractive feature of most Δ/Σ converters is that their entire system can be scaled in frequency (including internal filters) simply by changing frequency of the external sampling clock. At low sampling rates, Δ/Σ converters can achieve very high resolutions (20 and even 24 bit). This drastically simplifies data acquisition of small, low-level, low-frequency signals such as strain gauges, thermocouples, etc., by allowing direct connection of the sensor and A/D converter.

Unlike other types of A/D converters, which produce a single sample within a single sampling period, Δ/Σ converters produce a stream of samples that cannot be easily managed on a sample-by-sample basis. Internal digital filters of the Δ/Σ converters have storage and, therefore, introduce long delays that preclude quick switching of the analog signals on the input of the Δ/Σ converter. Multiplexing of multiple analog signals for digitization with a single converter is feasible only when such switching is not time-critical or the digitization channel's allowed settling time is longer than the internal delay of the Δ/Σ converter. Another disadvantage of high-speed Δ/Σ converters is that oversampling requires high-speed operation of the modulator. This leads to an increase in power consumption, compared to other methods, such as the successive approximation converters.

A typical commercial Δ/Σ converter (e.g., CS5510-AS 8-Pin device from Cirrus Logic) operates from a single +5-V power supply or various types of bipolar signals. It comes as a 16-bit device in eight-pin packages. Its linearity error is 0.0015% FS with a noise-free resolution up to 17 bits. The input ranges from 250 mV to 5 V with a 16- to 326-Hz output word rate. The power consumption is 2.5 mW in normal operating mode.

3.2.3 Digital-to-Analog Converters

Digital information requires D/A converters for the conversion of the binary words into proportional analog voltages. A typical D/A converter has three major elements: (1) resistive or capacitive networks, (2) switching circuitry, operational amplifiers, and latches, and (3) a reference voltage. Also, an important requirement is by some means the switching of the reference voltage from positive to negative for bipolar analog signal outputs.

Figure 3.16 shows a typical design of an n-bit D/A converter. The set of latches hold the binary number that is to be converted into the analog signal. The output of each latch controls a transistor switch, which is associated with a specific resistor in the resistor network. The reference voltage, which is connected to the resistor network, controls the output voltage range. The operational amplifier works as an adder.

The operation of the D/A converter is as follows: Switches are set to the digital word; i.e., switch is closed for logic ONE and switch is grounded for

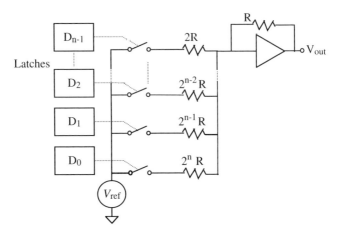

FIGURE 3.16
Basic D/A converter design.

logic ZERO. The weighted resistors determine the amount of voltage at the output. For each one of the resistors in the network, when its corresponding switch is on, the operational amplifier adds a voltage to V_{out} that is inversely proportional to the value of the resistor and directly proportional to the reference voltage and the feedback resistor (R). For example, for the $2R$ resistor, the operational amplifier adds the following value to V_{out}:

$$V_{out} = V_{ref} \frac{R}{2R} = \frac{V_{ref}}{2} \tag{3.9}$$

The total value of V_{out} can be calculated by the following formula:

$$V_{out} = V_{ref} \left(\frac{D_{n-1}}{2} + \frac{D_{n-2}}{4} + \dots + \frac{D_1}{2^{n-1}} + \frac{D_0}{2^n} \right) \tag{3.10}$$

where D_{n-1}, D_{n-2}, etc., represent the values of the bits of the binary number to be converted. In this formula, each active bit contributes to V_{out} with a value proportional to its weight in the binary number, and thus the output voltage generated by the D/A converter is proportional to the value of the input binary number.

3.2.3.1   Weighted Current D/A Converters

The weighted current D/A converters generate a weighted current for each bit containing logic ONE. This kind of D/A converter is useful when an analog current needs to be implemented directly. When voltage is needed, the output voltage can simply be determined by a known load resistor that

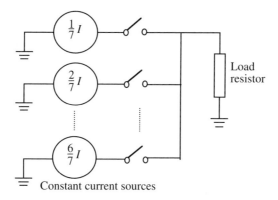

FIGURE 3.17
A weighted current D/A converter.

carries the current. The circuit diagram for the weighted current D/A converter is shown in Figure 3.17.

When the switches are closed, in the nonzero case, Kirchoff's current law applies. That is, the current leaving the junction is the addition of all the currents entering the junction; hence, the weighted sum of the current is equivalent to the analog value of the signal.

One of the advantages for this type of D/A converter is that it is easy to implement constant current drive circuits using weighted resistors to set the value of each current source for unipolar analog outputs. However, limitations exist, especially in the overall accuracy. There are leakage currents in the zero condition (all switches open) occurring particularly at higher temperatures. This type of D/A conversion is suitable for low–medium accuracy and can be used with high conversion rates.

As in the case of signal sampling, the minimum required conversion rate of the D/A converter is twice the desired signal's bandwidth (Nyquist rate). This is discussed in the mathematical treatments of signals in Section 3.3. The D/A converters should be followed by an interpolating filter that fills in the gaps between the samples. The interpolating filter bandwidth should be such that it passes only the desired part of the digital signal spectrum and rejects all other spectrum components. The complexity of the filter can be significantly reduced, if the conversion rate is much higher than the signal bandwidth. However, this requires generation of the oversampled digital signal (with a sampling rate much higher than the Nyquist rate) that needs correspondingly larger signal buffers to store it and faster D/A converters to convert it to the analog domain.

Transfer of digital signal data from the computer memory to the D/A converter is analogous to the task of moving A/D data to the computer memory. Periodic, sample rate movement of signal data from computer memory to the D/A converter may cause significant processor overhead; it may be best handled by using DMA peripherals.

A typical D/A converter (e.g., DAC14135 by National Semiconductors) is a monolithic 14-bit, 135-MSPS device. It can have many features, such as a proprietary segmented D/A converter core, differential current outputs, band-gap voltage references, and transistor–transistor logic (TTL)- or CMOS-compatible inputs. The device has a wide dynamic range, supplies differential current outputs, and operates on +3.3-V or +5-V power supplies. Normal operating temperature ranges from –40 to +85°C.

3.2.4   Data Acquisition Systems

Electronic portable instruments are at a crossroads, changing and improving constantly. One form of electronic portable instruments is the use of laptop and Palm computers as portable instruments. In this case, the computers are equipped with appropriate data acquisition cards (e.g., Personal Computer Memory Card International Association (PCMCIA)) and supporting software (e.g., LabView) to behave just like a portable instrument. In addition, electronic portable instruments frequently acquire and process signals from multiple sensors. In some advanced instruments, such as the portable mineral exploration systems or weather stations, there are many sensors and computers operating at the same time. In these systems each sensor signal is processed by its own dedicated circuitry, so in a sense, all channels operate independent of each other in the analog state. Once the analog signals are prepared for the digital conversion, the conversion takes place by using sample-and-hold devices and A/D converters as described above. In this way, all the sensors can be connected to a host microcontroller or computer by multiplexing the inputs. The integrated systems for carrying out all these operations of multiple inputs are called data acquisition systems. These systems can manage one or multiple analog signals. In the former case, they are called single-channel systems, and in the latter case, they are called multichannel systems. In the following subsections, the basic structure and the main functions carried out in single- and multichannel data acquisition systems are presented.

3.2.4.1   Single-Channel Systems

Figure 3.18 shows the typical structure of a single-channel data acquisition system. Signals generated by sensors that may not be suitable for A/D

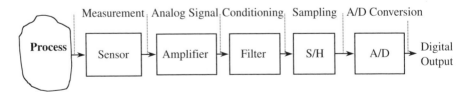

FIGURE 3.18
Basic structure and functions of a single-channel data acquisition system.

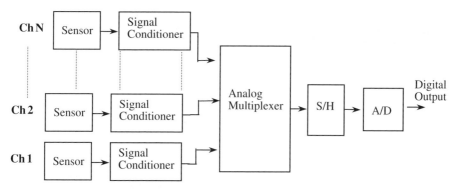

FIGURE 3.19
Multichannel data acquisition system with analog multiplexing.

converters have low amplitudes and are mixed with noise, which perturbs the quality of the measurement provided by the sensor. So these signals must be transformed into an appropriate form. In low-cost applications, the output generated by signal conditioning circuits can be directly introduced into an A/D converter. However, this approach is not valid for most applications, when higher precision is required; the conditioned analog signal must be sampled before the conversion takes place. The single-channel systems have been introduced implicitly in the previous subsections and therefore will not be further discussed here.

3.2.4.2 Multichannel Systems

When there are several acquisition channels, these channels can be multi-plexed. The multiplexing process can be carried out by two distinct methods: analog or digital multiplexing.

Analog multiplexing means that the multiplexing function is carried out on the analog signals. Figure 3.19 shows the most common configuration of an analog-multiplexed data acquisition system. The multiplexer alternately connects the analog channels of the system to a S/H circuit. This S/H circuit samples the signal and holds it, so that the A/D converter can carry out the conversion. The S/H circuit enables the multiplexer to switch to another channel if needed while the A/D is still carrying out the conversion.

3.2.4.2.1 Digital Multiplexing

When there are many acquisition channels, both the analog multiplexers and the shared follow-up resources must be very fast, even if the signals in the channels are slow. In this case, an alternative configuration to that shown in Figure 3.20 can be used. This alternative configuration is based on using one A/D converter for each channel of the acquisition system. The outputs of the converters are connected to a common bus by using interface devices. The bus is shared with the digital processor used to process the digital outputs of the converters. Using the bus, the digital processor can alternately

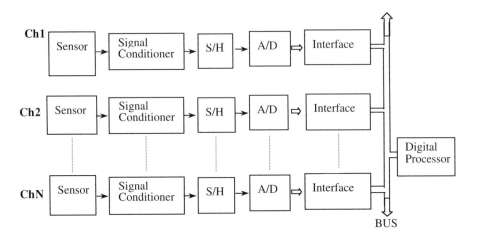

FIGURE 3.20
Multichannel data acquisition system with digital multiplexing.

access the outputs generated by each A/D converter, thus carrying out the digital multiplexing of the signals.

Irrespective of the type of multiplexing selected, multiplexers are usually under the control of a central processor, microcontroller, or computer. This approach provides small-size devices, good power consumption, and cost savings. However, multiplexing has a number of problems. For instance, multiplexing generates switching transients at the input of the A/D converter, so it is necessary to allow the signals to settle before conversion takes place. Transients can propagate through the entire chain of amplifiers and filters. Another undesirable effect of multiplexing is signal leakage (cross talk) between channels due to switching effects and stray capacitances in the multiplexer. These effects may drastically limit the rate of data acquisition. Finally, if each data acquisition channel requires different settings and parameters, such as gain, frequency response, etc., then these parameters must be switched on a per channel basis, which can be a difficult, slow, and expensive task.

3.2.5 Data Acquisition Boards

A data acquisition board is a plug-in board that allows the processing of analog signals by digital systems or computers. Data acquisition boards are gaining significant applications in electronic portable instruments, as many handheld, Palm, or laptop computers can be equipped with these boards. With the aid of virtual instrumentation software and advanced communication systems, these computers can be made to behave like oscilloscopes, spectrum analyzers, or any other portable instruments.

Data acquisition boards form one of the key elements of connecting the computer to processes for measuring, monitoring, and controlling purposes.

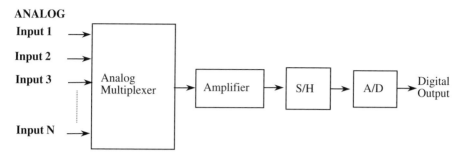

FIGURE 3.21
Input system of a data acquisition board.

These boards normally operate on analog conditioned signals, that is, signals that have already been filtered and amplified. However, some special boards can deal with the signals directly generated by sensors, and they include the necessary signal conditioning circuits on board.

The number of analog inputs normally managed by a data acquisition board is 4, 8, or 16 single-ended inputs. These inputs are treated by a system-like approach, as shown in Figure 3.21. This system is made up of an analog multiplexer, an amplifier with programmable gain, a sample-and-hold circuit, and an A/D converter. All the elements of this system are shared by all the analog input channels. Some data acquisition boards also offer analog outputs to carry out control operations. When available, the normal number of analog outputs is 2.

In addition to analog inputs and outputs, data acquisition boards also supply digital I/O lines and counters. Digital lines are used for process control, communication with peripheral devices, and other digital systems. Counters are used for applications such as counting the number of times an event occurs or generating a time base.

The summary of the features offered by a typical data acquisition board is as follows:

Analog Input

Analog inputs	12 bits, 2 or 4 inputs
Maximum sampling	5 MS/sec
Minimum sampling	1 KS/sec
Input range	±42, ±20, ±10, ±5, and ±1 V; ±500 mV
Input coupling	AC or DC
Over voltage	±42 V
Relative accuracy	±0.5 LSB
Data transfer	DMA, interrupts, programmed I/O
Dynamic range	200 mV–10 V, 75 dB; 20–50 V, 70 dB
Amplifiers	Currents; ±200-nA bias, ±100-pA offset; 1-MΩ input impedance, CMRR 60 dB (typical)
Bandwidth	5 MHz, 500 mV–50 V

Analog Output

No. of Channels	2
Resolution	16 bits, 1 in 65,536
Maximum update	4 MS/sec, one channel; 2.5 MS/sec, two channels
Relative accuracy	±4 LSB
Voltage output	±10 V
Output impedance	50 Ω ± 5%
Settling time	300 nsec ± 0.01%
Slew rate	300 V/μsec
Noise	1 mV$_{RMS}$, DC–5 MHz

Digital I/O

No. of channels	8 input, 8 output
Compatibility	5 V/TTL
Data transfer	Programmed I/O
Input, low voltage	0–0.8 V
Input, high voltage	2–5 V

Timing

Base clock	20 MHz and 100 kHz
Clock accuracy	±0.01%
Gate pulse duration	10 nsec, edge detect
Data transfer	Programmed I/O

Trigger

Analog trigger	Internal; external ±10 V; 8-bit resolution, 5 MHz
Digital trigger	Two triggers; start/stop gate with 10-nsec minimum pulse width; 5 V/TTL external with 10-kΩ impedance

Others

Warm time	15 min
Calibration interval	1 year
Temperature coefficient	±0.0 ppm/°C
Operating temperature	0–45°C
Storage temperature	−20 to 70°C
Relative humidity	10–90%
Bus interface	Master, slave
I/O connector	68 pins, male connector
Power requirements	4.65–5.25 VDC

However, many other boards with more inputs, higher resolution, and faster scanning speeds can be found in the enormous and highly competitive market of data acquisition systems.

3.3 Digital Signal Processing

Electronic portable instruments are designed to sense and process signals generated by physical phenomena. Signals are the natural emission of energy from physical objects or man-made information that manifests and characterizes the behavior of objects or groups of objects (systems). A wide range of signals are of practical importance for electronic portable instruments, such as the human voice, temperature, light emitted from substances, etc. A brief introduction has been given on signals and signal conditioning in Chapter 1. In this section, greater detail will be provided on digital signal processing since the majority of today's electronic portable instruments are essentially digital.

However, in many systems, signal transformations take place between the inputs and outputs. These transformations must be understood and carefully analyzed so that meaningful information about the system and its functions can be extracted. These transformations may be described by linear or nonlinear equations. In practice, the linear systems are very common. Even if the systems are nonlinear, the majority of them may be linearized by application of suitable mathematical techniques. This section will concentrate mainly on the linear systems.

Mathematically, signals are represented as functions of one or more independent variables. For example, time-varying voltage and current signals obtained from many sensors are functions of a single variable (i.e., time); the electromagnetic signals are functions of time and space in x, y, and z in Cartesian coordinates. Often, signals from sensors need to be analyzed and processed to extract useful information. Among many others, Fourier transforms and the Fourier series are two common methods that can be used in signal analysis. In these methods, the signal representation involves decomposition in terms of sinusoidal or complex exponential components. The decomposition is then said to be in frequency domain. The decomposition is determined by frequency analysis tools that provide a mathematical and pictorial representation of frequency components, which is referred to as the frequency spectrum. The frequency spectrum can be determined by frequency or spectral analysis techniques. The Fourier series is useful in representing continuous-time periodic signals, whereas Fourier transforms are useful in representing periodic signals as well as nonperiodic signals, which are also known as finite energy signals.

A number of important mathematical operations are performed before or during analysis of signals. Many of these operations involve transformation of the independent variable. Some important operations are time shifting, reflecting, and time scaling of signals. Time shifting of signals is often applied in radar, sonar, communication systems, and seismic signal processing. Reflection of signals is useful in examining symmetry properties that a signal may possess. Time scaling is useful for observing the behavior of signals in expanded or contracted times to observe the required details.

In this chapter, a brief introduction on signal processing is given in order to develop a good background knowledge for the understanding of signal processing practiced in electronic portable instruments. There are many volumes of books written on signals, analysis, and signal processing techniques. Detailed information on signal processing is beyond the scope of this book.

Interested readers can refer to books (Stanley et al., 1984; Proakis, 1992) listed in the Bibliography.

The signals generated by a physical phenomenon can be periodic, nonperiodic, or digital.

3.3.1 Continuous-Time Signals

One way of classifying a signal is according to the nature of the independent variable. If the independent variable is continuous, the corresponding signal is said to be a continuous-time signal and is defined for a continuum of values of the independent variable. An example of continuous signal is the sinusoidal signal; the sine wave can be used to explain the fundamentals behind continuous signal processing. A sinusoidal signal $x(t)$ can be described by

$$x(t) = A \sin(\omega t + \theta), \quad -\infty < t < \infty \tag{3.11}$$

where A is the amplitude, ω ($2\pi f$) is the frequency in rad/sec, and θ is the phase in radians.

By using Euler identity ($e^{\pm j\Phi} = \cos\Phi \pm j\sin\Phi$), the relationship in Equation 3.11 can be expressed in exponential form as

$$x(t) = A \sin(\omega t + \theta) = \frac{A}{2} e^{j(\omega t + \theta)} - \frac{A}{2} e^{-j(\omega t + \theta)} \tag{3.12}$$

This indicates that a sinusoidal signal can be obtained by adding two equal-amplitude complex–conjugate exponential signals. The use of complex exponential tools plays a major role in signal analysis.

In general, periodic signals can be represented as a sum of simpler signals. Linear system analysis considers each constituent signal to be independent of the others. In many applications, it is possible to represent an infinite variety of signals with a few simple signals; therefore, it is mathematically possible to write signals as continuous periodic functions in trigonometric series forms (i.e., Fourier series):

$$x(t) = a_0 + \sum_{k=1}^{k=\infty} a_k \cos(tk) + \sum_{k=1}^{k=\infty} b_k \sin(tk) \tag{3.13}$$

or using identities:

$$\cos(t) = \frac{1}{2}(e^{jt} + e^{-jt}) \quad \text{and} \quad \sin(t) = \frac{1}{2} j(e^{jt} - e^{-jt}) \tag{3.14}$$

Thus, the continuous periodic signal can be described by

$$x(t) = \sum_{k=-\infty}^{\infty} c_k e^{jkw_0 t} \qquad (3.15)$$

where

$$c_k = \frac{1}{T} \int_0^T x(t) \exp\left[\frac{-j\omega_0 t}{T}\right] dt \qquad (3.16)$$

where c_k is the Fourier coefficient for $k = 0, \pm 1, \pm 2, \ldots$, and ω_0 is the fundamental frequency in rad/sec ($\omega_0 = 2\pi f_0$, f_0 is in Hertz).

The plot of $|c_k|$ vs. $k\omega_0$ is called the magnitude spectrum and is often used in signal analysis to see the magnitudes of different frequency components in a complex signal. The plot of angle c_n vs. $k\omega_0$ is called the phase spectrum. The magnitude and phase spectrums of periodic signals constitute line spectrums.

The Fourier series is a powerful tool for treating periodic signals in that they can be represented by an infinite number of harmonics, based on a common period. Another powerful technique is Fourier transforms, which can describe both periodic and nonperiodic signals. Fourier transforms are often described in Fourier transform pair as:

$$x(t) = \frac{1}{2\pi} \int_{-\infty}^{\infty} X(\omega) \exp[j\omega t] d\omega \qquad (3.17)$$

and

$$X(\omega) = \int_{-\infty}^{\infty} x(t) \exp[-j\omega t] dt \qquad (3.18)$$

where $X(\omega)$ is the Fourier transform of $x(t)$ and has the same role as c_k in a periodic signal.

The magnitude of $X(\omega)$ plotted against ω is called the magnitude spectrum of $x(t)$, and $|X(\omega)|^2$ is called the energy spectrum. The angle of $X(\omega)$ plotted against ω is called the phase spectrum.

Essentially, the Fourier transform specifies the spectral content of a signal, thus providing a frequency domain description of the signal. The magnitude and phase spectrums of a nonperiodic signal are not line spectrums, as in the case of periodic signals, but instead occupy a continuum of frequencies.

The continuous-time Fourier transform and its discrete counterparts find extensive applications in electronic portable instruments, communication systems, control systems, and other engineering disciplines. They provide useful information for many operations such as amplitude modulations, frequency multiplexing, sampling theorem, design of filters, and so on.

3.3.2   Discrete-Time Signals

A discrete-time sinusoidal signal may be expressed in terms of sequences as

$$x(nT) = A \sin(\omega nT + \theta), \quad -\infty < t < \infty \tag{3.19}$$

where n is the integer variable termed as the sample number and T is the time interval between samples, known as sampling period.

The Fourier series representation of a periodic discrete-time signal $x(n)$ sequence is

$$x(n) = \sum_{k=0}^{N-1} c_k e^{j2\pi kn/N} \tag{3.20}$$

where N is the fundamental period.

The coefficient c_k is given by

$$c_k = \frac{1}{N} \sum_{n=0}^{N-1} x(n) \exp\left[\frac{-j2\pi kn}{N}\right] \tag{3.21}$$

The frequency domain of discrete-time signals that are not necessarily periodic can be expressed in discrete Fourier transform (DFT) as

$$X(\omega T) = \sum_{n=-\infty}^{\infty} x(n) \exp[-j\omega nT] \tag{3.22}$$

and

$$x(n) = \frac{1}{2\pi} \int_0^{2\pi} X(\omega T) \exp[-j\omega Tn] d(\omega T) \tag{3.23}$$

These equations are particularly useful in sampling theorem, analog-to-digital signal conversion, signal reconstruction, and digital filters.

3.3.3 Energy and Power Signals

In many cases, currents or voltages generated by the sensors can be expressed in the form of energy or power signals:

$$E = \int_{-L}^{L} |x(t)|^2 \, dt \qquad (3.24)$$

$$P = \lim_{L \to \infty} \left[\frac{1}{2L} \int_{-L}^{L} |x(t)|^2 \, dt \right] \qquad (3.25)$$

When the limit in Equation 3.24 exists and yields $0 \angle E \angle \infty$, the signal $x(t)$ is said to be an energy signal. On the other hand, when the limit in Equation 3.25 exists and yields $0 \angle P \angle \infty$, the signal $x(t)$ is said to be a power signal.

3.3.4 Digital Signals

Digital signals are information, arrays of numbers that are used in computational analysis of systems and signals. While analog signals are continuous in time and magnitude, sampled analog signals are discrete in time and continuous in magnitude. Digital signals are discrete in both time and magnitude. Digital signals can be either generated by calculations or directly derived from analog signals using A/D converters.

An analog signal is first sampled at discrete-time intervals to obtain a sequence of samples that are usually spaced uniformly in time. This is achieved ideally by multiplying a periodic train of Dirac delta functions spaced T seconds apart. T represents the sampling period, and its reciprocal, $f_s = 1/T$, is called the sampling rate. The signal obtained from this process of continuous-time uniform sampling is referred to as the ideal sampled signal (ideal because it is realistically impossible to implement a Dirac delta function). Due to the duality of the Fourier transform, by sampling a signal in the time domain (i.e., multiplication), the signal becomes periodic in the frequency domain.

After having sampled the analog signals, quantization is necessary to put the discrete-time signal into discrete values to be followed by coding of signals to express them as binary sequences.

3.3.5 Nonperiodic and Random Signals

Nonperiodic signals do not repeat in time and can be continuous or random. Nonperiodic signals can be characterized by the continuous complex spectral function $X(\omega)$. As can be seen from Equation 3.15 to Equation 3.17, most signals can be represented as a scaled sum (or integral) of complex sine waves $e^{j\omega}$. Consequently, linear transformation of any signal can completely be determined by just how it scales the complex sine waves as:

$$Y(\omega) = X(\omega)H(\omega) \qquad (3.26)$$

where $Y(\omega)$ and $X(\omega)$ are frequency spectrums of output and input signals and $H(\omega)$ is a system frequency response, constant complex function of frequency, or coefficient array that is defined by the properties of the system.

Random signals are met often in nature. Theoretically, an infinitely long time record is necessary to obtain a complete description of these signals. However, statistical methods and probability theory can be used for the analysis by taking representative samples of signals from the ensemble. Mathematical tools such as probability distributions, probability densities, frequency spectra, cross-correlations, autocorrelations, DFTs, FFTs, autospectral analysis, rms values, and digital filter analysis are some of the techniques that can be employed. Interested readers should refer to the Bibliography for further information.

Discrete Fourier transform is a powerful technique to analyze digital signals. In order to calculate the discrete Fourier transform of a nonperiodic signal $x(n)$, we can assume that over the finite range of N samples, $x(n)$ represents one period of a periodic signal $x_p(n)$ and determine the DFT of $x_p(n)$. Truncation of $x(n)$ introduces errors in the Fourier spectrum. This can be mitigated by tapering the ends of the truncated signal with a window function defined over the interval N; although reduced, the errors still remain and appear as loss of spectral resolution and spreading of peaks.

3.3.6 Sampling Process

One fundamental requirement of A/D conversions is that the digital signal obtained in a conversion process must be fully representative of the original analog signal. This means that the analog signal should be completely recoverable, through the reconstruction process, using the values of digital signal. To guarantee the complete recoverability of the digital signal, the sampling period used to carry out the conversion must fulfill the postulates of the sampling theorem. This theorem, which is of crucial importance in the theory of signal acquisition, states that "under certain conditions a continuous-time signal can completely be represented by and recoverable from knowledge of its instantaneous values of samples equally spaced in time." The conditions are as follows:

- The original analog signal must be band limited. This means that the Fourier analysis of the signal must determine that the frequency content of the signal ranges from 0 to a maximum value of ω_M (fm in Hz).

- The sampling frequency $\omega_s = 2\pi/T$ in rad/sec ($f_s = 1/T$ in Hz) must be equal or greater than twice the highest frequency present in the signal. That is, $\omega_s \geq 2\omega_M$.

To avoid aliasing of the out-of-band signals into the band of interest, the input signal must be filtered before A/D conversion takes place. Correct sampling frequency must be selected to avoid aliasing within the band of interest. The sampling theorem allows sampling of narrow-band signals with a sampling frequency that can be significantly lower than the signal itself.

However, in the real world it is very difficult to find band-limited signals, because sensors produce signals with an infinite number of frequency components. So the ideal conditions for the sampling theorem are not fulfilled. Therefore, there is no possibility of processing real signals without error. However, by carrying out an appropriate treatment of signals, this error can be minimized. One important action, which can be achieved, is the conversion of the real signal into a band-limited signal before the sampling process. This is carried out by using low-pass filters, which cause the information contained in the higher frequencies to be lost. However, as long as this information is of little importance, the complete process is valid while ω_s is greater than twice ω_M.

Using a continuous signal as an input, the sampling operation obtains a signal, which is discrete in the time domain. The sampling of a signal consists of multiplying itself by a train of impulses separated by the sampling period T. The sampled signal is also a train of impulses, modulated in amplitude. Figure 3.22 shows the sampling process of a signal $x(t)$, using a train of impulses $p(t)$, to obtain the sampled signal $X_s(t)$, which mathematically can be expressed as

$$x_s(t) = x(t)p(t) \tag{3.27}$$

where $p(t)$ is the periodic impulse train, which can be described by

$$p(t) = \sum_{-\infty}^{\infty} \delta(t - nT) \tag{3.28}$$

where t is time, T is a sampling interval, n is a sample number, and δt is a unit sample such that $\delta 0 = 1$ and $\delta t = 0$ for $t \neq 0$.

During the conversion stage, the A/D converter quantizes the sampled analog signal to one of the finite set of discrete numeric values available in the converter. The sampled analog signal is thus transformed into a digital signal $x(n)$. Once created, digital signals are just arrays of numbers stored in the computer memory, and their association with time (sampling rate at which they were acquired) is stored and maintained by the signal processor

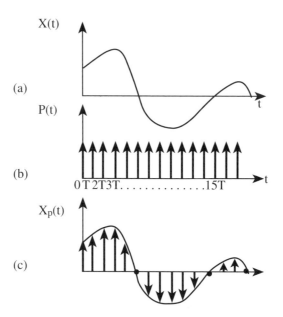

FIGURE 3.22
(a) Representation of an analog signal. (b) Train of impulses to sample the analog signal. (c) The sampled signal, which is a train of impulses modulated in amplitude.

independently of the signal. Each stage of the A/D conversion produces a different representation of the original analog signal.

The sampling theorem allows us to completely reconstruct a band-limited signal from instantaneous samples taken at a rate $\omega_s = 2\pi/T$. Using Equation 3.27 and Equation 3.28, the relationship between a continuous analog signal and a sampled analog is given by

$$x_s(t) = x(t) \sum_{-\infty}^{\infty} \delta(t - nT) \tag{3.29}$$

The Fourier transform of the sampled signal $x_s(t)$ becomes

$$X_s(\omega) = \sum_{-\infty}^{\infty} x(nT) \exp[-jn\omega T] \tag{3.30}$$

Because the lowering sampling rate saves the processor memory and computational time, there is a strong incentive to set the sampling frequency as low as possible, while maintaining faithful and accurate representation of the original signal. The sampling theorem of digital signal processing sets the lower limit on the sampling frequency of a band-limited signal and specifies a method for the recovery of the original analog signals.

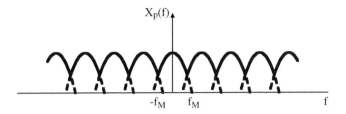

$X_P(f)$

$-f_M$ f_M

f

FIGURE 3.23
Representation of the aliasing phenomena.

Figure 3.23 shows the phenomenon that occurs when the sampling frequency (f_s) is less than twice the highest frequency (f_M) of the signal. This phenomenon, which is called aliasing, provokes overlapping of frequency patterns of the sampled signal. When aliasing occurs, it is impossible to recover the original signal.

Under certain easily satisfied conditions, the sampling uncertainty can be overcome and even exploited. It can be shown that sampling introduces periodicity in the frequency spectrum of the original analog signal. In fact, the sampling frequency f_s is the period of the sampled analog signal spectrum. If the bandwidth of the analog signal spectrum is limited to $f_s/2$, then the sampled analog spectrum consists of identical replicas of the analog spectrum. If, however, the analog spectrum bandwidth extends beyond $f_s/$ 2, then its replicas would overlap and produce new, distorted spectra. Therefore, to establish direct one-to-one correspondence between the analog signal and its sampled version, one has to know the maximum extent of the analog signal frequency spectrum and to make sure that the sampling frequency is high enough to produce exact replicas of the original signal without harmful overlapping.

Considerable reductions in the number of computation required by the DFT can be achieved by exploiting symmetry and periodicity of complex sine waves. The original Cooley–Tukey fast Fourier transform algorithm requires N to be a power of 2. This algorithm is computationally advantageous over the DFT whenever the number of spectral points to be computed is smaller than $\log_2 N$. The many existing variations of the FFT algorithms are those that optimize the number of required complex multiplications, amount of memory storage, order of execution, algorithms that allow N to be prime numbers, etc.

3.3.7 Reconstruction of Signals

The sampling theorem states that if a signal with a bandwidth, BW, is sampled at a rate f_s no lower than 2 BW, then this signal can completely be reconstructed from its samples. The sampling rate f_s is called the Nyquist rate. The sampling theorem also defines the method of how the reconstruction of a continuous-time analog signal from the sampled analog should be done:

$$x(t) = \sum_{n=-\infty}^{n=+\infty} x(nT) \frac{\sin[(\pi / T)(t - nT)]}{(\pi / T)(t - nT)} \qquad (3.31)$$

Equation 3.31 shows that reconstruction can be done by multiplying signal samples by $\sin(x)/x$ function and summing their product over all samples. Therefore, the D/A conversion must first produce a sampled analog signal and then apply mathematical operation to the analog samples. The reconstruction formula is a special type of operation known as convolution that can be described by

$$x(n) = \sum_{k=-\infty}^{k=+\infty} x(k)h(n-k) \quad \text{or} \quad x(n) = x(n) * h(n) \qquad (3.32)$$

where * is the convolution operator.

Convolution is a mathematical operation describing action of the linear system with an impulse response $h(n)$ on the input signal $x(n)$ that produces output signal $x(k)$. In a frequency domain, convolution corresponds to simple multiplication

$$Y(e^{j\omega n}) = X(e^{j\omega n})H(e^{j\omega n}) \qquad (3.33)$$

The terms $X(e^{j\omega})$ and $Y(e^{j\omega})$ are Fourier transforms of the input and output signals, and $H(e^{j\omega})$ is the system frequency response. Because multiplication is a much simpler operation than convolution, analysis of systems and signals is often much simpler in the frequency domain than in the time domain.

Continuous signals also can be recovered from the sampled signal using bandpass filters and interpolation filters. An ideal interpolation filter cannot be realized in practice, because it requires zero-width roll-off between the passband and stopband regions of its frequency response. A practical analog filter has a finite transition. So to recover an analog signal using such a filter, the original sampling frequency must be increased to account for the roll-off of the interpolation filter.

3.3.8   Quantization

Amplitude quantization is defined as the process by which a signal with a continuous range of amplitudes is transformed into a signal with discrete amplitudes. The quantization process used in practice is usually memoryless. That is, the quantizer does not depend on previous sample values to predict the next value. While this is not an optimal solution, it is commonly used in practice.

The precision of the quantizer is limited by the number of discrete amplitude levels available for the quantizer to approximate the continuous signal

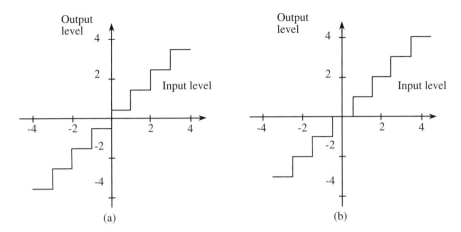

FIGURE 3.24
Types of uniform quantizers: (a) midrise and (b) midtread.

amplitudes. These levels are referred to as representation or reconstruction levels, and the spacing between the two representation levels is called the step size.

This type of quantization is known as uniform quantization. Uniform quantizers can be subdivided to be the midtread quantizer or the midrise quantizer, as shown in Figure 3.24.

The main characteristic of uniform quantization is that the absolute quantization error stays constant in the whole range. As a result, the relative error is variable and is maximum for the minimum amplitudes of the original analog signal.

Some applications require the relative quantization error to be held constant. In this case, another type of quantization, called nonuniform quantization, is used. Nonuniform quantization is based on varying the size of the intervals proportionally to the amplitudes of the analog input signal. For instance, the ranges of amplitudes covered by voice signals, from passages of loud talk contrasted with passages of weak talk, are in the order of around 1000 to 1. For this reason, in a nonuniform quantizer, the step size increases as amplitude increases. Therefore, infrequently occurring loud talk is catered for, but weak passages of talk are more favored with smaller step sizes.

The use of quantization introduces an error between the continuous amplitude input and the quantized output, which is referred to as quantization noise. Quantization noise is both nonlinear and signal dependent; thus, quantization noise cannot be alleviated by simple means. However, provided that the step size is sufficiently small, it is reasonable to assume that the quantization error is a uniformly distributed random variable, and therefore, the effect of quantization noise is similar to that of thermal noise.

By modeling the quantization noise as a uniformly distributed random variable, it is possible to ascertain a signal-to-noise ratio for the quantizer. As stated earlier, the precision of the quantizer is dependent on the number of representation levels available. This also corresponds to the number of bits per sample, R (the number of representation levels $L = 2^R$). The SNR is exponentially proportional to the number of bits per sample, R, so increasing the number of bits exponentially increases the SNR. However, increasing the number of bits also increases the bandwidth, so there is a trade-off between the two.

3.3.9 Coding

Coding is a process of representing the finite output states of a quantizer using sequences of n bits: a different sequence of bits is assigned to each output state. For n bits, 2^n different sequences can be generated, and consequently, 2^n output states can be coded. The codes formed using bits are normally known as binary codes, and they can be of two types: unipolar and bipolar.

3.3.9.1   *Unipolar Codes*

Unipolar codes are used to represent unipolar quantities, that is, quantities with a predefined sign (positive or negative). The most common unipolar codes are *natural binary, binary-coded decimal* (BCD), and the *Gray code*. In order to illustrate the coding process, the natural binary code is described below.

In the natural binary code, each bit of a number represents a successive power of 2 according to the position that the bit occupies in the number. Given a number of n digits (represented by D_{n-1}, D_{n-2}, ... D_1 and D_0), the value of the number is determined by the following expression:

$$D_{n-1}D_{n-2}...D_1D_0 = D_{n-1} \times 2^{n-1} + D_{n-2} \times 2^{n-2} + ... + D_1 \times 2^1 + D_0 \times 2^0 \quad (3.34)$$

D_{n-1} is the most significant bit (MSB) of the number, and D_0 is the least significant bit (LSB). For example, the sequence of bits 1010101 represents the following quantity:

$$1010101 = 1 \times 2^6 + 1 \times 2^4 + 1 \times 2^2 + 1 \times 2^0 = 85 \quad (3.35)$$

However, the output code of an A/D converter is normally interpreted as a fraction of its full-scale (FS) value, which is considered the unit. So the binary codes provided by an A/D converter in its output must be interpreted in fractional numbers, which is equivalent to dividing the value of each bit by 2^n. Taking this into account, the value of a sequence of n bits is:

$$D_{n-1}D_{n-2}...D_1D_0 = D_{n-1} \times 2^{-1} + D_{n-2} \times 2^{-2} + ... + D_1 \times 2^{1-n} + D_0 \times 2^{-n} \quad (3.36)$$

The bit sequence is interpreted by adding an implicit binary point at the left end of the number. Using this form of representation, the above bit sequence (1010101) actually represents the binary number 0.1010101, whose value is

$$1 \times 2^{-1} + 1 \times 2^{-3} + 1 \times 2^{-5} + 1 \times 2^{-7} = 0.6640625 \tag{3.37}$$

In this way, if the FS voltage value of a D/A converter is V_{FS}, the voltage V_o corresponding to an output word of n bits will be

$$V_o = V_{FS} \sum_{i=0}^{i=n-1} (D_i / 2^{n-i}) \tag{3.38}$$

For example, for a D/A converter of 8 bits and an FS value of 10 V, the output word 11010000 represents the following voltage value:

$$V_o = 10 \times (2^{-1} + 2^{-2} + 2^{-4}) = 8.125 \text{ V} \tag{3.39}$$

3.3.9.2 Bipolar Codes

Bipolar codes are used to represent bipolar quantities, that is, quantities that can be either positive or negative. The most common bipolar codes are 2's complement, 1's complement, sing-magnitude, and offset binary. These codes are used when there is a need to preserve the positive–negative excursion of the original analog signal after the A/D conversion.

3.3.10 Digital Filters

Filtering is probably the most commonly employed signal processing technique. It is used to remove unwanted signal components, such as distortions or noise, or alternatively, to select only the desired components, such as selection of a particular radio band. As a system, a filter can be described by several key characteristics, such as frequency, phase, and impulse responses.

All practical filters, whether analog or digital, have at least three regions: a "passband," the region of the frequency spectrum that is passed through the filter without significant change; a "stopband," the region that is not passed through; and a transition region between the passbands and stopbands. Existing terminology divides all filters into four basic types: the low-pass, high-pass, bandpass, and bandstop. Frequency responses of each filter are specified in terms of their cutoff frequencies, the required accuracy in passing desired frequencies, and the degree of attenuation of the unwanted ones.

The two most common digital filters are the FIR filter and the infinite impulse response (IIR) filter. Both of these filters can easily be implemented by microprocessor systems and DSPs. The FIR filter computes its output samples as a weighted sum of delayed input samples:

$$y(n) = \sum_{k=0}^{k=N} h(k)x(n-k) \qquad (3.40)$$

where $x(n)$ is the input signal and $h(k)$ is a set of coefficients that define the filter performance.

Computational implementation of the FIR filters is accomplished with a set of memory elements to store $x(n-k)$ samples and an add–multiply arithmetic unit. The FIR filter coefficients $h(k)$ also form its impulse response. When stimulated by an input impulse, an FIR filter produces an output signal that is a sequence of its coefficients (all multiplication products are zero, except the one at a point where the impulse is stored). The FIR filters are unique to the digital domain and can only be approximated by analog filters.

The IIR filters are also called recursive filters, because they use both input and previously calculated output values to compute the current output value. Like the FIR filters, the IIR filters also employ memory elements and perform add–multiply operations. The IIR filters are digital equivalents of analog filters and, for a comparable performance, they require fewer calculations than FIR filters. However, FIR filters have a very useful property that IIR filters do not have. FIR filters can be designed to have a linear phase response. The phase response of a system is derived from the complex frequency response $H(e^{j\omega})$:

$$\theta(j\omega) = \arg\left[H(j\omega)\right] \qquad (3.41)$$

Equation 3.41 characterizes the phase shift for each complex sine wave when it passes through the system. Phase response is important whenever the time waveform of a signal must be preserved, as in the case of digital communication. If the delay is constant and frequency independent, then its phase response must be linear; otherwise, the system would distort the phase relationship between the spectral components of the signal. Changes in the relative phases of spectral components will result in the distortion of the signal waveform.

3.3.11 Digital Signal Generation

Signal generation is used extensively in many applications. Examples include generation of sensor excitation and testing and performance measurements of complex signal processing devices and systems, such as transmission and communication systems and equipment. Signal generation is used in the component testing measurements of inductive and capacitive parameters, in eddy current instrumentation, and as a local oscillator in modulation and demodulation processing, among many other uses. Generation of periodic signals using analog techniques can be simple and effective in many inexpensive portable instruments, such as impedance meters. However, as most analog systems, the signal generators frequently suffer from relatively low accuracy and drift in frequency and amplitude, and are subject to temperature changes and other environmental factors.

Another method of analog signal generation, direct digital synthesis (DDS), is a digital technique that converts computer-generated digital signals to the analog form by using D/A converters. Although limited by the capabilities of digital hardware, digital signal generation now extends to the frequencies in hundreds of megahertz and offers a powerful and flexible method of signal generation that provides good frequency accuracy, stability, reasonably low noise, and harmonic distortion performances. For comparable costs, digital generators are simpler, precise, and, especially in terms of frequency stability, always more flexible.

The basic mechanism of digital signal generation is a mathematical computation that is easily achieved in a digital computer. Most existing microprocessors are fully capable of generating even the most complex signals given sufficient time. Also, most compilers include libraries of functions that include sine waves, logarithmic and exponential functions, square roots, etc. So in principle, it is simple to write a program that would generate many common signals, such as sine and square waves, triangular and sawtooth waves, amplitude and frequency modulated signals, polynomial signals, etc., and then send these signals to the D/A converter. For example, an L sample i-long sine wave signal can be computed using the following C code:

$$\text{for } (I = 0; i < L; I++)$$

$$x(i) = A * \sin\left(2 * \pi * \frac{f}{f_s} * i\right) \tag{3.42}$$

where A and f are the amplitude and frequency, respectively, of digital sine wave x, f_s is its sampling frequency, and L is the length of the signal array. Note that the frequency, f, gives the digital signal its analog meaning only in relation to the sampling frequency, f_s, such as, for example, an update rate of the D/A converter. Such limited lengths in digital signal can be precalculated, converted to the integer form, and then stored in the instrument's memory and retrieved as needed.

While generation of nonperiodic signals may require continuous computation of signal samples, generation of periodic signals requires only a single block of samples that can be used repeatedly. For example, in Equation 3.42, block length L of a periodic signal x must be such that

$$LT_s = NT \tag{3.43}$$

where $T_s = 1/f_s$, $T = 1/f$, and N is an integer number.

Under the conditions of Equation 3.43, a sine wave x must have integer number of periods within the length of L sampling intervals, so that $x(i) = x(L + i)$ for any i. It is also a consequence of Equation 3.43 that only sine waves with frequencies given by

$$f = \frac{Nf_s}{L} \qquad (3.44)$$

can be generated using a single L-long signal table. More generally, all periodic signals should satisfy the conditions in Equation 3.43 and Equation 3.44 in order to be fully represented by a sample block L.

Digital sine wave generation is a very common task in many instrumentation systems. The required digital hardware and D/A converter are frequently combined in the monolithic DDS integrated circuit. The DDS IC contains ROM with a sine wave table, an arithmetic hardware to calculate a variable pointer into the table, and a D/A converter to convert table data to the analog form. The common way to implement a digital signal generator is to maintain a pointer into the signal table and increment the pointer by a value derived from Equation 3.43 and Equation 3.44. If the increment value is an integer, then a discrete number of signals that differ by f_s/L can be generated from a given signal table. However, if a fractional increment is allowed, the sine wave of any frequency can be calculated. Of course, since the fractional increment would result in a fractional pointer, such a pointer would have to be rounded; otherwise, a small frequency inaccuracy (modulation) would result.

3.3.12 Data Storage

A frequently overlooked, but very important function is the effective delivery of the digital sample data into the memory, especially at high sampling rates. After the A/D converter finishes conversion, the instrument's computer must read the A/D output and store it in its memory. Although simple, this task, if not efficiently done, may require significant computational resources. The simplest way to implement data storage is to use a computer interrupt mechanism that is part of most existing microprocessors and microcontrollers. Whenever the A/D converter completes conversion, it asserts a "data ready" signal that is connected to the microprocessor interrupt line.

On the assertion of the interrupt, the computer immediately stops its current execution, stores its internal state, and branches to the part of the program that reads the A/D output, retrieves the current memory address, and writes the data to the memory. After having done this, the program restores its previously stored processor state and resumes the execution. This process must be repeated every time a data sample is acquired. It is also inefficient because most of the computing time is wasted in saving and restoring its state. Such inefficiency can be tolerated if interrupts come at a relatively low rate and a small proportion of the overall computer resources is wasted. More severe limitation to interrupt-based data acquisition exists in instruments that use some complex operating system software. Some of these systems limit user control of the available interrupts, frequently take a long time to respond to an interrupt, and therefore can guarantee only relatively low acquisition rates.

A more sophisticated method of data storage involves special direct memory access (DMA) computer peripherals that are often an integral part of the microprocessor device. A typical system may use more than one source of data and frequently requires multichannel DMA peripherals, each operating independently of the others according to a priority schedule set up by the system designer. Every DMA channel has several user-programmable registers that store addresses of the source (A/D) and destination (memory) end point of the transfer. As in the case of the interrupt-based systems, whenever a "data ready" signal from the A/D converter comes to the DMA peripheral, the processor stops its execution and DMA takes over the system bus. The A/D read and memory write operations are performed, the destination address is incremented in the anticipation of the next, and program execution is returned to the CPU. This is a highly efficient sequence that does not require storing and restoring the processor state. Once a programmed number of data samples have been transferred, the DMA peripheral stops and generates an interrupt that notifies the system that a required block of data has been stored. If the number of transferred samples is relatively large, then the DMA interrupt comes much less frequently than a sample interrupt and can be efficiently handled by most software systems. To repeat the DMA process, the user must reload DMA registers and restart the channel. This processing overhead usually is relatively small.

The method of storing data also can be important. For example, in a multiplexed system, samples may have to be acquired by quickly scanning all acquisition channels, so that samples are generated in a multiplexed fashion, channel after channel. However, often it is more convenient if data from each acquisition channel could be stored as a single monolithic block. Many DSPs are optimized for operations on the blocks of data and would process them much more efficiently if they were stored sequentially. The burden of de-multiplexing data can be accomplished by using more sophisticated DMA peripherals that allow programmable modification of the source and destination addresses. In a multichannel destination, memory address could be incremented by the length of the channel buffers to implement contiguous channel data storage.

Modern DMA peripherals, especially in DSPs, frequently implement other modes of operation, such as automatic restart of the DMA channel at the end of a block transfer, and complex increment of source and destination addresses, which could be especially useful in acquisition of two-dimensional and three-dimensional data.

Sometimes, there is no need to perform a special action at the end of the block transfer; rather, it is desirable to continue to repeat DMA operations indefinitely. This is frequently the case in signal generation, where a single period of a digital signal must be sent to the D/A converter. The DMA unit also can be programmed to branch at the end of the block transfer to a different memory location to fill another block of memory or repeat the previous block transfer continuously. The latter mode of operation is especially useful in signal generation where a digital signal must continuously write to the D/A converter.

3.3.13   Handling Small Signals

The A/D converter can realize its full accuracy and resolution only when an A/D input signal optimally matches the dynamic range of the A/D. Scaling signal up (gain) or down (attenuation) varies widely according to application. Small signal amplifiers typically have to provide large gains under conditions of low internal noise and immunity to the external interference. For example, biopotential amplifiers must provide very large gains of 1000 to 10,000 to bring signals in the hundreds microvolts to millivolts range to the typical A/D signal span of 3 to 5 V. Low-noise amplification requires selection of special low-noise components, especially in the first stages of the amplifier.

In small signal handling, filtering may be used that serves several purposes. As it was shown in this chapter, irrecoverable aliasing distortions may occur during sampling of analog signals. This distortion can be avoided by limiting signal bandwidth relative to the sampling frequency. Typically, an anti-aliasing filter is placed between the amplifier and A/D converter to eliminate undesirable out-of-band signals before they fold over into the band of interest.

A problem arises whenever a small signal is mixed with a large in-band signal such as power line interference. This problem is quite common, where high impedance sensors generate small signals and are connected to the instrument by a long cable. The interference must be removed before a sensor signal is amplified, so a bandstop or notch filter must be inserted in front of the amplifier. Another design option is to use a very high-resolution A/D converter that would not clip the interference but still provide enough resolution even for a small signal, and then to perform filtering digitally to remove interference. High-resolution 18- to 24-bit Δ/Σ converters are now available in low sampling frequency ranges (up to 5 kHz) and offer an attractive option in the design of many existing portable instruments. Another method of removing large interfering signals is to use special instrumentation amplifiers. These devices are designed to amplify the difference of two signals carried by the cable wires and reject their common components, such as the interference.

3.4   Instrument Communication and Telemetry

The utilization of digital processors and computers has improved the capabilities of electronic portable instruments and instrumentation systems extraordinarily. Computers, microprocessors, microcontrollers, and digital signal processors provide flexibility in the design of instrumentation systems and great computational power to process signals. The computational power has been growing constantly since its invention, and this allows the development of more challenging applications in electronic portable instruments. Improvements in such portable devices have been enhanced considerably by today's available communication technology.

In the past, electronic portable instruments were simple devices and they hardly had any communication facilities. If books on portable instruments were written, they would have had very little information the topic of instrument communication. Today, it is hard to see any portable instrument not being equipped with at least an RS-232 serial interface. Once an instrument has RS-232, it can easily be supported by radio frequency (RF) transmitters and receivers. This gives additional flexibility to portable instruments. Therefore, it can confidently be said that digital communication technology has been making revolutionary impacts on the development and use of electronic portable instruments. Elimination of the need for cables for communication purposes liberated them for greater uses. Most of these instruments are now capable of transferring and receiving information in all types of environments without any constraints. In order to appreciate recent developments in the portable instruments, it is necessary to introduce some fundamental concepts in communication systems.

3.4.1   Protocols

Information flow between the nodes, individual instruments, and computers is regulated by protocols. According to the IEEE, a protocol is defined as "a set of conventions or rules that must be adhered to by both communicating parties to ensure that information being exchanged between two parties is received and interpreted correctly." In general, communication protocols are defined by the following:

1. *Network topology*: star, ring, bus (tree), etc.
2. *ISO reference model layers*: physical, data link, network, transport, session, presentation, and application layers
3. *Data communication modes*: simplex, half-duplex, or duplex
4. *Signal types*: digital or analog

5. *Data transmission modes*: synchronous (include bit-oriented and character-oriented) or asynchronous

6. *Data rate supported*: several bits per second (bps) to several Gbps, it depends on both oscillator frequency and transmission medium

7. *Transmission medium supported*: twisted pair, coaxial cable, optical or microwave, etc.

8. *Medium access control methods*: carrier sense multiple access with collision detection (CSMA/CD) or control token

9. *Data format*: mainly based on data transmission modes and individual protocol specifications

10. *Types*: order of messages that are to be exchanged

11. *Error detection methods*: parity, block sum check, or CRC

12. *Error control methods*: echo checking, automatic repeat request (ARQ), sequence number, etc.

13. *Flow control methods*: X-ON/X-OFF, window mechanisms, etc.

The International Organization for Standardization (ISO) reference model layout for the protocols is shown in Table 3.3. The reference model has seven layers, each of which is an independent functional unit that uses functions

TABLE 3.3

ISO Reference Model

	Layer	Application	Protocols
1	Physical	Electrical, mechanical, process; functional control of data circuits	ISO/IEEE-802.4, phase coherent carrier, broadband (10 Mbs), etc.
2	Link	Transmission of data in LAN; establish, maintain, and release data links, error, and flow	IEEE-802.4 token bus, IEEE-802.2 type 1 connections
3	Network	Routing, segmenting, blocking, flow control, error recovery, addressing, and relaying	ISO DIS 8473, network services, ISO DAD 8073 (IS)
4	Transport	Data transfer, multiplexing, movement of data in network elements, and mapping	ISO transport, class 4; ISO 8073 (IS)
5	Session	Communication and transaction management, synchronization, administration of control	ISO session kernel, ISO 8237 (IS)
6	Presentation	Transformation of information such as file transfer, data interpretation, format and code transformation	Null/MAP transfer, ISO 8823 (DP)
7	Application	Common application service elements, manufacturing message services (MMSs), network management	ISO 8650/2 (DP), RS-511, ISO 8571 (DP), IEEE-802.1

from the layer below it and provides functions for the layer above it. The lowest three layers are network dependent, the highest three layers are network independent (application-oriented), and the middle layer (transport layer) is the interface between the two.

Each layer, except the physical layer, of the reference model has its own protocol control information (PCI), which specifies the services it provides. The PCI may vary from several bytes to hundreds of bytes, and it specifies the processing to be done to the data by the remote party's corresponding layer.

All digital instruments need to obey the ISO reference model when communicating with each other. In some applications, some of the seven layers of the ISO reference model may be omitted, as in the cases of many fieldbuses.

3.4.2   Communication Networks

In modern electronic portable instruments, local and global bus systems are used for the internal connection of components and for external connections to other digital systems. They may be a part of complex networks that may contain groups of instruments, controllers, microprocessors, and computers. The hardware architectures of a digital network system can be in centralized, decentralized, hierarchical, or distributed configurations. There are many different network topologies, as illustrated in Figure 3.25.

Since the instrumentation and sensors are the main contact between the process and controllers and computers, the networking of sensors and instruments becomes very important for efficient, reliable, and economical operations. Generally, instruments and sensors are grouped together to operate

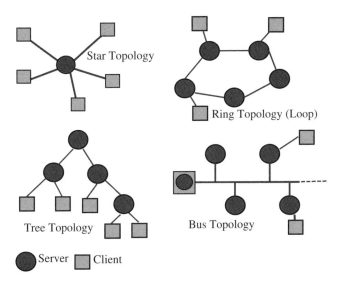

FIGURE 3.25
Examples of network topology.

in nodes by the help of the communication protocols, microcontrollers, and microprocessors. At the grassroots of this networking lies the fieldbus for industrial systems and other communication protocols developed by vendors to enable reliable and fast data communications among nodes, computers, and individual sensors. At the moment, manufacturers offer a bewildering range of buses and protocols.

It is worth noting that the grouping arrangements of digital systems result in many advantages. For example, in message sharing, because all the nodes are connected together by one single transmission medium, the nodes can demand information from other nodes to implement strategies, for setting and resetting alarms, etc. Also, due to the simple data transmission medium and the node configuration, any problem from sensors or transmissions can be isolated, thus giving rise to easy maintenance without affecting the performance of the network. Once the digital systems are grouped together to form a network, the burden of the computer can be relieved; the burden can be distributed among other units. Although the initial implementation of such networks is relatively more expensive than that of the traditional ones, the ease and benefits can outweigh the initial costs in the long run.

There are a number of digital systems that electronic portable instruments can easily be integrated with, such as the SCADA systems and various fieldbuses. SCADA systems will be explained next, and fieldbuses will also be discussed briefly.

3.4.2.1 SCADA Systems

SCADA stands for System Control and Data Acquisition. These systems are a typical example of instrumentation systems that all kind of instruments with communication capabilities can be integrated with, including electronic portable instruments. The term *SCADA* has been adopted by the process control industry to describe a collection of computers, sensors, and other equipment interfaced by telemetry technology in order to monitor and control processes. The uses of SCADA systems are almost endless, as they are only limited by the designer's imagination. These systems have been used to monitor the status of high-voltage power distribution lines and stations, RF energy delivered by broadcasting stations, water levels in reservoirs, and so on. Most of these applications use portable or portable-like instruments that are powered by batteries or some other means without the use of main power. As digital systems become more sophisticated, the use of remote telemetry and control systems becomes essential for adequate management of the resources.

One of the advantages of using SCADA systems is that the existing sensors in current applications can be incorporated easily into the overall system, thus decreasing the economic cost of migrating to the new technological solution. Doing so, the SCADA system adds a new level of intelligence and provides additional capabilities to current systems. Furthermore, system

TABLE 3.4

Common Fieldbus Protocol Layers

User layer
Application layer
Data link layer
Physical layer

control can be centralized; this is what makes the control of different remote locations available from a central commanding post.

3.4.2.2   *Fieldbuses*

As highlighted earlier, in instrumentation systems, all the signals from or to sensors and actuators need to be transmitted and processed suitably. One of the solutions for this problem comes up with the new emerging technology called fieldbus. Fieldbuses represent protocols using the lower communication layers of the ISO reference model, given in Table 3.3. The fieldbus protocols are all-digital, high-performance, multidrop communications protocols. They are taking more and more of a major role in instrumentation systems, automation systems, industrial embedded systems, etc. Some important fieldbuses include Profibus, DeviceNet, Foundation Fieldbus, CAN, Lightbus, Interbus, LonWorks, P-NET, Modbus, Bitbus, and many others.

Common to all fieldbuses, with few exceptions, LonWorks protocol layers 3 to 6 (network, transport, session, and presentation) of the ISO reference model are omitted to improve performance. Most fieldbuses contain an extra layer on top of the application layer, known as the user layer, as illustrated in Table 3.4. Many of the functions that would have been handled by the omitted layers are carried out by the data link layer or the fieldbus application layers.

The protocol implementations of fieldbuses usually involve hardware as well as software, with a trade-off between the two. Some fieldbuses have associated chips for implementing all or portions of protocol specifications. Some issues that cannot be implemented by the hardware have to be done by software. For example, in LonWorks protocol, all layers can be implemented by chips and associated software. The software defines all the objects and libraries that are readily available for the customer. But for CAN, the lower two layers are implemented by the protocol chip (82C200). All the software has to be done by the users so that it is suitable for their requirements.

In fieldbus systems, the management of the asynchronous information flow, such as alarms, is particularly critical. It is usually difficult to foresee the asynchronous information flow requirements (i.e., the bandwidth required for the transmission) *a priori*. Therefore, a policy for the management of the asynchronous information flow is needed in the beginning for feasible efficiency. For example, the Foundation Fieldbus, is composed of only three levels: application, data link, and physical layers. The data link layer provides policies for management of asynchronous information flow.

An important point in the fieldbus protocol is the servicing of the feedback control loops. In some cases, as in the case of fast systems, the communication delays may degrade the achievable performances of the feedback loops. To overcome such problems, there are a number of methods developed, such as the adapted synchronous model feedback systems (ASMFS) or priority systems. Therefore, in portable instruments, careful attention to the problems of efficient and effective use of fieldbuses is necessary where fast responses are required.

3.4.3 Telemetry

Telemetry is very important in the development and use of current and future electronic portable instruments. It is the science of gathering information at some remote location and transmitting the data to a convenient location to be examined and recorded. Telemetry can be done by different methods: optical, mechanical, hydraulic, electrical, etc. More recently, the use of optical fiber systems allows use of telemetry with broad bandwidths and high immunity to noise and interference. Other telemetry systems are based on ultrasound, capacitive or magnetic coupling, infrared radiation, and so on, although these methods are not routinely used.

The discussion in this section will be limited to the most used systems — telemetry that is based on electrical signals. Electrically based telemetry does not have limits regarding the distance between the measurement and the analysis areas, and it can easily be adapted and upgraded in already existing infrastructures. Telemetry methods can be divided depending on the transmission channel that they use — wired or wireless (or radio) telemetry. Wired telemetry is used when the transmission can make use of already existing infrastructures, for example, in electric power lines that are also used as wire telemetry carriers. Wireless telemetry is more complex, as it requires RF or microwave energy for transmission and reception of information. RF transmission of information is widely used in electronic portable instruments. Information can be transmitted over long distances. Wireless transmission is used in those applications where measurement areas are not normally accessible or the system transmitting or receiving the information is mobile, which suits portable instruments perfectly. Some wireless telemetry units have enough capacity to transmit several channels of information simultaneously.

Figure 3.26 displays a generic telemetry system where the main blocks and their functions are as follows:

1. Transducers — convert physical variables to be measured into electrical signals that can be easily processed
2. Conditioning circuits — amplify the low-level signal from the transducer, limit its bandwidth, and adapt impedance levels

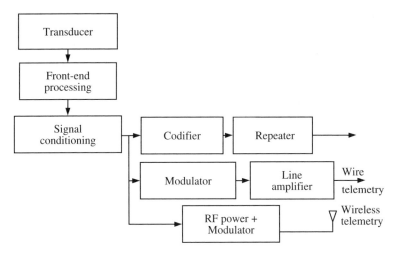

FIGURE 3.26
Basic components of a telemetry system.

3. Signal processing circuit — sometimes can be integrated into the previous ones

4. Subcarrier oscillator — its signal is modulated by the output of the different transducers once processed and adapted

5. Codifier circuit — can be a digital encoder, analog modulator, or digital modulator; adapts the signal to the characteristics of the transmission channel, that is, a wire or an antenna

6. Radio transmitter — in wireless telemetry, modulated by the composite signal

7. Impedance line adaptor — in case of wire transmission, to adapt the characteristic impedance of the line to the output impedance of the circuits connected to the adaptor

8. Transmitting antenna — for wireless communication; not all blocks need to be present in all types of telemetry systems

The receiver end consists of similar modules. For wireless telemetry, these modules are:

1. Receiving antenna designed for maximum efficiency in the RF band used

2. Radio receiver with a demodulation scheme compatible with the modulation one

3. Demodulation circuits for each of the transmitted channels

The transmission in telemetry systems, in particular wireless ones, is done by sending a signal whose analog variation in amplitude or frequency is a known function of the signals from the sensors or transducers. More recently, digital telemetry systems send data digitally as a finite set of symbols, each one representing one of the possible finite values of the composite signals at the time that it was sampled.

Effective communication distance in a wireless system is limited by the power radiated in the transmitting antenna, the sensitivity of the receiver, and the bandwidth of the RF signal. As the bandwidth increases, the contribution of noise to the total signal also increases, and consequently, more transmitted power is needed to maintain the same SNR. This is one of the principal limitations of wireless telemetry systems. In some applications, the transmission to the receiver is done on base-band, after the conditioning circuits. Base-band transmission is normally limited to only one channel. The advantage of base-band telemetry systems is their simplicity.

Almost all digital instruments are good candidates to be integrated in a network by a telemetry link. Telemetry is widely used in space applications for measurements of distant variables or controlling of the actuators. In most of these applications, for example, in space telemetry, it is very important to design the telemetry systems in order to minimize the consumption of power. Some vehicles and trains in transport systems use wireless telemetry. In clinical practice, the use of telemetry for patients increases their quality of life and their mobility, as the patients do not need to use fixed monitoring systems. Several medical applications are based on implanting a sensor in a patient and transmitting the data for further analysis and decision making.

The designers of a telemetry system also need to keep in mind the conditions in which the system will have to operate. In most of the applications, as in the case of electronic portable instruments, the telemetry systems must operate repeatedly without adjustment and calibration, in a wide range of temperatures. Finally, it is necessary to point out that as different telemetry systems are developed, there is a greater need of compatibility of transmitting, receiving, and signal processing equipment.

3.4.3.1 Multiple-Channel Telemetry

In many applications, multiple-channel telemetry is used that requires the measurement of many different processes and physical variables. As in the case of process industry and hospitals, the physical variables may be scattered spatially by being at different locations. In these multiple measurements, base-band telemetry is not an option, as it would require building a different system for each channel. Multiple channel telemetry is achieved by sharing a common resource (transmission channel), as shown in Figure 3.27. The sharing of the transmission channel by all the measurement channels is achieved by multiplexing.

There are two types of basic multiplexing techniques: frequency division multiplexing (FDM) and time division multiplexing (TDM). In FDM, different subcarrier frequencies are modulated by different measurement channel signals, which causes the information spectrum to shift from base-band to

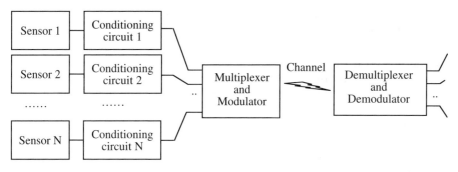

FIGURE 3.27
Block diagram of a multiple-channel telemetry system.

the subcarrier frequency. Then the subcarrier frequencies modulate the RF carrier signal, which allows the transmission of all desired measurement channels simultaneously. In TDM, the whole channel is entirely assigned to each measurement channel, although those channels may be active for a relative fraction of the time. TDM techniques use digital modulation to sample the different measurement channels at different times.

3.5 IEEE-1451 Standards for Sensors and Actuators: Smart Transducer Interface

There are new standards that are emerging for hardware architecture, software, and communications of modern intelligent (smart) sensors, thus making a revolutionary contribution to electronic portable instruments. As new sensors are developed complying with these standards, together with the wireless communication technology, we would expect a good growth in the variety of portable instruments being offered.

For example, IEEE-1451 is a set of standards that defines interface network-based data acquisition and control of smart sensors and transducers. These standards aim to make the creation of solutions easy by using existing networking technologies, standardized connections, and common software architectures. The standard allows application software, field network, and transducer decisions to be made independently. It offers flexibility to choose the products and vendors that are most appropriate for a particular application.

At the process connection level, IEEE-1451 provides standard ways of creating self-describing measurement and control devices. Sensors complying with these standards are expected to contain on-board information on the chip, such as serial numbers, calibration factors, accuracy specifications, and so on. During the installation, location information can also be loaded so that the system can have self-describing properties.

The IEEE-1451 standards come in four sections. The first section, 1451.1, is titled "Network Capable Application Processor (NCAP) Information Model." This section is concerned with the software architecture that moves the intelligence to the device level. Section 1451.1 approaches enable the benefits of objected-oriented technology. It creates flexible, natural software modules that allow engineers to think at the level of real-world systems, not at the level of programming languages. Object-oriented systems are built by assembly and produce systems that are much easier to adapt to new demands. The object-oriented technology makes open systems possible. This breakthrough could have the same liberating effect in software configuration as open systems had on hardware. In this way, flexibility can be achieved so that the systems can be assembled, reassembled, or modified quickly.

The second section, 1451.2, concerns transducers and is titled "Microprocessor Communication Protocols and Transducer Electronic Data Sheet (TEDS)." The IEEE-1451.2 standard defines the transducer data and electronic interface of digital information direct from sensors, thus creating a modular architecture. The modular architecture allows embedding of the module to any field networks automatically and transparently.

The third section, 1451.3, is titled "Digital Communication and Transducer Electronic Data Sheet (TEDS) Formats for Distributed Multidrop Systems."

The fourth section, 1451.4, is titled the "Mixed Mode Communication Protocols and Transducer Electronic Data Sheet (TEDS) Formats." The 1451.4 defines a standard interface that uses the connecting wires for analog signals or for digital communications.

There are positive signs that many companies, such as National Instruments, are taking notice of IEEE-1451 standards and producing devices that comply with them. For example, National Instruments is actively promoting and introducing plug-and-play sensors, complying with IEEE-1451.4. There are many smart sensors available in the marketplace that comply with these standards, for example, accelerometers and pressure transducers. More information on these sensors and IEEE-1451 will be given in Chapter 4.

3.6   Virtual Instruments and Portable Instruments

The applications of virtual instruments are gaining momentum in portable instruments and instrumentation systems. Nowadays, Palm and laptop computers are widely available, and they can easily be equipped with commercial data acquisition boards such as the PCMCIA interface cards, together with suitable interface cards and virtual instrumentation software, such as LabView. Equipped with the right sensors, these computers can easily function just like any other electronic portable instrument.

Traditional instrumentation systems are made up of multiple stand-alone instruments, which are interconnected to carry out a determined measure-

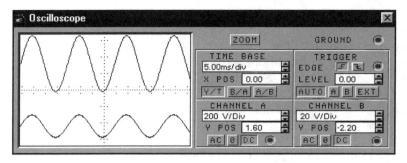

FIGURE 3.28
A virtual oscilloscope.

ment or control operation. Nowadays, however, the functionality of many of these stand-alone instruments can be implemented by using a computer (for example, a PC), a plug-in data acquisition board suitable for that computer, and some software routines implementing the functions of the system. The instrumentation systems implemented in this way are referred to as virtual instrumentation systems.

The major advantage of virtual instrumentation is the flexibility, because changing a function of an instrument simply requires reprogramming, which can easily be done with the use of convenient software tools. However, the same change in a traditional system may require adding new components, redesigning both hardware and software, or totally substituting that particular instrument with the new one equipped with the required features. This exercise may become difficult and expensive. These steps are eliminated in virtual instruments; for example, a new analysis function, such as Fourier analysis, can be easily added by the corresponding software to the library.

Virtual instruments also offer advantages in displaying and storing information. Computer displays can show more colors and allow users to quickly change the format of displaying the data that are received by instruments. Virtual displays can be programmed to resemble familiar instrument panels, including buttons and dials. Figure 3.28 shows an example of a virtual display corresponding to a two-channel oscilloscope. Computers also have more mass storage than stand-alone instruments, and consequently, virtual instruments offer more flexibility in storing measurements.

3.6.1   Data Acquisition Boards for Virtual Instruments

A data acquisition board is a plug-in board that allows the treatment of a set of analog signals by a computer. So these boards are the key elements to connect a computer to a process in order to measure or control it. Data acquisition boards normally operate on conditioned signals, that is, signals that have already been filtered and amplified. However, some special boards can deal with the signals directly generated by sensors. These boards include all the necessary signal conditioning circuits. In the case of virtual instruments, the information given on data acquisition boards in Section 3.2 is fully applicable; therefore, it will not be repeated here.

There is great variety of data acquisition cards (DAQs) available for portable computers, laptops, and handheld computers. Some laptops are available especially with custom design add-on units for data logging and analysis. The wireless communication capabilities of some of these devices enable remote data transmission and Internet connectivity. Plug-and-play cards are often used to collect data from field instruments. They can support various buses, such as PCI, PCMCIA, ISA, and IEEE-1394.

As an example, a laptop computer equipped with a suitable data acquisition card can support up to 16 single-ended inputs with sampling rates varying from 40 kilosamples/sec to 1.25 megasamples/sec. Common input resolution is 12 to 16 bits. The software support may come from various options such as LabView of National Instruments. Such software offers many features, such as analog and digital inputs and outputs, implementation of controllers, artificial intelligence and fuzzy logic capabilities, many different signal processing tools, data debugging tools, etc.

3.6.2   Software for Virtual Instruments

Virtual instruments can be programmed to fulfill many functions, such as:

- PC plug-in systems
- Oscilloscopes
- FFT analyzers
- Transient recorders
- Strip-chart recorders

There is a vast range of plug-and-play boards available commercially. The sensors such as temperature cards, sound, video, and network adaptors can be connected to portable computers to behave as portable instruments.

To develop the software of a virtual instrumentation system, a programming language or a special software package must be utilized. However, the option of using a traditional programming language (C, for example) can generate several problems. One of the problems may be the difficulty of programming the graphical displays and supporting elements. Another disadvantage may be the difficulty of learning and developing, requiring engineers designing virtual instruments to also be experienced programmers.

Nowadays, a more utilized option is the Microsoft Visual Basic programming language, which runs under the Windows operating system. Visual Basic has become quite popular in the development of virtual instrumentation systems, because it combines the simplicity of the Basic programming language with the graphical capabilities of Windows. Another important reason for its popularity is that it is an open language, meaning that it is easy for third-party vendors to develop products that engineers can use with their Visual Basic programs. Several companies now sell dynamic link libraries (DLLs), which add custom controls, displays, and analysis functions to Visual Basic. The controls and displays mimic similar functions found on stand-alone instruments, such as toggle switches, slide controls, meters, and LEDs. By combining various controls and displays, any virtual instrument can easily be programmed.

A third option for developing virtual instrumentation systems is to use a software package specifically designed for the development of these systems, such as LabView, PC600, and many others. The crucial advantage of these packages is that it is not necessary to be a Windows programmer to use them. Several packages of this type exist, but only one of them has reached great diffusion, LabView, developed by the National Instruments.

LabView is a graphical language, and it is based on two concepts: the front panel and the block diagram. The front panel is the face of a virtual instrument, and it is made up of controls and indicators. The display shown in Figure 3.28 is an example of a LabView front panel. The block diagram represents the "backside" or the front panel. It shows how all the controls and indicators fit together, as well as the hidden modules that operate on the data managed by the virtual instrument. Using these two concepts (front panel and block diagram), the development of complex virtual instrumentation systems can be easily achieved.

LabView is useful for graphical application development and integration with a wide variety of I/O. It offers multithreaded compiled performance and flexible and scalable standard software. The measurement, control, and automation applications can be created and tested using graphical development tools. An interactive user interface can be created to control a real system by specifying the system functionality in assembling block diagrams.

LabView was designed to be extendible, so new functional modules can be added to a program. For example, a manufacturer of an interface card may provide the user with a LabView driver that appears as a virtual instrument representing the card and its functionality in the LabView environment. LabView modules can also be written using general-purpose languages, such as C or C++. These languages provide great flexibility to program functions that perform complex numerical operations with the data received by a virtual instrument.

The PC600 data acquisition system is a typical example of the modern virtual instruments that can be purchased as a complete package. This system offers a range of plug-and-play boards suitable for laptop computers. The basic features of PC600 are:

- Intel microprocessor-based PC (selectable Pentium or Pentium II)
- MS-DOS, Windows 3.x, 95, or 98 operating systems

- Win600 software operating system: ready to run Win600 controls for digitizing setups, data capture, analysis, report writing, printing, and plotting
- 1.44 Mbyte floppy, 500+ Mbyte hard drive, 8 Mbyte RAM for printers and plotters
- One parallel and two serial ports, mouse or pointing device, keyboard
- Monitor or active matrix color display

3.6.3 Software for Electronic Portable Instruments

Electronic portable instruments are application-specific instruments that perform specific tasks. Therefore, there is no universal software available that can address the needs for all kinds of portable instruments. However, the software for portables is mainly determined by the type of microprocessor or microcontroller used. The software for instruments can be divided into two main components: (1) software for the internal operation of the system, together with the backing of diagnostic and troubleshooting, and (2) application-specific software for signal processing and external communications.

Today's portable instruments largely represent the classic embedded system environment, as they require maximum capability, and quality in a minimal processor and memory environment. Unlike extremely simple embedded systems, the processors are becoming much more sophisticated. The development of the RISC processor with variable instruction sets and implicit power reduction functionalities facilitate the use of higher-level language development of the applications. The increasing availability of larger RAM, ROM, and flash memory has also made life easier for software developers.

The major development languages used in electronic portable instruments are C++ and C, with some notable developments still being done in other languages such as EC++, Java, Ada, Basic, and Forth; however, these represent a small percentage of the current marketplace. The need for standardization of these languages has become much more evident as the flexibility of target processor selection has become more common.

If C programming language is selected, to get full functionality out of C, one must be able to easily convert standard C input/output functions to work under the often limited display interface conditions provided by the target system. A good way to check this is to investigate the implementation of the standard C function "printf." A fully functional printf is a significant piece of code, but by limiting the formatting capabilities, variations can be provided that reduce the memory demands dramatically. However, the programmer needs to check and modify the software termed "fundamental character I/O driver" to support "printf" code, thus avoiding modification and customization of the system hardware.

It is better to write a carefully developed and efficient application specific software. To understand the differences that result from writing good embedded C, compared to a bad one, consider the following example. Create a function with the following prototype:

```
void frequencystring(unsigned int period, char str[])
```

The definition of the function is as follows: The function is to take a 16-bit value representing the period of a 50-Hz signal in 0.5-μsec increments and return a four-character string showing the frequency to one decimal place. The normal operational range for the frequency may be between 49.8 and 50.2 Hz, but the conversion is to be within 0.05 Hz over a range from 48 to 52 Hz.

A typical C programmer would write a function similar to the following:

```
* 0.5microseconds per cycle gives 2000000 per second */
#define countspersec 2000000.
void frequencystring(unsigned int period, char str[])
{ float freq;
int error;
freq = countspersec/period;
error = sprintf(str,"%4.1f",freq);
}
```

But, writing short programs for portable instruments may result in hazards. For example, if this short piece of code, given above, was compiled for the 68HC11 microcontroller, the workbench product code would take 79 bytes of ROM, but the functionality would use library routines, which would add another 2091 bytes to the final executable code. Another example is as follows.

Alternatively, a piece of code that looks much more complex is shown below but compiles in the same environment to 219 bytes and does not need any library support.

```
#define FiftyHz 40040
#define OneHz 800
#define PointOneHZ 80
void frequencystring(unsigned int period, char str[])
{ int count;
count = period - FiftyHz;
if (count > 0) {/* Below 50.0 HZ */
str[0] = '4';str[1] = '9';str[2] = '.';str[3] =
'9';str[4] = 0;
while ((count - = OneHz) > = 0) {
```

```
    str[1] - = 1;
    }
    count + = OneHz;
    while ((count - = PointOneHz) > = 0) {
    str[3] - = 1;
    }
    }
    else {/* Above 50.0 Hz */
    str[0] = '5';str[1] = '0';str[2] = '.';str[3] =
    '0';str[4] = 0;
    count = -count;
    while ((count - = OneHz) > = 0) {
    str[1] + = 1;
    }
    count + = OneHz;
    while ((count - = PointOneHz) > = 0) {
    str[3] + = 1;
    }
    }
    }
```

The second solution takes 10.1% of the original code and works just as well. Incidentally, an assembly language version of the same code takes only 93 bytes.

There are more than 50 suppliers of compilers and language products to the embedded marketplace, including Avocet, Green Hills, Hi-Tech, IAR Systems, Metrowerks, Microsoft, Microware, Tasking Inc., and so on.

In selecting a language supplier, it is necessary to look at the ease of use of the development environment. The major PC-based development environments are integrated development environments (IDEs), with the emphasis on the commonality of functionality and user interface across a wide range of processors so that conversion from one processor to another is simplified and the only rewriting necessary is where a low-level assembly language driver has been necessary to achieve specific functionality.

In addition to language selection, the system developer has now to consider the operating system and supplementary packages that will be utilized in the target processor. Historically, the complete system would be developed as a single entity with a "hand-crafted" real-time kernel when necessary. The rapid proliferation of these systems together with the market competition and increased functionality has seen a significant growth in the use of standard modular kernels and the reuse of functional modules.

For example, typical real-time kernels in use are by Wind River Systems, QNX, or Microware and come from the embedded system's marketplace developing over the last 10 years or so, but there is rapidly growing competition from Microsoft, with Windows CE, and from Linux.

At least 70 vendors support this marketplace, including Accelerated Technology, Green Hills, Kadak, LynuxWorks, Microsoft, Microware, QNX, Red Hat, TimeSys, etc.

The capability of incorporating high-reliability operating systems into portable systems adds the capability of attaching to LANs for the download of results and upload of updates if needed.

In this context, developers can now purchase preprogrammed TCP/IP stacks to incorporate into their products, as well as full World Wide Web support by the provision of file transfer and http server facilities. The rapidly growing availability of Bluetooth networking will also quickly impact the portable instrumentation field; more information in this area is provided in Chapter 4.

4

Design and Construction of Electronic Portable Instruments

Introduction

In this chapter, all the information given in Chapters 1, 2, and 3 is brought together to explain design and construction of portable instruments. Additional information that is particularly relevant to electronic portable instruments has been highlighted.

Electronic portable instruments differ in many aspects from bench-top and fixed instruments. Although they have many points in common with the conventional instruments, they have many subtle differences, too. Therefore, they need special attention in design, prototyping, construction, and manufacturing. At the design and construction of portable instruments, there are many factors that must be considered carefully; some of these factors are:

1. Limited power supply from batteries.
2. Limited voltage levels.
3. Wide range in voltage supply from the batteries during the operation period. The instrument might have an overvoltage charge in the initial stages of the battery and a low voltage after some time in operation.
4. Wide variations in the environmental conditions, such as temperature and humidity.
5. Possibility of being subjected to mechanical shocks, vibration, etc.
6. Possibility of being operated by unskilled users.

During design and manufacturing of portable instruments, there will be a variety of requirements that must be taken into account, such as reasonably good performance, interoperability, reliability, and maintainability; immunity to noise and interference; easy and effective user interface and other human factors; safety and failure modes; variations in operational environments; support services once in the market; and market position and cost.

For the good performance of an instrument, other factors that need attention such as range, error rates, operation speed, size, weight, power consumption, efficiency, noise and interference, shock and vibration, temperature effects, etc. For reliability and maintainability, the failure rates, mean times between failures, and maintenance downtimes must be analyzed carefully.

In electronic portable instruments, human factors and user interface are important issues. Important factors are ease of use, intuitive operation, expertise required, and response latency. Safety and failure modes that are related to human factors must be handled with care.

Portable instruments are likely to be used in different environments by a diverse range of operators; therefore, the duty cycles, temperature extremes, stress ranges, etc., will play an important role in long-term operations. The support of instruments with proper documentation can provide distinct advantages for the product over those of competitors in the marketplace with similar functions. After all, quality and image of the product in the eyes of the consumers are what ultimately decide the success of the instrument in the market. Nevertheless, different markets may expect different characteristics from an instrument. For example, it is known that the consumer market may be more conscious about the cost, the medical market about safety, and the military market about reliability.

Adhering to standards ensures the quality of electronic portable instruments and guarantees some or all of the interoperability issues. In addition to standards and national or international guidelines that may exist for a specific instrument (particularly medical instruments), there are international quality standards, such as the ISO 9000. These are sets of standards to certify the design and production processes. ISO 9000 provides a road map for the series of standards ISO 9001 to ISO 9004. ISO 9001 covers design, manufacturing, installation, and servicing. ISO 9002 covers production and installation for commodities involving little design. ISO 9003 covers mainly distribution, whereas ISO 9004 gives guidelines for managing quality.

4.1 Design Considerations

There are distinct steps for designing and constructing electronic portable instruments:

1. *Conceptualization of the instrument*: This is the phase where the abstract ideas turn into a list of initial specifications. The specifications of the instrument should be drafted at this stage. The designers

and engineers should have some clear objectives about the functionality of the entire instrument and generate some ideas about how the instrument should look and what it should perform.

2. *Identification of functional blocks*: Once the specifications are drafted, the functional components need to be identified in blocks. These functional blocks may be the selection of suitable sensors, signal processing circuits, appropriate digital components, means of display, and the power supplies and associated circuits. The relationship between the functional blocks and their compatibility with each other needs to be considered carefully. The block diagrams should clearly indicate hardware and software needs. At this stage, the project can officially be deemed to have started since clear objectives are set.

3. *Feasibility study*: By this time, sufficient ideas have already been generated and the overall functionality of the instruments has been identified. Since the functional blocks have been determined, the scientific and technological risks can be assessed. The market potential of the instrument can be estimated. The market potential depends on the existence of competing instruments in the marketplace offering similar functionalities. This leads to the identification of the market position for the instrument. The selling price per item can be estimated, and the possible financial costs for the volume production of the instrument can be forecasted. Naturally, at this stage, estimated market size plays an important role in the price and financial investment necessary. Also, the life cycle of the instrument in the potential market needs to be assessed carefully.

4. *The first prototype*: A prototype can be built by construction of the previously identified electronic blocks in a simplistic manner, perhaps on a wire-rap breadboard. A breadboard facilitates quick connection of components and allows checking of the operations. However, circuits on the wire-rap breadboards may be limited in operation due to cross talk, unwanted impedances, and noise. If breadboards are used, each block can be built separately to be integrated together later, once their functionalities are proven. The sound performance of each block in the expected manner is essential. Once the blocks are integrated, the overall performance of the first prototype can carefully be assessed. This prototype can give fairly good insight into the realism of the specifications that were set in the previous stages. From the information gathered on the prototype, the necessary changes in the design and construction can be made in order to achieve the target specifications. The experience gained from the first prototype is valuable for reviewing the entire design to suit the target specifications and reassess the objectives and specifications if necessary. In many cases, specifications can be reviewed and upgraded without too much deviation from the main objectives.

At this stage, the use of computers for simulations can enhance the design and construction of the prototype.

In some cases, evaluation boards from manufacturers can be used. Evaluation boards contain particular components such as analog-to-digital (A/D) converters, microcontrollers, or microprocessors. They can connect to other circuits to emulate the final system. The evaluation boards can reduce the system cost and save time compared to custom design. They offer additional hardware for expansion, supporting software, debugging, and library facilities.

5. *Final prototyping*: The prototyping, testing, and reassessing of the overall approach should be repeated as many times as necessary. At this level, hopefully, each prototype leads to an improved version of the instrument to meet the targeted specifications. In the final prototype stages, some of the blocks may perform as expected, whereas others may need to be modified or even completely replaced. In some cases, only some of the components may need to be upgraded. In the final prototyping, the printed circuits boards (PCBs) of the instrument need to be designed, constructed, and tested. The PCB design and testing may have to be repeated several times until a satisfactory product is achieved. Once the PCB is constructed, it needs to be populated with the components and tested. This may have to be repeated a number of times.

6. *Volume manufacturing*: Once a working prototype is achieved with satisfactory performance of the targeted specifications, the manufacturing process can start for mass volume production. The manufacturing is usually an entirely separate process; therefore, it may need a totally new set of engineering teams. The smooth transition between final prototyping and volume production is very important. One needs to make sure that there is good information flow from one team to the next. It may be a good idea to involve some members of the design team in the initial stages of production. This is particularly important when the design teams are from research centers and universities. In many cases, research teams in centers that are not affiliated with the manufacturers design the instruments.

7. *Marketing, maintenance, and upgrading*: This stage is very important. If we look at the marketplace, the most successful companies are good implementers of this stage. As consumers use the instruments, these instruments are likely to be used under conditions that the companies never even dreamed of when designing them. Different users expose different expectations from the instruments (perhaps some of us remember that attempts were made by some consumers to dry living pets in microwave ovens when they were first introduced). Hence, feedback received from consumers on the functionality, operations, durability, and performance of the instrument is extremely useful to improve the instrument for the next level. At

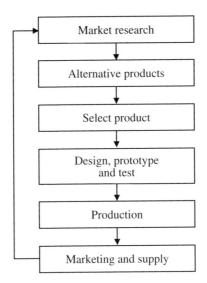

FIGURE 4.1
Market research and production stages of instruments.

this stage, it is worth mentioning that no matter how good and experienced the design teams may be, there will always be something that has been overlooked or has not been considered at all. In many cases, as a result of customer feedback, it may become apparent that the instrument needs to be upgraded right from the specifications. This stage is particularly important, as technology changes constantly; hence, upgrading becomes a "must" process for improvements, cost savings, and maintenance of the market share. Needless to say, the market forces constantly change, thus urging the manufacturer to evaluate market position and instrument specifications frequently.

A general design and marketing process of electronic portable instruments is given in Figure 4.1.

4.1.1   Reliability

The reliability of many electronic portable instruments is one of the major concerns. The success of an instrument in the marketplace will generally be governed by its reliability and durability. The instrument must work trouble-free for a reasonable length of time; therefore, it should perform the required functions under stated conditions for the expected period of time.

Reliability is expressed in statistical terms that an instrument will function without failure over a specified time. The failure could be temporary or permanent. Reliability of an instrument is determined by accelerated qualification tests. Accelerated qualification tests are generally conducted within some standards (such as MIL-STD-883) and yield statistical results on the *mean time between failures* (MTBFs).

During the design and construction of portable instruments, the following checklist for reliability may be useful:

1. All possible causes for the failure of an instrument must be investigated. The application of specific failure mode analysis can be very useful in the early stages of the design. The users may experience unpredicted causes of failures; if these are brought to the attention of the manufacturers, remedies against failures can be worked out.

2. The soldering of the components is very important in determining the life cycle of an instrument; therefore, proper soldering must be carried out by abiding good soldering rules.

3. It is important that all electronic portable instruments should provide fail-safe features. The frequency of fail-safe features in portable instruments can be expected to be higher than that in nonportable counterparts.

4. High-quality components, connectors, and display units must be selected.

5. Self-test features, such as monitoring of low battery voltages and excessive current consumption, can increase the cost of an instrument. But self-test features will also increase the reliability considerably.

6. As stated earlier, electronic portable instruments are more likely to be used by unskilled people. Therefore, in the design stages, developing a user-proof philosophy and estimating possible conditions of use by trained and untrained operators can increase the reliability and expected lifetime of an instrument.

4.1.2   Failure Mode Analysis

In order to achieve the specified operation features of an electronic portable instrument, it is important to conduct a comprehensive failure mode analysis. These analyses can be conducted at different levels. In the design stages, each electronic working block needs to be tested in its own right for many reasons, for example, sudden supply failures, reverse voltage connections, and under- and overvoltage levels. These failure modes can occur in the normal course of use, and the knowledge gained at this stage helps the reasons of failure of the assembled instrument. Failure mode analysis of an assembled instrument can be conducted by subjecting it to different operational conditions in different environments, such as environmental test chambers.

The environmental test chambers are devices in which most reliability tests are conducted. They can be set at steady temperature and humidity levels for any length of time. In addition, drop and ingress progress (IP) tests must be conducted to assess failure modes. IP tests are designed to test the instruments to decide their durability when they are exposed to various levels of environmental conditions such as dust and water.

4.1.3 System Voltage

The correct choice of system voltage for electronic portable instruments is very important. The different components may need different voltage levels. For example, most digital devices will operate at a 5-V supply or less, whereas analog devices are traditionally manufactured to operate at ±15 V. A modern portable instrument may be expected to contain both digital and analog components, thus requiring different voltage levels. Clearly, too many different voltages increase the system complexity, reliability, and cost. Also, multiple power and voltage supplies introduce different failure modes that need to be taken care of separately for each supply.

In some applications, multiple power supplies may be inevitable. In others, a single supply may be sufficient. In the case of single supplies, it is convenient to use a single set of batteries and hence save from the battery size and weight. But in order to compensate the need for multiple-level voltages, extensive voltage regulations may need to be included.

Basically, in electronic portable instruments, there are two types of power distribution systems: centralized and distributed supply systems. The centralized supply converts, distributes, and transmits regulated power from a single location. The distributed power system, on the other hand, supplies unregulated power to each subsystem and the local regulators, as shown in Figure 4.2, and regulates the power within each system. Ideally, an electronic portable instrument should be supplied by a single power source, since supplying with a single power source is easier, avoiding the need for complex regulating systems. At this point, it is worth mentioning that digital systems are more flexible in accepting variations in supply voltages than their analog counterparts.

However, there are some advantages in supplying multiple voltages in an instrument. If multiple voltage levels are available, the choice in the selection of suitable components will increase. Also, the effect of the internally gen-

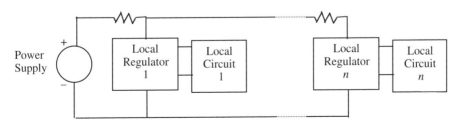

FIGURE 4.2
A distributed power system.

erated noise, due to step-up operations, can be minimized particularly at high voltages. Complementary metal oxide semiconductor (CMOS) circuits are more suitable for high-voltage operations and run faster.

4.1.4 Temperature

Electronic portable instruments are more likely to be used in conditions with wide temperature variations; therefore, they should be built strong enough to withstand extreme temperatures. It is well known that the performances of most electronic components (such as semiconductors, capacitors, inductors, and resistors) are temperature dependent.

The effects of higher temperatures on the instruments, compared to lower temperatures, must be considered separately. Generally, increased temperatures may lead to unexpected performances more quickly than low-temperature operations. This is because most of the components can deviate substantially at high temperatures from their expected or linear operations. Also, the increased temperatures can change the values of components, thus leading to undesirable secondary effects. One secondary effect can be the thermal runaway, where the heating increases the conduction to levels where permanent damage can occur. In conclusion, the instrument is very likely to be damaged by higher temperatures rather than lower ones.

Defining the expected operational temperature ranges of an electronic portable instrument in the early design stages is a good practice since it leads to the selection of appropriate components. The temperature range of an instrument is limited by the maximum and minimum boundaries of the electronic components of the device. The temperature range of most of the commonly available electronic components is usually from 0 to 70°C. The temperature range for military-grade electronic components is far better, typically between –55 and 125°C. Unfortunately, military-grade electronic components are much more expensive and not readily available.

Apart from the deviations from the expected linear operations, the temperature vs. frequency characteristics of the components may be very significant too. At high temperatures, the operating frequencies may deviate severely with the increasing temperature from their normal designed values. As explained above, it is normal to expect electronic portable instruments to be subjected to extreme temperatures; therefore, in the design and construction stages, the necessary precautions must be taken to maintain normal operations under all possible conditions of use.

4.1.5 Humidity and Moisture

If electronic portable instruments are used outdoors, it is normal to expect that they will get wet one way or another. Therefore, the effect of moisture and water on the performance of the instruments must be assessed carefully. Conducting ingress progress (IP) tests can be very helpful. It is possible that most consumers expect instruments to have passed higher levels of the IP test. More information on the IP test is provided in Section 4.7. Nevertheless, in many cases, the effect of moisture may have less limiting effects than those of temperature. Moisture usually introduces low-impedance paths across components and tracks, thus causing short circuits.

Counteracting the effects of moisture on the instrument may be absolutely necessary. There are a number of ways of eliminating these effects. In some cases, the printed circuit boards inside the casing are sealed with silicone, epoxy, or other water-resistant coatings. In extreme situations, the whole circuit may be immersed in a plastic resin. If such measures are taken, possible effects of resin on the components and circuit connections must be considered carefully, and various tests must be conducted in the early design stages to assess possible deviations from the expected performances.

4.1.6 Corrosive Fumes and Aerosols

Atmospheric pollutants, salts, and other chemicals exist in the natural environment. Also, in some industrial situations, the level of atmospheric pollution and aerosols may be very high. These adverse conditions can affect the instruments, causing malfunctioning and physical damages. One possible effect may be the corrosion of the components of the instrument. In some cases, the damage-causing fumes can originate within the instrument itself, i.e., acid leak from batteries, leaks from capacitors, and overheated resistors.

4.2 Power Supplies and Energy Sources

Electric power can be alternating current (AC) or direct current (DC). Electronic portable instruments are powered by DC supplies ranging from 1.2 V to hundreds of volts. Some standard voltages used in portable instruments are shown in Table 4.1.

Table 4.1 indicates that there may be more than one voltage level needed in a complex instrument. This may require the conversion of power from one level to another. During the power conversion, inductive and capacitive

TABLE 4.1

Possible Voltages Used in Portable Instruments

Voltage (V)	Components
+2.4 to +3.3	Low-power ICs, CMOS, and bipolar CMOS
+5	TTL, NMOS, CMOS
±12 to ±15	Analog circuit chips, CMOS
−5.2	Emitter-coupled logic circuits

effects of switching must be taken care of. In addition, the power supplies need to maintain the voltages and currents at required levels at all times. They also need to respond to load variations and irregular power demands during the start-up and shutdown.

In many aspects, the design and construction rules of electronic portable instruments differ considerably from those of their conventional counterparts, for example, in weight, size, battery power, versatility, and need for sturdiness and robustness due to the possibility of their use in wide-ranging and harsh environments. Electronic components must be selected in such a way that the instrument uses and wastes minimum power. In this respect, recently, the CMOSs have been the primary contributors in the development of portable instruments technology. But there are many other low-power draining devices that support portable instruments, such as monolithic sensors, intelligent and low-power LCDs, lithium batteries, etc.

In electronic portable instruments, effective power regulation and a well-balanced power distribution system are necessary. Using a common power supply can cause severe internal noise spilling problems from digital circuits to analog circuits. It is known that switching of digital circuits can generate strong spikes in the power bus of the instrument. Some sensitive circuits (particularly analog) can be affected if rejection of these noises is not conducted properly. Some of these spikes may easily be suppressed simply by using capacitive circuits.

The limited power availability in electronic portable instruments is one of the main characteristics of such devices. Therefore, careful planning of the power requirements and the design and selection of appropriate components and power supplies are absolutely necessary. Some of the concerns for power requirements may be listed as follows:

- Average and peak current consumption during routine and unusual operations
- Voltage levels that are necessary to drive various circuits
- Minimum allowable voltage that the device can operate at full capacity
- How frequently the batteries need to be recharged or replaced
- Recharge time of the batteries; could be very critical in applications where continuity in measurements is essential
- Use of interchangeable battery packs
- Shelf life of the instrument, which is the expected time that the instrument may wait unused for a long period of time, but still be able to perform fully within the specifications
- External temperature ranges of normal expected operations and response of the instrument to extreme temperature conditions, hence associated power supply fluctuations
- Weight, size, and cost restrictions

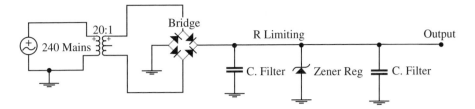

FIGURE 4.3
A typical AC/DC converter with voltage regulator.

- Integration of the battery charger within the instrument and its cost implications
- Prediction of unusual circumstances that may affect the performance of the instrument, such as strong light and strong electric, magnetic, or electromagnetic fields

In some applications, such as domestic and laboratory environments, the portable instrument may be operating near an AC source so that these instruments may be charged when necessary. Also, under normal operations, it may be convenient to use the AC supply whenever possible to conserve battery energy. When the main power is used, it is necessary to convert AC into DC form. In some cases, external converters convert the AC power and DC is plugged into the instrument, whereas in others, the AC/DC conversion unit is integrated within the device. The choice depends on the market position of the instrument, its weight and size, and how often it is expected to be powered from an AC source. There are many different AC/DC conversion methods, and also, there is a wide range of converters available in the marketplace. Figure 4.3 illustrates a typical general-purpose converter with a regulator and capacitive filter integrated together with it.

The main components of such converters are:

- Transformer, to reduce the AC to acceptable levels and isolate from power lines
- Rectifier, to convert the AC into DC
- Filter, to minimize or eliminate ripples after rectification
- Regulator, to maintain a constant level of voltage against any fluctuations

Electronic portable instruments that use distributed power systems often use DC/DC converters for local regulation. Figure 4.4 illustrates a 6-W DC/DC converter with an input voltage ranging from 3.5 to 16 V. The output is fixed at level 5, 12, or 15 V.

Many of these DC/DC converters allow a wide range of input voltages and incorporate overvoltage and overcurrent protection. They provide isolation from each other so that if one or more regulators are down, others

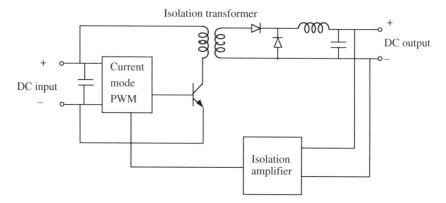

FIGURE 4.4
A typical DC/DC converter.

will still be operational. The DC/DC converters can be classified according to their power handling capacity. The power can range from 0.5 W to hundreds of watts. As a typical example, a 1-W, 5-V nominal input converter can handle input voltages between 4.5 and 5.5 V. The output voltages can be 5, 6, 6.5, 7, 9, 12, 15, or 24 V for an output current ranging from 40 to 150 mA, depending on the output voltage. Some of these converters can be trimmed to give nonconventional outputs such as 5.6 or 13.3 V. There are many other devices in the market to meet a diverse range of requirements.

In some applications, regulators may be necessary for voltages greater than 6 V. There are many different types of regulators, ranging from a single Zener diode to highly sophisticated regulators manufactured in chip forms. Some examples of regulators are Zener diodes, monolithic regulators for CMOS circuits, or active regulators for analog and other sensitive circuits. Typical examples of such regulators are the 7800 series of three terminal regulators. The 7805, 7806, 7809, 7812, and 7815 are available depending on the voltage requirement, such as 5, 6, 9, 12, or 15 V. There are corresponding 7900 regulators for regulating negative voltages. For variable-voltage supplies, an LM723 regulator can be used to generate positive or negative power. Some of the companies offering a variety of regulators together with DC/DC converters are listed in Table 4.3.

4.2.1 Batteries

Batteries are electrochemical devices that store chemical energy and turn it into electrical form. For electronic portable instruments, the selection of correct batteries is very important since they are the only power source within the instruments. The batteries for portable instruments have several implications:

- Batteries have a limited lifetime, and lifetime in an instrument is determined by frequency of its use.

- Batteries provide good isolation and offer noise-free operations.
- Output power, the capacity, and the terminal voltages of batteries are temperature dependent — they increase as the temperature gets higher.
- Terminal voltage drops when heavy currents are drawn from the battery. This is due to internal resistance of the battery.
- Output voltage deteriorates by aging, which can be attributed to the deterioration of chemicals and electrodes inside the battery.
- Large currents drawn from the battery can cause heating. When large currents are drained, some batteries such as lithium or nickel–cadmium types can be damaged permanently.
- Magnitudes of the internal resistances of batteries differ; while lithium or nickel–cadmium types have low internal resistances, zinc–carbon types have high resistances.

Batteries are classified to be either primary (nonrechargeable) or secondary (rechargeable) batteries. In electronic portable instrument applications, both primary and secondary battery power has distinct advantages and equally significant disadvantages and limitations. In this respect, capacity, energy density, discharge service life, shelf life, and discharge rates of batteries bear particular importance.

The *capacity* of a battery signifies the amount of available energy, which is usually expressed in ampere-hours. The capacity is an important factor in the design of battery packs and the recharging systems. *Shelf life* of a battery is the length of time that it can be stored without losing a portion of its charge and still is able to operate normally. The portion of charge is regarded to be typically 50% of the full capacity. Since all batteries have internal leakage currents, their voltages slowly decrease in time. This is particularly severe in the case of rechargeable batteries. The *energy density* of a battery is the amount of energy in watt-hours per unit volume or weight. Clearly, energy density is dependent on the size and weight of the battery.

The *continuous* and *momentary discharge rates* of batteries indicate how much current a battery can supply under various conditions. The discharge rates may be limited by cell polarization. The cell polarization causes the accumulation of chemical products such as gas bubbles. This interferes with the reaction in the electrodes and appears as an increase in the internal resistance. In some cases, the heat produced, when a battery supplies excessive currents, may limit the momentary discharge rates.

The types of batteries that are suitable for portable instruments are discussed below.

4.2.1.1 Carbon–Zinc Batteries

Carbon–zinc batteries are cost-effective batteries, and they are used in many disposable instruments. They are available in many sizes, a variety of shapes, and at various terminal voltages. However, they have limited energy storage capacity, which gives them a short period of usage and a short shelf life. They also have large drops in terminal voltages after a relatively short usage. High internal resistances limit the maximum currents that they can supply. Carbon–zinc batteries also have a tendency to leak corrosive liquids; therefore, they may need replacing often.

4.2.1.2   Alkaline Manganese Dioxide Batteries

Alkaline manganese dioxide batteries can be used in almost all situations, except in those instruments requiring strict voltage regulations. They have a relatively longer shelf life and a higher capacity, but higher leakage resistances. They are easily available and come in all the standard sizes, ranging from button cells to large lantern batteries.

4.2.1.3   Nickel–Cadmium Batteries

Nickel–cadmium batteries are the most popular rechargeable batteries that are used, particularly in small instruments. Special nickel–cadmium batteries are available for applications such as high temperatures and when quick charge rates are needed. They are suitable in applications that require low internal battery resistances; hence, low voltage drops at high currents. The power densities of nickel–cadmium batteries are comparable to those of zinc–carbon batteries. Their output voltage tends to remain constant with discharge, which indicates that they have very low internal resistance (in the region of 10 mΩ). These batteries are known to perform well in all high- and low-temperature ranges. The charge–discharge life cycle of a well-manufactured quality nickel–cadmium battery can be up to 1000 cycles.

The nickel–cadmium batteries need to be fully charged regularly. Discharging of the battery to extremely low levels can cause changes in the polarization characteristics of the weak cells, thus leading to early failures. They have a relatively shorter shelf life and lose capacity quickly on long-term storage, typically 2 to 3 months. They need a few good charge–discharge cycles to maintain the energy storage capability at the maximum levels.

Similar to nickel-cadmium, nickel metal hydride (NiMH) batteries are available in various sizes, shapes, and power densities. They are available at 1.2, 3, and 6 V.

4.2.1.4   Lead–Acid Batteries

Modern lead–acid batteries are relatively much heavier than other equivalents, and they have low-power densities. The power density can be as low as one fifth of a zinc–carbon battery. The problem with lead–acid batteries is that their cells cannot easily be miniaturized, but they have very low internal resistances and can deliver high currents. Therefore, lead–acid batteries are suitable in applications that demand large currents and quick rechargeable needs. They are known to be very reliable and also almost maintenance-free.

4.2.1.5   Lithium Batteries

Lithium batteries are relatively more expensive, but they have much higher power densities than other types. As a comparison, lithium batteries can contain up to 30 times the energy density of zinc–carbon batteries. Lithium batteries are known to be very robust and have very long shelf lives, up to 10 years. Many varieties of lithium batteries are available, including the sulfur dioxide and thionyl chloride types. These batteries find extensive applications in memory devices. They can also be miniaturized and built into integrated circuits (ICs) as backup power sources.

Currently, in electronic portable instruments, lithium ion (LiIon) batteries are extensively used. It is expected that the next generation of lithium polymer batteries will be common and will offer many advantages in energy density, weight, shape, and size.

4.2.1.6   Mercury and Silver Oxide Batteries

Mercury and silver oxide batteries have many useful properties such as constant voltages and longer shelf lives. But, these batteries do not have high-current properties as do lithium batteries. They are very expensive, simply because their basic materials, both silver and mercury, are rare and expensive commodities. They are manufactured in small sizes. They are most suitable in applications requiring low-current drains, such as watches, cameras, and hearing aids.

The properties of batteries that are suitable in electronic portable instruments are summarized in Table 4.2.

4.2.1.7   Miniature Power Supplies

Ultrathin (less than 5 mm and 2 cm^2) miniature high-voltage power supplies delivering 10 to 90 V are available for special uses such as powering avalanche photodiodes. The input voltage can be between 2.7 and 6.7 V.

4.2.2   Batteries and Power Management

In electronic portable instruments, the battery packs are constructed in such a way that they can easily be interchanged. The battery holders should be robust, and the materials used should usually be strong and noncorrosive. The batteries are firmly anchored to the case to give them extra strength in case the device is subjected to unduly forces and accidental drops. Usually, a fuse in the power supply circuit is incorporated to protect the instrument against any short circuits or unexpected excessive currents.

Many portable instruments have battery monitors to give advance warning about the battery condition. Also, there is a possibility that the battery terminals may be wrongly connected by reverse installations; therefore, the instruments are protected against the reverse polarities. The protection against the reverse polarity may simply be provided by diode circuits, as

TABLE 4.2

Batteries for Portable Instruments

	Primary					Secondary			
	Carbon–Zinc	Alkaline Manganese	Mercury or Silver	Zinc–Air	Lithium	Lead–Acid	Nickel–Cadmium	Silver–Zinc	Nickel Hydride
Volts/cell (V)	1.2	1.25	1.35	1.2	2.75	2.0	1.0	1.3	1.2
Energy density (W-h/kg)	55–75	66–100	99–126	155–330	260–330	18–35	18–45	60–210	56–70
Operating temperature (°C)	–7 to 54	–30 to 54	0 to 54	–30 to 54	–40 to 60	–40 to 60	–40 to 70	–40 to 74	–18 to 30
Capacity D (A-h)	2.1	4.8	14	—	20	—	—	—	—
Low-temperature performance	Poor	Good	Poor	—	Good	Good	Good	Good	Medium
Charging method	—	—	—	—	—	Taper current	Constant current	Constant current	Constant current
Cost	Lowest	Low	Medium	Expensive	Expensive	Medium	Expensive	Expensive	Expensive

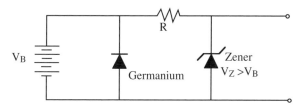

FIGURE 4.5
A protection circuit against reverse-polarity insertion of batteries.

shown in Figure 4.5. In this figure, the low forward voltage drop characteristics of the germanium diode protect the circuits against the reverse voltages
if the batteries are inserted wrongly. The breakdown voltage of the Zener
diode is set slightly higher than the correct battery voltage, thus providing
protection against excessive voltages.

4.2.2.1 Switching Regulation

Regulation has briefly been introduced above; because of its importance for
portable instruments, further detail will be given here. In many instruments,
some suitable switching circuits regulate the power of the device. For the
purpose of regulation, transistors can be used as the switching elements.
The regulation process is different in the step-up voltages compared to the
step-down operations. An example circuit for a step-down regulator is
illustrated in Figure 4.6(a), and a typical step-up circuit is given in Figure
4.6(b).

In the case of step-down of voltages, the output voltage is controlled by
the time that a switch is kept on or off. A negative feedback loop may be
necessary between the output voltage and the variable duty cycle to realize
the required regulation. Simply a square-wave oscillator can provide the
duty cycle. When the square waves are used, appropriate filters are necessary

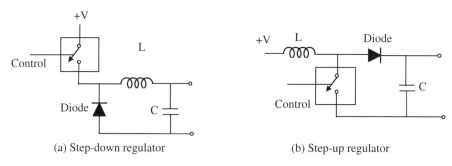

(a) Step-down regulator (b) Step-up regulator

FIGURE 4.6
Examples of battery voltage regulators. (a) A step-down regulator, and (b) A step-up regulator.

to smooth the voltages. In the step-up regulators, some special filters and feedback circuits are necessary.

4.2.2.2 Programmable Power Management Chipsets

Programmable chipsets are available for low-voltage, high-current multiphase synchronous DC/DC converters. They contain programmable, pulse width modulation (PWM) controller ICs with on-board drivers to control power metal oxide semiconductor field-effect transistors (MOSFETs). Some of these units offer programmable output voltages from 1.075 V up in steps of 25 mV, with 1% output voltage accuracy. The controller IC offers many protection and monitoring features, such as undervoltage lockout, overvoltage protection, externally programmable soft start, and so on.

In Table 4.3, a list of battery manufacturers and suppliers with telephone numbers is provided. Only a few names are given, but the list is by no means exhaustive. Many companies offer more than one type or all types of products. However, the repetition of company names has been avoided whenever possible.

TABLE 4.3

Battery Manufacturers and Suppliers

Type	Manufacturers and Dealers (Phone No.)	
Alkaline batteries	Centurion Wireless Technologies (402-467-4491)	Panasonic Batteries (877-726-2228)
	Mouser Electronics (800-346-6873)	Rayovac Corp. (608-275-3340)
	Multiplier Industries Co. (914-241-9510)	SelfCHARGE Inc. (425-881-9199)
Lead–acid batteries	Advanced Battery System (800-634-8132)	Nexergy Inc. (800-575-2191)
	Alpha Technologies (360-647-2360)	Plainview Batteries (800-642-2354)
	JBRO Batteries (630-281-1900)	Power-Sonic Corp. (650-364-5001)
LiIon batteries	BYD America Corp. (847-690-9999)	PolyStor Corp. (925-245-7000)
	Electrovaya (905-855-4610)	SAFT (800-399-7238)
	Hutton Communications (877-648-8866)	Toshiba America (949-455-2000)
	Motorola Energy Systems (770-338-3742)	Ultralife Batteries (315-332-7100)
NiCd batteries	AF CommSupply (800-255-6222)	PowerBurst Batteries (781-929-6242)
	Freemann & Wolf (410-557-4352)	Tri-M Systems Inc. (604-527-1100)
	Moltech Power Systems (904-462-3911)	W & W Manufacturing (800-221-0732)
NiMH batteries	BYD America Corp. (847-690-9999)	RBRC (678-419-9990)
	Dataman Programmers Ltd. (904-774-7785)	Shokia Far East Ltd. (914-736-3500)
	Freemann & Wolf (410-557-4352)	W & W Manufacturing (800-221-0732)
Other batteries	AER Energy Resources (770-433-2127)	Target Enterprises (978-664-5303)
	Cadex Electronics (604-231-7777)	TDI Batteries (630-679-8200)
	Panasonic Batteries (877-726-2228)	Ultralife Batteries (315-332-7100)
Smart battery systems	AF CommSupply (800-255-6222)	Elpac Power Systems (949-476-6070)
	Bourns Inc. Multifuse (909-781-5005)	Galaxy Power Inc. (610-676-0188)
	Cadex Electronics (604-231-7777)	PowerSmart Inc. (203-925-1340)
	E-one Moli (Canada) (604-466-6654)	Soft Tools Technology (408-973-7828)
Regulators, DC/DC converters	AAK Corp. (603-295-7200)	National Semiconductor (408-721-5000)
	Alpha Technologies (360-647-2360)	SelfCHARGE Inc. (425-881-9199)
	Datel Inc. (800-233-2765)	TK Power Inc. (510-770-1114)
	International Rectifiers (310-252-7019)	WinTron Technologies (814-625-2720)

Source: Portable Design, March 2002, PennWell.

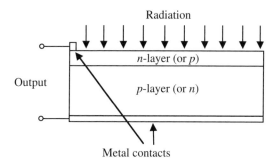

FIGURE 4.7
A typical silicon photovoltaic cell.

4.2.3 Unconventional Power Sources for Portable Instruments

In some applications, electronic portable instruments are required to operate in remote areas without human interrupts for long periods. In such cases, an alternative power supply to the batteries may be necessary. The alternative power can be obtained from various sources, such as solar power, fuel cells, wind, electromagnetic energy, and thermoelectricity. Undoubtedly, the most frequently used alternative power source in portable instruments is solar energy, but fuel cells are very likely to be a strong alternative to battery power in the very near future. Solar energy and fuel cells will be explained next.

4.2.3.1 Solar Cells

The power density of sunlight on an average sunny day is about 1000 W/m^2. Photovoltaic cells are normally used to convert solar power into electrical energy. The most common types of photovoltaic cells, as illustrated in Figure 4.7, are the silicon types. In the silicon photovoltaic cells, the surface of a *p*-type wafer is doped to make a thin transparent *pn* junction. In this junction, electrons are promoted by light falling on the junction to be trapped in the *n*-layer. A suitable electrode arrangement conducts the current away from the junction for external use.

4.2.3.2 Fuel Cells

A fuel cell is an electrochemical device that combines hydrogen fuel with oxygen to produce electric power. Oxygen is drawn from the air and hydrogen is carried as fuel in a pressurized container. As alternative fuel, methanol, propane, butane, and natural gas can be used.

In a type of fuel cells, proton exchange membrane (PEM) technology is used to dilute methanol or ethanol for the chemical reaction. Platinum–ruthenium on one side and platinum on the other side are used as membranes. Platinum acts as a catalyst that makes the chemical reaction occur. The methanol is converted into protons and electrons. The combination of protons and electrons with air creates water, thus releasing energy.

However, fuel cells have many inherent problems. For example, use of water for dilution and creation of water as a result of reaction necessitate extensive water handling mechanisms. Therefore, miniaturization of fuel cells is not possible at this time. Nevertheless, recent research on the topic indicates that fuel cells may be available for portable instruments in the near future. Alternative methods such as the use of liquid electrolytes appear to be promising.

4.3 Passive Circuit Components

In portable instruments, the available power is limited; therefore, special care must be taken to conserve power. The conservation of power can be done in a number of ways, for example:

- By minimizing the electric current consumption, which can be achieved by building circuits with low-power ICs. To do this, the use of CMOS components is desirable since this technology is based on voltage control rather than current control. Virtually all types of analog and digital components are available in CMOS. In addition, some low-power bipolar devices are also available in CMOS, such as LM308 and LM324. This makes the selection CMOS components very attractive.

- By designing and constructing proper grounding of the entire circuits to avoid floating inputs. Floating inputs can cause unduly excessive currents.

- By avoiding unnecessary currents that may be generated due to harmonics. If the waveforms contain distortions, they can generate additional harmonic voltages; therefore, the waveforms must be as clean as possible, such as the undistorted square waveforms.

- By selecting low-power-consuming passive components such as good capacitors with low leakage. Low-tolerance passive components can help the conservation of power.

- By avoiding high-power-consuming devices such as Zener diodes. Devices such as the Zener diode can drain substantial currents before they are turned on.

- By replacing high-power-consuming passive components with their active counterparts.

- By synchronizing all the signals. This can avoid losses due to unnecessary and untimely switching operations.

In addition to the necessity of selecting good-quality ICs, in electronic portable instruments, the proper selection of passive component resistors, capacitors, and inductors is significant too. For example, if proper care is not taken, the resistors may introduce nonlinear effects in normally linearly operating circuits. Leaky capacitors can introduce unaccounted resistances, which may have significant effects, particularly at high frequencies, and so on. The characteristics of passive components are discussed in detail next.

4.3.1 Resistors

In electronic circuits, the primary function of a resistor is to restrict current, but resistors find many other uses, such as pull-up, pull-down, impedance matching, voltage division, gain control, bias set, etc. Resistances of a resistor vary with manufacturing tolerances, temperature, humidity, and voltage stress. Manufacturing tolerances arise due to manufacturing techniques, trimming to remove material for the desired value, packaging material, and soldering of leads to resistive conductor. The percent variation of resistances quoted by the manufacturers can be 20, 10, 5, and 1%. A precision resistor can have a tolerance of less than 1%.

The effect of temperature on resistance can be significant. The temperature gradients of a resistor can deviate from the expected value due to self-heating and soldering heating during the assembly process. Humidity also can alter the resistance; the change in resistance against humidity is measured in parts per million. Resistors also exhibit inductive and capacitive properties that can have significant effects at certain frequencies. They can also contribute significant amounts of thermal noise and other noises. The noise can be regarded as proportional to resistance, temperature, and bandwidth.

Resistors are made from different materials. In portable instruments, four types of resistors are of particular concern: carbon composition, carbon-film, thin-film, metal-film, wire-wound, and foil resistors. The properties of these are given in Table 4.4.

4.3.1.1 Carbon Composition Resistors

Carbon composition resistors are made from graphite, ceramic, and resin compressed at high temperatures. These resistors have poor temperature coefficients, stability, over-temperature, and shelf life. Use of these resistors needs to be avoided if the portable instrument has large temperature swings and excessive humidity and moisture variations. Another variation is the carbon-thin resistor, which is made from vacuum-deposited carbon films onto a ceramic core. These resistors have better temperature, humidity, and shelf life properties. Carbon-film resistors generally are used in low-power and higher-resistance applications, above 1 MΩ. They have low-noise properties.

4.3.1.2 Metal-Film Resistors

Metal-film resistors are formed by sputtering metal onto ceramic. They have good accuracy and low variability, and therefore are most suitable for long-

TABLE 4.4

Characteristics of Resistors Used in Electronic Portable Instruments

	Carbon Composition	Carbon Film	Metal Film	Thin Film	Wire Wound	Foil
Accuracy (±ppm)	10,000	5000	100	50	20	5
Maximum resistance (MΩ)	22	50	10	1	1	0.25
Temperature coefficient (±ppm/°C)	1200	1000	100	100	10	2
Maximum voltage (V)	500	500	350	200	1000	500
Humidity effects (%)	15	3	0.5	0.5	0.5	0.02
Power range	Low	Low	Low	Low	Any	Low
Noise	Low	Low	High	Low	Low	Low
Long-term stability	Poor	Low	Fair	Fair	Excellent	Excellent
Price	Very low	Low	Moderate	Moderate	High	High

term and continual use. They have a relatively high degree of accuracy, but tend to be expensive. Their noise figures are much lower than those of carbon composition resistors. Metal-film resistors have very low temperature coefficients and are suitable for linear circuits.

4.3.1.3   Thin-Film Resistors

Thin-film resistors are made from tantalum nitrate sputtered and etched on a silicon substrate. They have low temperature coefficients and are suitable for resistance networks and surface-mount technology.

4.3.1.4   Wire-Wound Resistors

Wire-wound resistors are made from nickel–chromium wire wound on a beryllium oxide core. They are relatively expensive. They have excellent temperature coefficients. Wire-wound resistors find applications particularly in high-power circuits where accuracy is the prime importance. One drawback of wire-wound resistors is that they may have substantial inductances that can make them unsuitable in high-frequency applications (say above 50 kHz).

4.3.1.5   Foil Resistors

Foil resistors are made from nickel–chromium bonded onto a ceramic substrate and laser-trimmed for accuracy. They have the best temperature coefficients and very high precision. Foil resistors have low inductive and capacitive properties. They are good for high-frequency applications.

General-purpose, high-quality resistors are available in dual in-line packages (DIPs), which look like ICs. They are used in many applications, particularly in digital circuits as pull-up or pull-down resistors.

4.3.1.6   *Variable Resistors, Trimmers, and Potentiometers*

Variable resistors, trimmers, and potentiometers find frequent use in all types of electronic circuits for trimming operational amplifiers (op-amps), tapping reference voltages, setting trip points, and adjusting voltages, currents, and frequencies. They are available in carbon composition and wire-wound type resistors. The variable resistors tend to have good temperature stability. If applied in electronic portable instruments, the variable resistors need to be high quality, particularly the multiturn types, so that they do not wear out easily over time. Nevertheless, variable resistors can be a source of externally induced noise; therefore, it is advisable to provide a good shielding against possible electromagnetic interference (EMI).

In general, variable resistors can be sensitive to moisture, dirt, and wear and tear and should be avoided whenever possible; fixed resistors with the desired settings should be selected instead. The variable resistors basically are mechanical devices with inertia, friction, and the associated limitations. The sliding electrical contacts can also be prone to oxidization.

4.3.2   Capacitors

Capacitors are important components of all types of electronic circuits. Their primary function is to store charge. They have a diverse range of applications in filters, couplers and decouplers, tuning circuits, compensators, isolators, energy storage, noise suppression, and so on. Commonly used capacitors are constructed with air, paper, mica, polymers, and ceramic dielectric materials. Variable capacitors are generally made with air or ceramic dielectric materials. Capacitors are manufactured in greater varieties than resistors.

Basically, capacitors are made from parallel plates as shown in Figure 2.9. The capacitance of a capacitor is affected by three parameters: the separation between plates, d; effective area of plates, A; and relative dielectric constant of material between plates, ε_r. The relation between these values can be expressed by

$$C = \frac{\varepsilon_0 \varepsilon_r A}{d} \tag{4.1}$$

where ε_0 is the permittivity of free space (8.85×10^{-12} F/m).

Experience shows that a substantial part of component failures in electronic equipment is due to capacitors; therefore, in this book more space is allocated to capacitors. They can fail in several ways, such as excessive voltage, excessive current, temperature, and sudden charging and discharging. The most common failure modes are open circuit, short circuit, arcing between plates, and leakage failure. An arc between plates can evaporate the metal contacts opening the circuit, or can damage the dielectric and thus short-circuit the capacitor.

The major cause of failure in circuits can be attributed to the improper selection and inappropriate applications of capacitors. The following factors are therefore important criteria in the selection of capacitors in many circuit applications.

The capacitance *values and tolerances* are determined by the operating frequencies or by the value required for timing, energy storage, phase shifting, coupling, or decoupling. The *voltages* are determined by the type and nature of the source, AC, DC, transient, surges, and ripples. The *stability* is determined by operating conditions like temperature, humidity, shock, vibration, and life expectancy. The *electrical properties* are determined by life expectancy, leakage current, dissipation factor, impedance, and self-resonant frequency. The *mechanical properties* are determined by the types and construction, e.g., size, configuration, and packaging. The *cost* is determined by the types and physical dimensions of capacitors and the required tolerance.

The capacitors available for electronic portable instruments are discussed next.

4.3.2.1   Ceramic and Glass Capacitors

Ceramic capacitors are very common in all types of electronic circuits. They are used in bypass and coupling circuits. The dielectric of the capacitor is a ceramic material with deposited metals. They are usually rod or disc shaped. Ceramic capacitors have good temperature characteristics and are suitable in many high-frequency applications. There are many different types of ceramic capacitors, for example: low K ceramic, high K ceramic, and miniature ceramic. *Low K ceramic capacitors* are made with materials that contain a large fraction of titanium dioxide (TiO_2). The relative permittivities of these materials vary from 10 to 500, with a negative temperature coefficient. The dielectric is TiO_2 + MgO + SiO_2, suitable in high-frequency applications in filters, tuned circuits, coupling, and bypass circuits. The dielectric of *high K ceramic capacitors* contains a large fraction of barium titanate ($BaTiO_3$) mixed with $PbTiO_3$ or $PbZrO_3$, giving a relative permeability of 250 to 10,000. They have high losses and high voltage time dependence with poor stability. *Miniature ceramic capacitors* are used in critical high-frequency applications. They are made in the ranges of 0.25 pF to 1 nF.

In some dielectric ceramic capacitors, the material is a semiconducting ceramic with deposited metals on both sides. This arrangement results in two depletion layers that make up the very thin dielectric. In this way, high capacitances can be obtained. Due to thin depletion layers, only small DC voltages are allowed. Ceramic capacitors have good temperature stability and are used in small and lightweight equipment such as hearing aids.

Glass capacitors are made with glass dielectric materials. The properties of glass dielectrics are similar to those of ceramic materials.

4.3.2.2   Paper Capacitors

Usually, paper capacitors are made with thin (5- to 50-μm thick) wood pulp. A number of sheets are used together to eliminate possible chemical and fibrous defects that may exist in each sheet. The paper sheets are placed between thin aluminum foils and convolutely wound. The losses and self-inductance of paper capacitors are sizable and frequency dependent. The applications are usually restricted to low frequency and high voltages.

4.3.2.3 Electrolytic Capacitors

These are many capacitors in which the dielectric layer is formed by an electrolytic method. Electrolytic capacitors in dry foil form may be similar in construction to paper-film capacitors; that is, two foil layers separated by an impregnated electrolyte paper spacer are rolled together. In this case, one of the plates is formed using metallization of one side of a flexible dielectric film. A foil, e.g., aluminum, is used as the other plate. The capacitor is then hermetically sealed in an aluminum or plastic can. These capacitors can be divided into two main subgroups: tantalum electrolytic capacitors and aluminum electrolytic capacitors.

4.3.2.3.1 Tantalum Electrolytic Capacitors

In tantalum electrolytic capacitors, the anode consists of sintered tantalum powder and the dielectric is Ta_2O_5, which has a high value of ε_r. A semiconductor layer MnO_2 surrounds the dielectric. The cathode made from graphite is deposited around MnO_2 before the capacitor is sealed. The form of a tantalum electrolytic capacitor includes a porous anode slug to obtain a large active surface. These capacitors are highly stable and reliable, have good temperature ranges, and are suitable for high-frequency applications.

4.3.2.3.2 Aluminum Electrolytic Capacitors

Foil–foil is oxidized on one side as Al_2O_3. The oxide layer is the dielectric having a thickness of about 0.1 µm and a high electric field strength (7×10^5 V mm^{-1}). A second layer acting as a cathode made from etched Al–foil is inserted. The two layers are separated by a spacer, and then the layers are rolled and mounted.

Electrolytic capacitors are extensively applied in filters and power supplies due to their wide capacitance value ranges with reasonable costs. However, they have severe limitations due to internal heating by increased leakage currents at high temperatures.

Electrolytic capacitors must be handled with *caution*, since the electrolyte in these capacitors is polarized. That is, the anode should always be positive with respect to the cathode. If not connected correctly, hydrogen gas will form and damage the dielectric layer, causing high leakage currents or blowup. These capacitors may be manufactured at values up to 1 F. They are used in not-so-critical applications, such as coupling, bypass, filtering, etc. However, they are not very useful at frequencies above 1 kHz or so.

4.3.2.4 Mica Capacitors

A thin layer of mica, usually muscovite mica (≥0.003 mm), is stapled with Cu–foil or coated with a layer of deposited silver. It is then vacuum-impregnated and coated with epoxy. The field strength of these capacitors is very high (10^5 Vmm^{-1}) and resistivity $\rho = 10^6$ to 10^{15} Ωm. These capacitors are available in values from 1.0 pF to several microfarads for high-voltage (from 100 to 2000 V) and high-frequency applications. They have tolerances between ±0.5 and ±20%.

4.3.2.5 Metallized-Film Capacitors

Polymer, Mylar, polypropylene, polycarbonate, and Teflon capacitors are made from alternating layers of metal foil and flexible dielectric materials. Various polymers such as polycarbonate, polystyrol, polystyrene, polyethylene, and polypropylene are used as the dielectric. The construction is similar to that of paper capacitors. Polystyrene capacitors in particular are very stable and are virtually frequency independent. They have a low voltage rating and are used in transistorized applications as tuning capacitors and capacitance standards. Generally, metallized-film capacitors exhibit good thermal stability. Although their sizes tend to be large, they are used mostly in high-power applications.

4.3.2.6 Variable and Air Core Capacitors

Variable capacitors usually have air as the dielectric and consist of two assemblies of spaced plates positioned together by insulation members such that one set of plates can be rotated. A typical example of variable capacitors is given in Figure 4.8. Their main use is in adjustment of resonance frequency of tuned circuits in receivers and transmitters, filters, etc. By shaping the plates, various types of capacitances can be obtained, such as *linear capacitance*, in which capacitance changes as a linear function of rotation, and *logarithmic capacitance*.

Variable capacitors may be grouped as precision types, general-purpose types, transmitter types, trimmer types, and the special types such as phase shifters. The precision type variable capacitors are used in bridges, resonant circuits, and many other instrumentation systems. The capacitance swing

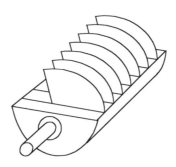

FIGURE 4.8
A typical variable capacitor.

can be from 100 to 5000 pF. They have excellent long-term stability with very tight tolerances.

General-purpose type variable capacitors are used as tuning capacitors in radio and other broadcasting devices. The normal capacitance swing is from 400 to 500 pF. In some cases, a swing of 10 to 600 pF is available.

Transmitter type variable capacitors are similar to general-purpose variable capacitors, but they are specially designed for high-voltage operations. The vanes are rounded and spaced wider to avoid flashover and excessive current leakages. The swing of these capacitors can go from a few picofarads up to 1000 pF. In some cases, oil filling or compressed gases are used to increase operating voltages and capacitances.

Trimmer capacitors are used for coil trimming at intermediate radio frequencies. They can be air-spaced rotary types (2 to 100 pF), compression types (1.5 to 2000 pF), ceramic dielectric rotary types (5 to 100 pF), and tubular types (up to 3 pF).

Sometimes, special type variable capacitors are produced for particular applications such as differential and phase shift capacitors in radar systems. They are used for accurate measurement of time intervals, high-speed scanning circuits, transmitters and receivers, and so on.

4.3.2.7   Integrated Circuit Capacitors

Integrated circuit capacitors have been adapted for use in microelectronic circuits. They include some miniature ceramic capacitors, tantalum oxide solid capacitors, and tantalum electrolyte solid capacitors. The ceramic and tantalum oxide chips are unencapsulated and fitted with end caps for direct-surface mounting onto the circuit board. The beam-leaded tantalum electrolytic chips are usually attached by pressure bonding. Integrated circuit capacitors made mostly within MOS integrated circuits are monolayer capacitors containing tantalum or other suitable deposits. The plates of the capacitors of the integrated circuits are generally formed by two heavily doped polysilicon layers formed on a thick layer of oxide. The dielectric is usually made from a thin layer of silicon oxide. Further information on integrated circuit capacitors can be found in Section 2.4.1.2.

4.3.2.8   Voltage-Variable Capacitors

Voltage-variable capacitors make use of the capacitive effect of the reversed-biased *pn*-junction diodes. By applying different reverse bias voltages to the diode, the capacitance can be changed. Hence, the name varicap or varactor diodes is given to theses devices. Varactors are designed to provide various capacitance ranges from a few picofarads to more than 100 pF. It is also possible to make use of high-speed switching silicon diodes as voltage-variable capacitors. However, they are limited by the very low maximum capacitance available. Typical applications of these varactor diodes are in the tuning circuits in radio frequency receivers. Present-day varactor diodes operate into the microwave part of the spectrum. These devices are quite efficient as frequency multipliers at power levels as great as 25 W. The efficiency of a correctly designed varactor multiplier can exceed 50% in most instances. It is also worth noting that some Zener diodes and selected silicon power supply rectifier diodes can work effectively as varactors at frequencies as high as 144 MHz. In the case of the Zener diode, it should be operated below its reverse breakdown voltage.

4.3.2.9 Characteristics of Capacitors

Capacitors used in all types of electronic circuits can be classified as low-loss, medium-loss, and high-tolerance capacitors.

Low-loss capacitors such as mica, glass, low-loss ceramic, and low-loss plastic-film capacitors generally have good stability. These capacitors are expensive and selected in precision applications, e.g., telecommunication filters.

Medium-loss capacitors have medium stability in a wide range of AC and DC applications. These are paper, plastic-film, and medium-loss ceramic capacitors. Their applications include coupling, decoupling, bypass, energy storage, and some power electronic applications, e.g., motor starters, lighting, power line applications, and interference suppressions.

High-tolerance capacitors such as aluminum and tantalum electrolytic capacitors deliver high capacitances. Although these capacitors are relatively larger in dimension, they are reliable and have a longer service life. They are used in polarized voltage applications, radios, televisions, and other consumer goods, as well as in military equipment and harsh industrial environments.

Capacitors are characterized by dielectric properties, breakdown voltages, temperature coefficients, insulation resistances, frequency and impedances, power dissipation and quality factors, reliability, and aging properties. The dielectric properties of capacitors are important, and the dielectric materials can be polar or nonpolar. Polar materials have dipolar characteristics; that is, they consist of molecules whose ends are oppositely charged. This polarization causes oscillations of the dipoles at certain frequencies, resulting in high losses.

The temperature characteristics of capacitors are largely dependent on the temperature properties of the dielectric materials used, as given in Figure 4.9. The temperature coefficients of glass, Teflon, mica, and polycarbonate are very small, whereas in ceramic capacitors, they can be very high.

If the capacitor is subjected to high operating voltages, the electric field in the dielectric exceeds the breakdown value that damages the dielectric permanently. The dielectric strength, which is the ability to withstand high voltages without changing properties, depends on the temperature, frequency, and applied voltage. It is commonly known that the use of capacitors below their rated values increases the reliability and expected lifetime. The standard voltage ratings of most capacitors are quoted by manufacturers as 50, 100, 200, 400, and 600 V. Tantalum and electrolytic capacitors have ratings of 6, 10, 12, 15, 20, 25, 35, 50, 75, and 100 V and higher.

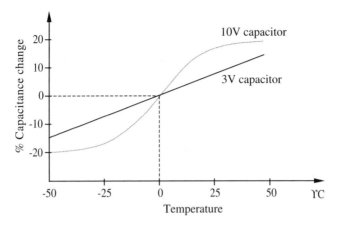

FIGURE 4.9
The typical temperature characteristics of capacitors.

The losses of capacitors depend on the frequency and impedances. In practical capacitors, losses are very high at low frequencies and high at high frequencies. At low frequencies, the circuit becomes entirely resistive and the DC leakage current becomes effective. At very high frequencies, the current passes through the capacitance and the dielectric losses become important. An ideal capacitor should have an entirely negative reactance, but losses and inherent inductance prevent ideal operation. Depending on the construction, capacitors will resonate at certain frequencies due to unavoidable construction-based inductances. A typical impedance characteristic of a capacitor is depicted in Figure 4.10.

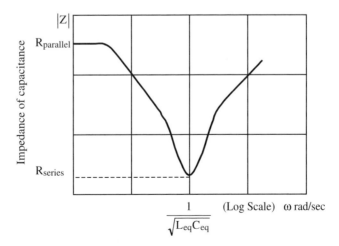

FIGURE 4.10
The typical frequency vs. impedance characteristics of capacitors.

Some of the common causes of capacitor failure are due to voltage and current overloads, high temperature and humidity, shock, vibration pressure, frequency effects, and aging. The voltage overload produces an excessive electric field in the dielectric that results in the breakdown and destruction of the dielectric material. The current overload is caused by rapid voltage variations, resulting in current transients. If these currents are of sufficient amplitude and duration, the dielectric can be deformed or damaged, resulting in drastic changes in capacitance values, and thus leading to equipment malfunction. High temperatures are mainly due to voltage and current overloads. Overheating and high temperatures accelerate the dielectric aging. This causes the plastic film to be brittle and also introduces cracks in hermetic seals. Moisture and humidity due to a severe operating environment cause corrosion, reduce the dielectric strength, and lower insulation resistances. The mechanical effects are mainly the pressure, variation, shock, and stress, which can cause mechanical damages of seals that result in electrical failures. Aging deteriorates the insulation resistance and affects the dielectric strength. Aging is usually determined by shelf life, and information about aging is supplied by the manufacturers.

In electronic portable instruments, all types of capacitors can be used, but mostly electrolytic capacitors are selected. These capacitors are made from two metal plates separated by a liquid electrolyte. The electrolytic capacitor can have high inductive properties, noticeable particularly at high frequencies. Also, the resistance at high frequencies can become significant; hence, the power dissipation at high frequencies increases. The applications of electrolytic capacitors should be restricted to lower frequencies unless they are specially manufactured for high-frequency operations.

The typical properties of capacitors commonly used in electronic portable instruments are summarized in Table 4.5. The most important properties are physical size, voltage rating, temperature coefficient, and availability within the required ranges. The voltage rating is the point at which the dielectric material breaks down. As the physical size gets smaller, the voltage rating becomes less.

For the selection of capacitors in electronic portable instruments, the following are recommended:

- Plain ceramic disc capacitors are suitable particularly for decoupling and despiking type of circuits. The voltage rating of the capacitor should be about twice that of the normal working voltage.
- Aluminum electrolytic capacitors are suitable for filtering and switching of the power supplies.
- Tantalum electrolytic capacitors are much smaller in dimension.
- There are electrolytic capacitors with low inductance and effective series resistances, but they tend to be expensive.
- Nonpolarized electrolytic capacitors must be selected if the polarity of charge of the capacitor changes.

TABLE 4.5

Characteristics of Capacitors Suitable for Electronic Portable Instruments

	High K Ceramic	Low K Ceramic	Mylar	Mica	Polycarbonate	Tantalum	Aluminum
Capacitance range	100 pF–100 nF	10–220 pF	1 nF–50 μF	2.2 pF–100 nF	100 pF–30 μF	100 nF–500 μF	100 nF–1.6 F
Maximum voltage (V)	20,000	100	750	600	600	100	600
Frequency	10^9	10^9	10^9	10^9	10^9	10^7	10^7
Dielectric constant	Up to 6000	35	3	7	2.7	11	7
Temperature coefficient	High	Small	Medium	Low	Low	Low–negative	Very high
Leakage	Medium	Medium	Low	Low	Low	Medium	Highest

- In applications where leakage currents are important, such as sample-and-hold (S/H) devices the selection of Mylar, polystyrene, or polycarbonate capacitors is more appropriate.
- If a low temperature coefficient and low leakages are important, such as in integrators and differentiators, medium K and low K ceramic capacitors are appropriate.
- At high frequencies, mica capacitors are a better choice than ceramic ones.

4.3.3 Inductors

Inductors are made from wires wound on a ferromagnetic or air core. The value of inductance depends on the geometric arrangements of the structure, number of turns, and relative permeability of the core material. They are nonlinear devices due to hysteresis, saturation, and eddy currents. Many inductors are custom produced or available in IC form. Their role in many circuits is significant but used less frequently than capacitors and resistors.

In Table 4.6, a list of RLC component manufacturers and suppliers with telephone numbers is provided. Only a few names are given; the list is by no means exhaustive. Many firms supply more than one or all types of products. Repetition is avoided whenever possible.

4.4   Active Components and Circuit Design

Circuits for electronic portable instruments are designed and constructed in accordance with the required functions, regulations, operational conditions, and environment. At the initial stages of the design, it is better to start with broad functions such as response time, data rates, input/output (I/O)

TABLE 4.6

RLC Component Manufacturers and Suppliers

Type	Manufacturers and Dealers (Phone No.)	
Fixed capacitors	AVX Corp. (843-448-9411)	KEMET Electronics (864-963-6300)
	California Micro Devices (408-263-3214)	Rohm Electronics (858-625-3600)
	EVOX RIFA (847-948-9511)	Schaffner EMC Inc. (732-225-9533)
Variable capacitors	Galco Industrial Electronics (848-542-9090)	Schuster Electronics (513-489-1400)
	IET Labs Inc. (516-334-5959)	TDK Corp. of America (847-803-6100)
	KAO Speer Electronics (814-362-5536)	Voltronics Corp. (973-586-8585)
Chip inductors	API Delevan Inc. (716-652-3600)	Spraque-Goodman (516-334-8700)
	CirQon Technologies (847-639-6400)	Taiyo Yuden (USA) (800-348-2496)
	Gowanda Electronics (716-532-2234)	TT Electronics (714-447-2300)
Coil inductors	Coilcraft (847-639-6400)	Schuster Electronics (513-489-1400)
	Intertechnologies LLC (914-347-2474)	TDK Corp. of America (847-803-6100)
	PICO Electronics (914-738-1400)	TT Electronics (714-447-2300)
Passive networks	California Micro Devices (408-263-3214)	Schuster Electronics (513-489-1400)
	EVOX RIFA (847-948-9511)	State of Art Inc. (874-355-8004)
	Rohm Electronics (858-625-3600)	TT Electronics IRC (828-264-8861)
Resistors	AVX Corp. (843-448-9411)	Schuster Electronics (513-489-1400)
	California Micro Devices (408-263-3214)	State of Art Inc. (874-355-8004)
	Isotek Corp. (508-673-2900)	TT Electronics IRC (828-264-8861)

Source: Portable Design, March 2002, PennWell.

requirements, and so on. As the design progresses, these functions are refined with operational concerns. Operational concerns lead to determination of specifications and other achievable targets. As progress is made, selections can be made for appropriate technology, such as application-specific integrated circuits or commonly available components. Commonly available components may be selected from various technologies such as transistor–transistor logic (TTL), CMOS, programmable logic arrays (PLAs), etc. In the case of microprocessor-based systems, once an appropriate processor is selected, the supporting components can readily be available for memory, I/O components, timers, coprocessors, and so on, perhaps offered by the same vendor.

Selection of IC packages is most appropriate since they provide both known mechanical and environmental properties, suitable for handling, service, and assembly. The selection and use of ICs for portable instruments is important mainly because of the reliability in and limited power availability. Therefore, the base selection criterion must be the reliability of operation of the ICs and low-power consumption. Also, there are other factors such as suitable operating voltage ranges, resistance to electrical damage, immunity to electromagnetic noise, availability, etc.

In the last few decades, microelectronics has become one of the greatest driving forces in all types of electronic devices. Microelectronic devices find applications in both analog and digital systems, but particularly digital systems, such as the personal computers (PCs), as well as today's modern electronic portable instruments. They are practically finding applications in all parts of modern life, from personal safety and health equipment to all types of scientific applications.

There are many semiconductor devices available to perform specific tasks, but only a few may be suitable to use in electronic portable instruments. In portable instruments, most of the electronic devices may be voltage-controlled (field-effect) types, especially CMOS due to low power consumption, and current-controlled devices, such as the TTLs. In electronic portable instruments, the TTLs should only be used when no substitute can be found, or when they are available for low-power use.

In the selection of semiconductor devices, comprehensive data should be gathered on the available types from the manufacturers' data books. Once a comprehensive list of available devices is prepared, the best ones should be selected for the project. In the selection process, the following criteria need to be considered:

- *Operating voltage*: CMOS circuits that have "C" suffixed after their numbers operate on voltages of 4 to 6 V. Standard metal-gate CMOSs can operate with 10 to 15 V. But most memory devices operate at a 5-V supply voltage. Analog devices usually operate at supplies of –15 and 15 V or –12 and 12 V.

- *Latch up resistances*: Some CMOS circuits can accidentally be subjected to voltages greater than 0.7 to 0.8 V even when the V_{CC} is off, which is the switching voltage of *pnp*- (or *npn*-) junctions, the inherent parts of CMOS construction. The affected chip may suddenly start drawing heavy currents to destroy itself. This can occur if the instrument is supplied from separate power supplies. Also, ringing and overshoots in signal lines might drive inputs momentarily higher than V_{CC}.

- *Power consumption*: Low power consumption per circuit is necessary, particularly when the instrument is complex and has many components. Some bipolar operational amplifiers such as LM308 and LM324 are specially designed for battery operation, and current consumption is within a few milliamperes.

- *Speed*: The behavior of circuits may be affected by the long propagation time of some integrated circuits. This type of behavior is common in microprocessor and memory circuits. Most IC makers offer high-speed versions of CMOS chips, usually indicated by the infix HC or HCT in the chip identification numbers.

- *Price*: In many situations, the price bears the least consideration. Integrated circuits may be small portions of the total investment on an instrument. Usually, the use of cheap parts leads to problems that can cost time and money to develop and maintain.

- *Availability*: As the integrated circuits have become more complex, they have also become susceptible to small changes in the manufacturing process. Sometimes a commonly known chip may suddenly disappear. Therefore, selecting chips that are offered by various manufacturers may ensure some degree of continuity.

Integrated circuits come in several physical forms. The most common of these is the DIP, which refers to the two rows of connecting pins on either side. The pins of DIPs are numbered in an industry-wide standard manner. The packages may be made of plastic or ceramic. Ceramic packages tend to be of better quality but more expensive. Less common integrated circuit packages are the "flat packs." They are good for space saving but may be difficult to use on electronic boards. Flat-package ICs are common in high-performance analog circuits such as operational amplifiers and voltage regulators.

Due to the wide variety and availability of analog integrated circuits, such as the operational and instrumentation amplifiers, the design of most analog circuits may be relatively simple.

4.4.1   Active Analog Components

Analog signal processing is an essential part in the majority of electronic portable instruments. Analog signal processing components, procedures, and governing mathematical approach are major topics for many books. Due to limited space, analog signal processing will not be dealt with in detail in this book. Some common signal processing techniques, together with associated components, are introduced briefly; detailed discussions can be found in Chapter 1 and in appropriate sections of Chapter 3.

A purely analog system measures, transmits, displays, and stores data in analog form. Signal conditioning is usually made by integrating many functional blocks such as bridges, amplifiers, filters, oscillators, modulators, offsets and level converters, buffers, and the like, as explained in Chapter 1. In this section, brief explanations on analog signal processing are given with relevance to electronic portable instruments.

4.4.1.1   *Operational Amplifiers and Circuits*

In electronic portable instruments, operational amplifiers (known as op-amps) and charge amplifiers are often used as common amplifiers, filters, oscillators, adders, summers, controllers, buffers, and so on. Some of the popular operational amplifiers in electronic portable instruments are listed in Table 4.7, but this list is by no means exhaustive. Instrumentation amplifiers are selected in situations where operational amplifiers do not meet the electrical requirements such as low noise and good common mode rejection ratio (CMRR).

In electronic portable instrument applications, there are a number of op-amp characteristics that must be watched; these are offset voltage, offset

TABLE 4.7

Popular Operational Amplifiers

Part No.	V_o Maximum	I_B Ambient	T_c of V_o (μv/°C)	A_o (V/mV) at DC	Noise (nV/Hz$^{1/2}$)	Type	Manufacturer
OP-27E	25 μV	10 nA	0.2	1800	6	Low noise	TI, Motorola
LT1008C	120 μV	0.1 nA	1.5	200	30	Low power	LTC
LM11	300 μV	17 pA	1	1200	150	Low power	National
TL051ACP	800 μV	4 pA	8	100	75	Precision	Motorola
MC34071	3 mV	100 nA	10	100	55	Low drift	Motorola
MB47833	5 mV	500 nA		110 dB	10	Low noise	Fujitsu
TLC271BCP	2 mV	0.6 pA	2	23	100	Programmable	TI, Thomson
AD546	1 mV	1 pA	20	1000	90	Low cost	AD

current, dynamic loop gain, CMRR, power supply rejection ratio, slew rate, and output ratio.

The offset voltage is a small input voltage that makes output go to zero, particularly in DC applications. It results in internal circuit imbalances during the IC manufacturing process. Offset voltages can be compensated by external components. The offset current, on the other hand, is the sum of the inverting and noninverting input currents when the output of the op-amp is zero at DC applications. The resulting output offset voltages due to current can be taken care off by techniques such as the use of integrators.

Loop gain can be a function of frequency that can affect stability. The CMRR defines how well an op-amp rejects the common mode voltage when the same voltage is applied to inverting and noninverting inputs. In electronic portable instruments, the use of good-quality op-amps with high CMRRs is desirable since compensating circuits for inferior CMRRs require additional circuits, space, and power.

The power supply rejection ratio defines how well the operational amplifier rejects the variations in voltage. The slew rate is the limiting rate of change in output for an input pulse that is faster than the specifications of an op-amp. Operational amplifiers have output noise due to thermal effects and offset voltages (e.g., popcorn noise).

Operational amplifiers are useful in eliminating the nonlinearity of the physical variables. The operation amplifiers may be used as inverting and noninverting amplifiers; and by connecting suitable external components, they can be configured to perform many other functions, such as voltage followers, limiters, oscillators, drivers, voltage-to-frequency converters, frequency-to-voltage converters, differentiators, integrators, summers, etc. Some of these op-amp based circuits are discussed next.

4.4.1.2   *Instrumentation Amplifiers*

Instrumentation amplifiers are operational amplifiers with stringent characteristics. An example of an instrumentation amplifier is given in Figure 4.11. This amplifier has an active feedback with variable gain.

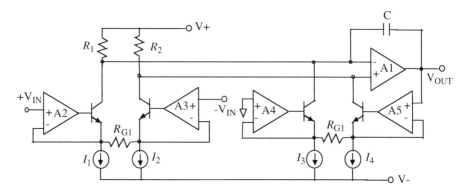

FIGURE 4.11
A typical instrumentation amplifier.

The gains of instrumentation amplifiers are usually limited to less than 100. Both inverting and noninverting inputs have high-input resistances such that they can be used in true differential forms.

4.4.1.3 Isolators

Isolators are used at various stages in the circuit designs. Isolation cuts the direct current path and removes the DC bias from the signal. It uses three physical principles:

1. *Optoisolation* — can provide up to 2500 V of isolation. It has a non-linear transfer function but is still suitable for digital signals and switched inputs.
2. *Capacitive isolation* — has a linear transfer function and is used in low-current applications.
3. *Transformer isolation* — has a linear transformation. It is not used in electronic portable instruments since it tends to be bulky. It also exhibits saturation problems of the magnetic components.

4.4.1.4 Circuit Protection ICs

There are a number of chips available for circuit protection and power distribution applications. They employ current regulation that eliminates high inrush currents, e.g., in highly capacitive loads. The current-limiting circuitry incorporates a programmable overcurrent filter that eliminates false tripping due to surge currents. Overvoltage lockout with a start-up delay eliminates high-current transients. Typical circuit protection ICs can be supplied by 2.3- to 13.6-V inputs, and they can support 2.5- to 12-V applications.

TABLE 4.8

Active Component Manufacturers and Suppliers

Type	Manufacturers and Dealers (Phone No.)	
Circuit protection	California Micro Devices (408-263-3214) Galco Industrial Electronics (248-542-9090) Keystone Electronics (800-221-5510)	Schuster Electronics (513-489-1400) TDK Corp. (847-803-6100) Thermometrics (732-287-2870)
DC amplifiers	Alpha Industries (978-241-7700) Entran Devices Inc. (973-227-1002) National Semiconductor (408-721-5000)	Pentek (201-818-5900) Sensotec Inc. (614-850-5000) Toshiba America (949-455-2000)
Differential amplifiers	Analog Devices Inc. (800-262-5643) Elantec Semiconductors (408-945-1323) Goal Semiconductor Inc. (800-943-4625)	Insight Electronics (800-677-7716) Linear Technology (408-432-1900) Precision Filters Inc. (607-277-3550)
Operational amplifiers	Advanced Linear Devices (408-747-1155) Conxant Systems Inc. (949-483-4600) Galco Industrial Electronics (248-542-9090)	Intersil (888-468-3774) Microchip Technology (480-792-7200) ON Semiconductors (602-244-6600)
Wideband amplifiers	Avtech Electrosystems (800-265-6681) Linx Technologies (541-471-6256) Mouser Electronics (800-346-6873)	Physical Acoustic Corp. (609-716-4000) Pro-Comm Inc. (732-206-0660) RF Micro Devices (336-664-1233)
Opto-couplers	Agilent Technologies (408-654-8675) Lumex Inc. (847-359-2790) Microsemi Corp. (949-221-7100)	Sealevel Systems (864-843-4343) Sharp Microelectronics (360-834-2500) Vishay Infrared Corp. (408-988-8000)
Opto-isolators	Agilent Technologies (408-654-8675) Galco Industrial Electronics (248-542-9090) Lumex Inc. (847-359-2790)	Schuster Electronics (513-489-1400) Sealevel Systems (864-843-4343) TDK Corp. (847-803-6100)
Oscillators	AKM Semiconductor Inc. (408-436-8580) Bomar Crystal Co. (800-526-3935) Epson Electronics (800-228-3964)	Fox Electronics (941-693-0099) Micro Crystal (847-818-9825) Vectron International (203-853-4433)
Programmable oscillators	Cardinal Components (973-785-1333) Comclock Inc. (800-333-9825) JFW Industries (317-887-1340)	Synergy Microwave (973-881-8800) Tellurian Technologies (847-934-4141) TLSI (631-755-7005)
VCOs	Agilent Technologies (408-654-8675) InterQuip Ltd. (913-397-0686) KS Electronics (602-971-3301)	Micro Networks (508-852-5400) Microsemi Corp. (949-221-7100) Z-Communications (858-621-2700)

Source: Portable Design, March 2002, PennWell.

4.4.1.5 Single-Chip Active and Passive Device ICs

Integrated passive and active devices (IPADs) are available to eliminate the problem of combining active and passive devices externally. They are most suitable for portable instruments, saving space and reducing assembly costs. IPAD technology implements a double side wafer diffusion to add integrated passive components such as resistor networks, capacitors, and inductors, together with high-performance devices. An IPAD can contain bias, pull-up, pull-down, and serial resistors ranging from 1 Ω to 100 kΩ; coupling, decoupling, and filtering capacitors ranging from 5 to 500 pF; diodes from 6 to 14 V; radio frequency (RF) detection signal schottkys; bipolar signal transistors; filters; and choke inductors.

In Table 4.8, a list of manufacturers and suppliers of active components is provided. For each component, only a few names are given to avoid repetition; the list is not exhaustive. Many firms supply more than one or all types of components.

4.4.2   Actuators and Controllers

Actuators are generally used in the output stages to control a process or an external phenomenon, such as mechanical movements. Actuators are used either inside of an electronic portable instrument to control some operations of the instrument, such as opening and closing valves, or outside to control the operation of an external process. In the latter case, the actuator is located on the external process and the portable device sends control signals. A typical example is the remote controllers for many devices, such televisions, air conditioners, garage door openers, and so on. In these applications, some kind of communication technique must be integrated between the portable and controlled devices.

Actuators can consume a considerable amount of power. Bearing in mind the limited amount of available power in portable instruments, there are many different types of actuators that can be selected that may be suitable for a particular process, depending on the characteristics of the internal circuits. Actuators may be mechanical relays, on–off switches, or solenoids. These will be explained briefly.

4.4.2.1   Mechanical Relays

A typical example of a mechanical relay is the reed relay. Reed relays have low-power requirements and can operate trouble-free for longer times. They are available in IC packages that make them suitable to be used in portable instruments.

4.4.2.2   On–Off Switches

On–off switches are used often in processes. On–off switches can be mechanical or electrical. Examples of electrical types include Darlington switches, switching power controllers, power MOSFETS, and solid-state relays.

A typical *Darlington* configuration is illustrated in Figure 4.12. These transistors can be cascaded and can deliver several amperes. However, when cascaded, each stage reduces the voltage, and hence the power wastage increases.

Switching power controllers are used to control power into the external devices simply by providing a switching action to the voltage supply of the device.

Power MOSFETS are also used for switching purposes and voltage control actions.

There are many other types of *solid-state relays*, most consisting of some kind of isolated drivers and a triac. Triacs are useful devices for switching both AC and DC voltages. These relays may have significant on-resistances, and hence greater heat dissipation.

4.4.2.3   Solenoids

Solenoids are also used in electronic portable instruments. They are made from a coil and a moving iron core. Usually the core is connected to the

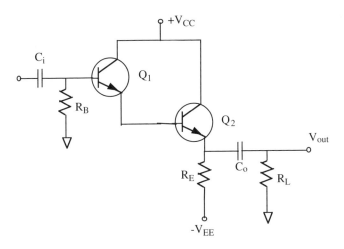

FIGURE 4.12
A Darlington configuration.

moving part, for example, camera shutter or valve stem. The recoiling action is provided by springs. Solenoids are available from a large number of manufacturers, and they come in many sizes, shapes, and voltage and power ratings. Most of the solenoids in portable instruments are used to operate valves and detectors.

In some cases, *small DC motors* are used. DC motors are more efficient than solenoids. The speed of the motor can easily be controlled (e.g., pulse width modulation, pulse frequency modulation) to suit the application needs.

If a precise control of rotary motions is required, the *stepping motors* offer good choices. Nevertheless, they may be inefficient in some electronic portable instrumentation applications, and they require complex control circuitry.

4.4.3 Fundamental Analog Circuits

4.4.3.1 *Amplifiers*

Operational amplifiers are used as inverting or noninverting amplifiers by suitable connection of some external components. Figure 4.13(a) illustrates an operational amplifier-based a non-inverting amplifier. The resistors R_1 and R_2 define the feedback. The resulting gain is

$$A = 1 + \frac{R_2}{R_1} \qquad (4.2)$$

The amplification, A, is constant over a wide frequency range. At high frequencies, where the gain-bandwidth product (GBW) is 1 at frequency f_{gb}, the amplifier cannot amplify. Therefore, f_{gb} becomes a limiting factor regardless of the feedback. The feedback improves the linearity, gain stability,

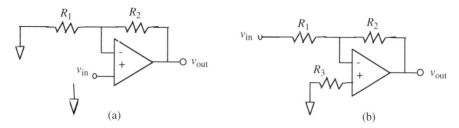

FIGURE 4.13
Operational amplifier as (a) noninverting amplifiers and (b) inverting amplifiers.

output impedance, and gain accuracy. As a general rule, for a moderate gain, the open-loop gain of an operational amplifier should be about 100 times greater than the closed-loop gain. The ratio of open-loop to closed-loop gain must be much higher (1000 or more) for better accuracies.

In Equation 4.2, when R_1 is set to infinity, the amplification becomes unity (e.g., a voltage follower). In this case, a capacitor parallel to R_2 is used to limit the bandwidth for the noise reduction. The 3-dB level may be estimated from the formula as

$$f_{3dB} = \frac{0.16}{R_2 C} \tag{4.3}$$

Operational amplifiers also can be configured as an *inverting amplifier*, as illustrated in Figure 4.13(b). The gain is $A = R_2/R_1$. The resistor R_3 is called the *ballast resistor* to equalize the voltage drop between the inverting and noninverting inputs. The ballast resistor is essential particularly if the bias current of the noninverting input is close to that of the inverting input. Due to common mode rejection, in phase voltage, drops at both inputs are cancelled out. In electronic portable applications, it should be noted that low-input resistors R_1 might result in excessive loading of sensors. In such cases, voltage followers may be used between the sensor and inverting amplifier.

4.4.3.2   Charge Amplifiers

Charge amplifiers are used to convert voltage signals generated from devices that generate very low charges or currents, such as active capacitive, piezo-electric, and pyroelectric sensors. They are designed to have extremely low bias currents. A typical charge amplifier, which acts as a charge-to-voltage converter, is depicted in Figure 4.14.

In Figure 4.14, the capacitor, C, is connected as the feedback element. The leakage resistance of the capacitor must be significantly higher than the capacitive impedance at the frequency of operation. At high frequencies, the operational amplifier operates near open-loop gain, thus causing oscillations.

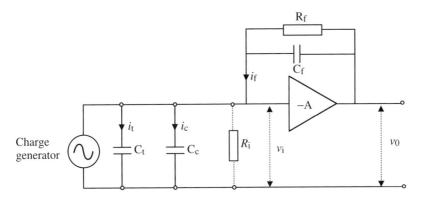

FIGURE 4.14
A typical charge amplifier based on an operational amplifier.

Various modifications of this circuit can be made to improve stability, noise, and bias current and to offset voltage improvements.

4.4.3.3 Voltage Followers

Voltage followers are often used for interfacing sensors to loads. They act as buffers by converting high-level impedances to low levels. A typical voltage follower, as in Figure 4.15, has a high-input impedance, high-input resistance, low-input capacitance, and low-output resistance. Voltage followers have unity voltage gains and high current gains, thus acting as a current amplifier and making them suitable to interface low-current sensors to the subsequent current-demanding signal processing circuits.

If voltage followers are used to interface current-generating sensors, the output current of the sensors must be at least 100 times greater than the input bias current of the operational amplifier. The input offset voltage must be low and trimmable if necessary.

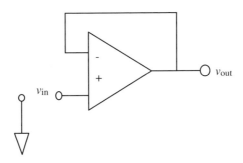

FIGURE 4.15
A voltage follower.

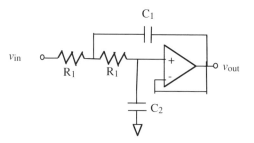

FIGURE 4.16
A typical low-pass filter.

4.4.3.4 *Filters*

Analog filters are essential for electronic portable instruments. They are used for the elimination of external and internal noise that interferes with the normal operation of an instrument. They are also used to extract useful information from the sensors, as most of the sensors are known to generate broadband signals containing various forms of noise. Filters can be analog or digital.

As far as the components and constructions of filters are concerned, there are two basic types: passive filters and active filters. Passive filters are made from only the passive components, such as resistors, capacitors, and inductors. Active filters contain active elements such as operational amplifiers, in addition to passive components. The filters can be configured as low-pass filters, high-pass filters, bandpass and bandstop filters, and notch filters. The subject of filters is a vast topic and cannot be discussed in detail here. However, in portable instruments, the active filters are more frequently used; therefore, they will be discussed in more detail and some examples will be given.

4.4.3.4.1 *Low-Pass Active Filters*

A typical example of a low-pass filter is given in Figure 4.16. In the process of an active low-pass filter design, the cutoff frequency of the filter must be determined first. The cutoff frequency is the frequency at which the gain of the filter is about −3 dB.

The choice of resistors in filters is critical. As a rule of thumb, resistors R_1 and R_2 are selected as, say, 25 kΩ for a 10-kHz filter or 250 kΩ for a 100-Hz filter. The values of capacitors can be calculated for the selected frequency by substituting the selected resistance values in the governing equations. The calculations are repeated to obtain the nearest available capacitors by changing the resistor values.

4.4.3.4.2 *High-Pass Active Filters*

A typical example of a high-pass filter is shown in Figure 4.17. They are analogous to low-pass filters.

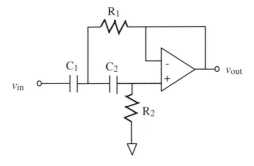

FIGURE 4.17
A typical high-pass filter.

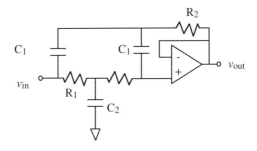

FIGURE 4.18
A typical bandpass filter.

4.4.3.4.3   Bandpass or Bandstop Active Filters

A straightforward way of building a bandpass or bandstop filter is to use a combination of low-pass and high-pass filters. A typical example of a bandpass filter is illustrated in Figure 4.18.

4.4.3.4.4   Notch Filters

These filters are usually used to suppress single sharp frequencies such as 50 Hz. The best performance is obtained by adjusting component values while monitoring the circuit performance on an oscilloscope.

There are various design techniques that are used to configure analog filters. Some of these design techniques include Butterworth, Chebyshev, Bessel–Thomson filters, etc. The complexity of filters increases as the order of filter (number of poles) increases. A typical complex active filter is given in Figure 4.19.

4.4.3.5   Oscillators

Oscillator circuits are the generators of electrical signals, and they are necessary in many applications. They are essentially unstable circuits that comprise of some gain, nonlinear features and positive feedback. There are three

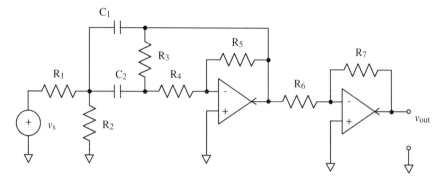

FIGURE 4.19
A typical high-pass filter.

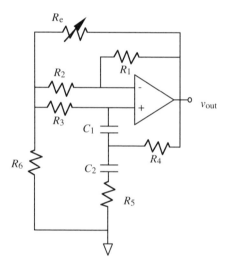

FIGURE 4.20
A sine-wave oscillator.

types of oscillators: resistive-capacitive (RC), inductive-capacitive (LC), and the *crystal oscillators*. In the case of *RC* and *LC* oscillators, the operating frequencies are determined by the values of resistive, inductive, and capacitive components. In the crystal oscillators, the frequency is defined by mechanical resonant properties of quartz crystals. As a typical example, an *RC* sine-wave oscillator is illustrated in Figure 4.20. This *RC* oscillator is useful for generating low-frequency sine waves.

In the circuit in Figure 4.20, the frequency can be controlled by a single variable resistance R_e. It can be shown that oscillation takes place when

$$R_5 \frac{C_2}{C_1} + R_4 \left(1 + \frac{C_2}{C_1} \right) = \frac{R_1 R_3}{R_2} \tag{4.4}$$

and the frequency of oscillation is

$$f = \frac{1}{2\pi\sqrt{C_1 C_2 R_3 R_4}} \sqrt{\frac{R_e / R_6}{1 + (R_1 / R_2)}} \tag{4.5}$$

There are many different types of oscillator circuits that can be found in books on operational amplifiers, analog and digital systems. Many oscillators are based on operational amplifiers, while others use *npn*-transistors as amplifiers and *LC* networks to obtain variable frequencies. Also, various types of multivibrators, together with the logic circuits, constitute the basics of many oscillators.

In many electronic portable instruments, improvised versions of bridges are used together with oscillatory circuits. Usually, the sensor is configured as a part of the oscillator that causes changes in the frequency of the oscillations. A suitable AC bridge, such as the Wien bridge or Schering bridge, is used to process the signals as part of the oscillator circuit. In these types of applications, particularly the capacitive sensors exhibit favorable characteristics with good frequency responses. The frequency, which is a direct indication of the physical variable, can be read by appropriate output devices such as electronic counters.

4.4.3.6 *Pulse Width Modulators and Linearizers*

In electronic portable instruments, pulse width modulators are often used to process the sensor output in the form of amplitude-modulated signals. Typical examples of such sensors are the capacitive vibrational and displacement sensors. The output of the sensor may be pulse width modulated to directly be interfaced with digital circuits.

Operational amplifiers are also used as linearizers to linearize the outputs of many nonlinear sensors. Usually, feedback linearization techniques are applied by using a feedback system that adjusts the currents or voltages to stay at some reference values. This is usually accomplished by obtaining a DC signal proportional to the current from a demodulator, comparing this current with the reference current, and adjusting the voltage amplitude of the system excitation until the two currents agree. Another linearization technique is square-wave linearization, which is often applied in capacitive accelerometers.

4.4.3.7 *Bridges*

In electronic portable instruments, bridges are commonly used to measure basic electrical quantities such as resistance, capacitance, inductance, impedance, and admittance. There are two groups of bridges, the AC bridge and the DC bridge. Also, in each type, there are many different types, such as

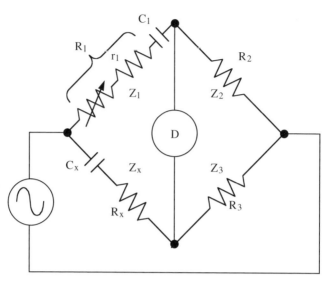

FIGURE 4.21
A series RC bridge mainly used for capacitance measurements.

Wheatstone and Kelvin DC bridges, and Schering, Maxwell, Hay, and Owen AC bridges.

Bridges are essentially two-port networks in which the component to be measured is connected to one of the branches of the network. In a particular instrument, the selection of the bridge to be employed and the determination of values and tolerances of its components are very important. It is not our intention to cover all bridges here; however, as a typical example of an AC bridge, a series RC bridge is given in Figure 4.21. We also offer some analysis briefly to illustrate their typical principles of operation.

Taking Figure 4.21, at balance,

$$Z_1 Z_3 = Z_2 Z_x \tag{4.6}$$

Substitution of impedance values gives

$$(R_1 - j / \omega C_1)R_3 = (R_x - j / \omega C_x)R_2 \tag{4.7}$$

Equating the real and imaginary terms gives the values of unknown components as

$$R_x = \frac{R_1 R_3}{R_2} \tag{4.8}$$

and

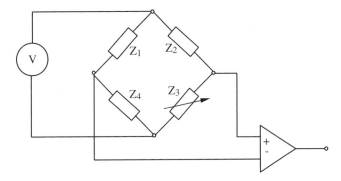

FIGURE 4.22
A disbalanced bridge.

$$C_x = \frac{C_1 R_2}{R_3}$$ (4.9)

There are two types of bridges commonly used in electronic portable instruments: disbalanced bridges and null-balanced bridges. A disbalanced bridge, known as the *deflection bridge*, is based on detecting voltage across the bridge diagonal arms, as illustrated in Figure 4.22. This figure also symbolizes a typical bridge amplifier. There are many other configurations of bridge amplifiers possible, depending on the grounding and availability of grounded or floating reference voltages.

In Figure 4.22, the output of the bridge is a nonlinear function of disbalance. In small ranges of measurements, the nonlinearity may be considered to be quasi-linear, but must be linearized by suitable techniques for large ranges.

The null bridge should be balanced at all times. This is achieved by adjusting one of the arms in response to variations in the other arm, which carries the sensor. Null bridges have superior linearization characteristics compared to disbalanced bridges.

In Table 4.9, a list of active electronic circuit manufacturers and suppliers with telephone numbers is provided. For each component, only a few names are given to avoid repetition. Many firms supply more than one or all types of components.

4.5 Digital Aspects, Communication, and Networks

4.5.1 Digital Aspects

Nowadays, digital portable instruments are very common. All the digital components discussed in Chapter 3 can be applied in portable instruments. Before we concentrate on the characteristics of digital components, it is worth

TABLE 4.9

Active Circuit Manufacturers and Suppliers

Type	Manufacturers and Dealers (Phone No.)	
Generators (audio, function, programmable, RF, pulse)	Alfa Electronics (609-897-1135) Analogic Corp. (978-977-3000) Avtech Electrosystems (800-265-6681) California Instruments (858-677-9040) Conxant Systems Inc. (949-483-4600)	DATEL Inc. (800-233-2765) JFW Industries (317-887-1340) MX-COM Inc. (336-744-5050) Rhom Electronics (858-625-3600) Sunshine Instruments (512-416-8500)
DC/DC converters	AF CommSupply (800-255-6222) Fujitsu Microelectronics (800-866-8608) IFR (316-522-4981) Intersil (888-468-3774)	Motorola Energy Systems (770-338-3742) National Semiconductor (408-721-5000) Texas Instruments (www.ti.com) Wilmore Electronics (919-732-9351)
Filters (high/ low bandpass/ bandstop, tunable)	Ace Technology Inc. (818-718-1534) Hutton Communications (877-648-8866) Insight Electronics (800-677-7716) KS Electronics (602-971-3301) Micro Electronic Manufacturing (949-206-9960) Mouser Electronics (800-346-6873)	Piezo Technology Inc. (407-298-2000) RF Micro Devices (336-664-1233) Schuster Electronics (513-489-1400) Tellurian Technology (847-934-4141) TLSI (631-755-7005) World Technologies (800-900-9825)
Mixers	Agilent Technologies (408-654-8675) Analog Devices (800-262-5643) Fujitsu Microelectronics (800-866-8608)	Intersil (888-468-3774) Maxim Integrated Products (408-737-7600) Synergy Microwave (973-881-8800)
Frequency multipliers	Frequency Electronics (516-794-4500) Insight Electronics (800-677-7716) Micro Networks (508-852-5400)	RF Micro Devices (336-664-1233) Synergy Microwave (973-881-8800) Toshiba America (949-455-2000)
Mixed-signal ICs	Advanced Linear Devices (408-747-1155) AKM Semiconductors (408-436-8580) Aura Communications (978-527-4100) Cirrus Logic (512-445-7222)	Fairchild Semiconductors (408-820-2200) Hitachi Semiconductors (408-433-1990) National Semiconductor (408-721-5000) Supertex Inc. (408-744-0100)
Demodulators	Analog Devices (800-262-5643) Insight Electronics (800-677-7716) Intersil (888-468-3774)	MX-COM Inc. (336-744-5050) Sharp Microelectronics (360-834-2500) Silicon Wave Inc. (858-453-9100)

Source: Portable Design, March 2002, PennWell.

highlighting some of the problems commonly associated with digital circuits. Hence, the uses of digital circuits in instruments introduce their own problems. For example, when the clock frequency exceeds 1 MHz in a system, the digital signals on the buses exhibit transmission lines type of behavior. This behavior of circuits as transmission lines requires careful design considerations, including bandwidth limitations, decoupling, ground bounce, cross talk, impedance mismatching, timing skew, and delays.

Limiting the bandwidth of the signals is an effective way of reducing noise, EMI, and other problems associated with the transmission lines. The bandwidth can be limited by increasing the rise and fall times of signal edges or by reducing the clock frequency. Reducing the clock frequency may not be a feasible option; however, edge rates may be controlled by selecting suitable logic families. For example, the rise time for a GaAs logic family is about 5 nsec, whereas for 5-V CMOS families, the rise time is about 100 nsec. Clearly, the CMOS family results in superior connection properties so far as transmission line behavior is concerned.

The switching of digital logic also causes voltage and current transients. These transients can generate noise in the rest of the circuits. If transients are severe, coupling capacitors may be used for suppression.

Ground bounce in a digital system is a voltage surge that couples through the ground leads of a component into nonswitching outputs, thus introducing glitches onto signal lines. Ground bounce can be severe in asynchronous signals, and it can be cumulative. A large-voltage surge on the ground circuits can reference the outputs above the input thresholds of the driven logic. The ground bounce can be minimized by several methods, i.e., reducing the loop inductance, reducing the input gate capacitance, or selecting appropriate logic families with slower fall-times.

Cross talk is the electromagnetic coupling between active and passive lines. The coupling can be capacitive or inductive. Cross talk can be controlled by line spacing and length of lines, characteristics impedance, and signal rise times. A good layout of bus and other connection lines can minimize cross talk problems.

Impedance matching can be a serious issue if not treated properly. A mismatched arrangement can cause ringing, undershoots, and overshoots in the signal pulses. Impedance matching can be done by making the source and termination impedances equal to the characteristic impedance of the transmission line. There are various other techniques (e.g., using capacitors, resistors, transistors, and other active circuits) that can be employed under severe mismatching conditions.

Timing can become a problem as the clock frequency increases, since the propagation delays, timing skew, and phase jitter can have serious effects on system operation that may need to be treated carefully.

In portable instruments, low-power design is desirable not only for limited available power, but also for isolation and low heat dissipation. In CMOS components, power is a function of frequency, f; load capacitance, C; and the DC supply voltage, V, which can be expressed by

$$p = fCV^2 \qquad (4.10)$$

The power consumption can be reduced by reducing the supply voltage, load capacitance, or frequency. As can be seen, power is proportional to the square of the supply voltage, and nowadays, low-voltage CMOS circuits are extensively used in portable instruments. There are many other options to minimize the power consumption, such as shutting down unused circuits, putting the controller into sleep mode when the device is not in use, terminating the unused inputs, and avoiding slow signal transition.

4.5.1.1 Low-Power Digital Components

Low-power microcontrollers form the basis of many portable systems. The latest products continue to add features that make the drives easier to program and more simple to integrate into a system. Some examples are provided below.

For example, Intel offers its Mobile Pentium line of processors. The Pentium III Processor M runs at 1.2 GHz and features 512 Kbytes of cache memory. It has a deep sleep mode. Pentium 4-M is a 1.6-GHz and 1.7-GHz microprocessor operating with an average power of less than 2 W. The central processing unit (CPU) dynamically ranges its operating voltage, running at 1.4 V in full-speed mode. It can operate with voltages down to 1 V. There is a range of ICs supporting such microprocessors.

Amtel's 8-bit 8051 family microprocessor operates on 1.8 V. Another product, NEC's PD789881 microprocessor, runs on 18 μA in operating mode and 0.9 μA at standby. The TC2000 microcontroller contains 64 Kbytes of standby random access memory (SRAM) and 4 Mb of flash memory.

Despite the availability of low-power microprocessors, other components in a digital system still require different voltages, such as:

Leading-edge microprocessors	1.2 V
PCMCIA interface chips	3.3 V
Memory	2.5 V
Display and backlit	9 to 22 V

This requires the support of different voltages in a system as discussed in Section 4.2.

4.5.1.2 Back-Plane Buses

Back-plane buses connect the circuit boards together in an instrument. There are a number of back-plane buses that are suitable for physical sizes and data transfer rates. For example, STD-32 is based on TTL logic circuits and has a data transfer rate on 20 Mbytes/sec. Other back-plane buses are G-64, suitable for the single Eurocard; EISA, Micro Channel, and NuBus, suitable for instrumentation and desktop computers; Multibus and VXIBus, suitable for complex instruments; and VMEbus and Futurebus, suitable for computers.

4.5.2 Signal Conversion

One of the main features of modern electronic portable instruments is the capability of digital operations. The recent progress in digital electronics has opened new dimensions in the development and usage of these instruments. However, most portable instruments contain sensors that provide analog information in voltage or current forms. These signals are converted to their equivalents to be processed further in the digital environment. For this purpose, A/D converters are used to transform the signals from the continuous analog domain to the digital discrete domain. There are many different types of A/D converters, which are discussed in detail in Chapter 3, but further information relevant to portable instruments will be supplied here. The correct choice of A/D converter for a particular electronic portable instrument is necessary, since it will set the limit on resolution of the system. Greater resolution translates to more bits of digital data, higher accuracy, and a wider dynamic range, but probably slower conversion speed and requires more computation.

In this chapter it will be sufficient to say that when building nonportable instruments, there are some limitations in the types of converters that can be employed. Generally, fast 12-bit or higher A/D and digital-to-analog (D/A) converters are used. However, in portable instruments, one would expect severe limitations in the operational characteristics of converters due to the limited availability of power and space within the instrument. The price and availability of suitable CMOS devices may be one of the reasons for the limitations; the other is the relatively poor reliability of CMOS converters. Also, the availability and performance characteristics of other major supporting components of converters, such as the multiplexers, sample-and-hold amplifiers, and active filters, may cause some problems.

Converters have many inherent errors — quantization, nonlinearity, missing code, aperture jitter, input noise, and settling errors — since the tendency in portable instruments is to use low-bit converters (8 and 12 bits) to save space and memory.

4.5.2.1   Digital-to-Analog Converters

Digital-to-analog converters take the binary numbers as the input and output currents or voltages proportional to the value of the binary number. D/A converters operate on the principle of summing currents using R-2R or C-2C ladder networks, as explained in Chapter 3. Basically, the current generated in each step of the ladder is proportional to 2^n. Since the output is in current form, the conversion of the current output to the voltage equivalent is left optional to the user.

Many D/A converters, particularly 8-bit converters, are fairly easy to incorporate into the circuits of electronic portable instruments. In many cases, the total current consumption of the converter is less than 2 mA, which makes them ideal for portable instruments. Nevertheless, an important point is that the settling time and transients of D/A converters may be very significant and can have negative effects on some circuits. In some cases, the filters may be useful, and in others, sample–hold amplifiers may be employed to hold the previous information a little bit longer while the D/A converter is settling.

4.5.2.2   Analog-to-Digital Converters

Analog-to-digital converters are much more complex in operation principles, design, and construction. Therefore, this is a much wider subject to discuss, mainly due to the availability of many different types. There are a few CMOS A/D converters available on the market, such as the AD7570. There are four types of A/D converters: integrating, successive approximation, delta-sigma, and flash converter types that cover slow, medium, and fast conversion speeds. The architecture of A/D converters differs significantly; each configuration has inherent capabilities that make it suitable for a particular application.

In portable instruments, all types of A/D converters are used, depending on the applications. For example, portable oscilloscopes use flash converters, and delta-sigma converters are finding a diverse range of applications. Most of these converters contain latching circuits, comparators, and counters on board the chip.

In some electronic portable instruments, the speed of conversion may not be particularly important. If this is the case, the best choice of converters may be the integration types, since they are inherently noise resistant and inexpensive. Slower A/D converters are frequently used, such as the dual-slope and quad-slope converters, where conversion speeds are usually less than 200 samples/sec. Integration time A/Ds are most suitable for low-frequency applications, such as monitoring of temperature and pressure. Integration type A/D converters are available in wide ranges that are offered by different manufacturers.

Successive approximation A/D converters are much more complex in circuits and operations; their typical conversion times are about 1000 times better than those of integration types. These converters work on a trial-and-error basis, needing only 8 trials for an 8-bit converter, 12 trials for a 12-bit converter, and so on. They are extremely prone to noise. A spike of noise in the input signal can set the comparator wrongly in the conversion process. A successive approximation converter counts a reference signal generated by an internal D/A converter. The converter selects the count or digital code that matches the analog input. They need an S/H circuit to present a stationary analog value to the comparator for each conversion. However, successful approximation converters are low-cost devices and find wide applications in all types of electronic portable instruments.

Successive approximation converters can operate at high speeds. In electronic portable instruments, if these converters are used, adequate measures must be taken to deal with the noise by means of some additional circuits, such as sample–hold amplifiers. For fast conversion rates, flash converters may be selected at the expense of higher-power consumption levels.

Flash converters use the principle of the 2-bit converters. The input signal is fed to a number of comparators, say 16 or 32, at a time. Decoding logic determines which is the highest converter that is being triggered by the input signal, and it outputs the corresponding binary number. Flash converters can run as fast as 10 nsec per conversion. They find applications in communication and radar systems and digital oscilloscopes.

In portable instruments, the output is usually required in the binary-coded decimal (BCD) format. This will require additional circuits in the instruments.

4.5.2.3 Multiplexers

Multiplexers are basically a collection of high-performance analog switches that are used to interface more than one analog input to the instruments. Apart from multiplexing the analog signals, multiplexers can also be used for other purposes, such as scaling and turning the output devices on. Multiplexers are available in CMOS technology, such as AD7501 from Analog Devices. Two important features of multiplexers are the on-resistances and settling times. The on-resistance can be as low as 100 Ω or as high as several thousand ohms. Most multiplexers have relatively large gates, and therefore large capacitances and large settling times, which can impose severe limitations to conversion rates.

4.5.2.4 Sample–Hold Devices

Sample–hold devices, also known as sample–hold amplifiers (SHAs), contain an analog switch, a high-performance capacitor, a high-input impedance amplifier, and some external capacitors. The S/H devices are used just before the A/D converters. They are an essential part of converters, particularly in successive approximation types, to hold the input voltage steady while the conversion is taking place. If a very fast signal needs to be held for several seconds, two or more S/H units can be cascaded. The first S/H has a small capacitor for fast acquisition and holds the signal long enough for the subsequent ones to sample and hold. The second and third S/Hs can have larger capacitances.

4.5.2.5 Software to Drive A/D Converters

In addition to hardware, extensive software is necessary to drive A/D converters. The complexity of the software depends on the complexity of the hardware and the extent of data handling. Some of the steps involved in developing the software are:

1. Selection of channels for the analog signals and the delivery of the corresponding binary numbers to the memory locations
2. Taking care of polarity of the input voltage and associated binary equivalents
3. Autoscaling for testing overflow and underflow conditions
4. Applying appropriate timing to eliminate settling times, aperture times, offsets, and other delays, thus making sure conversions are completed without loss of useful information

4.5.3 Intelligent Sensors

The emergence of satellite systems and advances in RF and microwave communications are greatly impacting the development and use of electronic portable instruments. The progress in hardware, as in semiconductors, intelligent sensors, and microsensors, and in software, as in the case of virtual instruments, is adding remarkable features and opening up many potential areas for applications.

Nowadays, many portable instruments use intelligent sensors, which contain the sensor, signal processors, and intelligence capabilities in a single chip. They are appearing in the marketplace as pressure sensors and accelerometers, biosensors, chemical sensors, optical sensors, magnetic sensors, etc. Some of these sensors are manufactured with the neural network and other sophisticated intelligence techniques on board the chip. Many different types of intelligent sensors are appearing, such as neural processors, intelligent vision systems, and intelligent parallel processors.

For example, the NC3002 is an intelligent sensor-produced "neuricam," which is based on the digital very large scale integration (VLSI) parallel processing technique. These find applications in machine learning and image recognition supported by artificial neural networks in quality inspection applications. The architecture of the sensor is structured in a way to implement the Reactive Tabu Search learning algorithm, a competitive alternative to back-propagation. Such sensors are suitable for fast parallel number-crunching purposes intended for operation with standard CPUs in either single-chip or multiple-chip configurations.

In parallel to rapid developments in hardware aspects of modern sensors, some new standards are emerging for hardware architecture, software, and the communications of modern smart sensors, thus making a revolutionary contribution to electronic portable instruments. Although not specifically addressing portable instruments, IEEE-1451 is an example of such standards that define interface network-based data acquisition and control of smart sensors and transducers.

The IEEE-1451 standard aims to make it easy to create solutions using existing networking technologies, standardized connections, and common software architectures. The standards allow application software, field network, and transducer decisions to be made independently. They offer flexibility to choose the products and vendors that are most appropriate for a particular application. As the new sensors are developed, coupling these standards together with wireless communication technology, we would expect good growth in the number and variety of portable instruments.

Sections 1451.3 and 1451.4 of the Institute of Electrical and Electronics Engineers (IEEE) standard are developed for communication purposes. The former is dedicated to the digital communication and transducer electronic data sheet (TEDS) formats for distributed multidrop systems, whereas the latter is for mixed-mode communication protocols and TEDS formats. IEEE-1451.4 is particularly important for portable instruments since it defines the mechanism for adding self-describing behavior to transducers with an analog signal interface. Rather than defining yet another network or bus, it defines a low-level transducer interface and information structure that can work across any type of measurement interface or network.

There are positive signs that many companies are taking notice of the IEEE-1451 standard and producing devices that comply with it. For example, National Instruments is actively promoting and introducing plug-and-play sensors that comply with IEEE-1451.4. There are many smart sensors available in the marketplace, accelerometers and pressure transducers, that comply with these standards.

4.5.4 Communications

The communications capability of portable instruments is enhancing their usability in many different types of applications. There are two levels of communications: (1) communication at the sensor level, which is mainly realized by intelligent sensors, and (2) communication at the device level. At the device level, the communication takes place in either wired form, infrared (IR), RF, or microwave techniques.

A typical example, IR communication technology is applied in a number of dosimeters. The information exchange between an electronic reader and the dosimeter is provided through the IR communication port operating through the RS-232 port. When the dosimeter is placed on the reader window, an IR emitter forms signals under the microprocessor control according to an exchange protocol. By using suitable software, the operator can:

- Register the dosimeter
- Enter the user's personal information
- Create a database for the registered dosimeter
- Transmit the dose accumulation history from the dosimeter memory to a PC and store this history in the database
- View the history stored in the database
- Transmit date, time, and threshold values
- Format the dose history storage to the dosimeter

Data acquisition, signal conditioning, instrumentation, and controller or interface modules have options for various computer buses, such as PCM-CIA, parallel, USB, Fire Wire, and GPIB. In a typical example, a portable multiple-sensor toxic gas detector may be equipped with a wireless RF modem that allows the unit to communicate and transmit readings and other information on a real-time basis with a remotely located base controller. The real-time data transmission takes place between the instrument and a base controller located a couple of kilometers away. Any personal computer equipped with communication facilities can be used as the base station.

Portable instruments are making a leap forward due to the implementation of modern RF communication systems. RF communication modules are usually integrated in I/O modules that eliminate cable connections between devices. In some industrial applications, two or more digital signals are generated say between the frequencies of 902 and 928 MHz using spread spectrum frequency hopping technology to guarantee an interference-free link between remote devices. Typically, they have 1-W power consumption to transmit signals for a 300-m range.

The spread spectrum RF technology offers an alternative to wire, reliability, superior jam resistance, structure penetration, signal sensitivity, and a good external signal range. Spread spectrum is available for nonmilitary use and operates at 915 MHz, 2.46 GHz, and 5.8 GHz. The transceivers complying with spread spectrum capabilities are available for $2 to $10 from companies such as Axonn–New Orleans. These transceivers operate at low voltages (1.8 to 5 V), and hence have long battery lives (3 to 10 years). They find extensive applications in distributed systems, e.g., System Control and Data Acquisition (SCADA). Another example is RCM2200 by Zworld Engineering, Davis, CA. This single-board computer has an Internet interface end-embedded TCP/IP that allows high-speed network data transfer. The firmware of RCM2200 can be programmed to support multiple-data output formats, including HTML for direct web interface.

4.5.5 Protocols

In the implementation of wireless communication features, in today's portable instruments, Bluetooth technology is gaining wide acceptance. Bluetooth is considered to be a low-cost and short-range wireless technology that provides communication functionalities, ranging from wire replacements to simple personal area networks. Initially, it was aimed to bring short-distance wireless interfaces mainly to consumer products on a large scale. As the number of products incorporating Bluetooth wireless technology increases, the development of various types of instruments for a diverse range of applications becomes more important.

However, Bluetooth has not escaped competition. ExtremeTech reports that Sony and Philips have agreed to jointly work on Near-Field Communication, a potential competitor to Bluetooth in the short-range personal area network (PAN) market. Also, the IEEE just recently adopted and approved the Bluetooth protocol under its wide PAN (WPAN) standards, the IEEE-802 series of protocols.

There are many Bluetooth products appearing in the marketplace, from the chip level to sophisticated devices. On the chip level, a typical example of Bluetooth products is the $5 chipset from Texas Instruments. It provides up to 1 megabit per second (Mbps) for fast data transmission. On the device level, Sony has introduced in Europe and Japan the ultracompact fixed-lens DSC-FX77 camera with built-in Bluetooth communication protocol. Many chip-level products target Bluetooth specifications. They exclude antenna and point-to-multipoint communications consisting of a base-band controller with flash memory, a reference crystal, and an RFCMOS transceiver. In some cases, the two-chip approach is used to ensure a good level of performance and reliability in RF-intensive environments. The architecture is based on independent silicon optimization with digital parts in standard CMOS and analog parts in BiCMOS or RFCMOS, enabling cost and size reduction.

Bluetooth is gaining wide applications in wireless smart transducer networks. Interfacing of IEEE-1451 smart transducer nodes to the Bluetooth network is gaining momentum. This involves a detailed study of network communication models specified for smart transducer communication as well as the Bluetooth protocol stack. Recently, the interface between IEEE-1451 and Bluetooth communication hardware has been researched extensively.

IEEE-802.11b is a high-bandwidth standard for the transfer of large amounts of data. It handles spread spectrum and high data bursts easily. The IEEE-802.11b standard is supported by three chips, compared to the two-chip or single-chip solution of Bluetooth. IEEE-802.11b is designed as a communication channel to host the processor-running TCP/IP. The encryption length is 64 bits.

Bluetooth offers 128-bit encryption. The single-chip solution of Bluetooth enables it to be used in unusual applications such as headsets and luggage security tags, and so on. It also finds extensive applications in phones, cameras, portable games, and linking devices to PCs. It is in wide use by companies such as IBM, Sony, Compaq, 3Com, NEC, Fujitsu, and TDK.

For the designers of portable instruments with remote communication facilities, there are many single-chip RF transceivers. A typical example is the LMX3162 from National Instruments. This transceiver is a 48-pin $7 \times 7 \times 1.4$ mm monolithic device operating at 2.45-GHz wireless systems. It contains phase locked loop (PLL) transmit and receive functions. The 1.3-GHz PLL is shared between transmit and receive sections. The transmitter includes a frequency doubler and a high-frequency buffer. The receiver consists of a 2.5-GHz low-noise mixer, an intermediate frequency (IF) amplifier, a high-gain limiting amplifier, a frequency discriminator, a received signal strength indicator (RSSI), and an analog DC compensation loop. The circuit features on-chip voltage regulation to allow supply voltages ranging from 3.0 to 5.5 V. Such chips are extensively used in personal wireless communication systems/networks (PCSs/PCNs) and wireless local area networks (WLANs).

If the data need to be transmitted for long distances in which transmitter and receivers are out of range, wireless bridges are employed. They provide long-range point-to-point or point-to-multipoint links. Some of these devices use direct sequence spread spectrum (DSSS) radio technology operating at typical frequencies (i.e., 2.4 GHz). Data may be transmitted at speeds up to 11 Mbps. These bridges comply with standards such as IEEE-802.11b, connecting one or more remote sites to a central server or Internet connection.

It is necessary to mention that in some cases, portable data loggers are employed as smart instruments to collect information on physical variables such as relative humidity, temperature, pressure, electrical current, pulse signals, etc. Information stored in the logger's memory is then transferred to computers.

Industrial plug-and-play or long-range wireless communication devices are offered by many companies, for example, Microwave Data Systems, Inc.

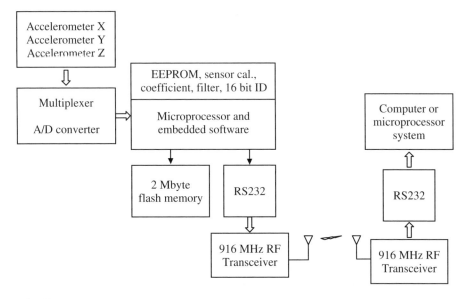

FIGURE 4.23
A typical wireless data-logging system.

Typical products provide point-to-point and point-to-multipoint communication of data with frequencies from 200 MHz to 2.4 GHz and speeds up to 8 Mbps. Such devices are used for health monitoring and other civil, military, and industrial applications. Networks of low-cost wireless sensors enable monitoring of large civil structures with a large number of sensing nodes.

A typical single-unit data-logging transceiver is a wireless data communications device that can serve multichannel sensor arrays to a remote data acquisition system hosted by a computer or a microprocessor. The frequency of transmission is 916 MHz, narrowband. The RF communication link operates with 19,200 baud, and the device is capable of triggering a sample to be logged (typically from 30 m) or requested data to be transmitted. A typical application of wireless data logging on a triaxial micro-electro-mechanical (MEM) accelerometer is illustrated in Figure 4.23. It has flash memory (typically 2 to 8 Mbytes). The nodes in the wireless network may be assigned with 16 bits, hence being able to address thousands of multichannel sensor clusters. They are powered by 3.1- to 9-V lithium ion AA size batteries. They draw about 10 mA. They contain 10- or 12-bit A/D converters. The dimensions of the total package are $25 \times 40 \times 7$ mm.

4.5.5.1　Built-in-Tests

RF data communication is extensively applied in built-in-tests (BITs) in which the sensors are embedded in operational systems such as rotating machinery and industrial systems, or even implanted in living organisms. The information gathered from the sensors is transmitted by a built-in RF transmitter to a nearby receiver. The quality of the BIT is improving steadily through the availability of intelligent and other sophisticated sensors, instrument software, interoperability, and self-test effectiveness. Applications such as built-in-tests are typical examples of recent progress taking place in electronic portable instruments.

4.5.6 Networks

Geographically distributed portable instruments can be networked as a distributed sensor network (DSN). The networking requires intelligent sensors that obey some form of hierarchical structure. The development of networked microsensors is progressing in several directions:

1. Development of low-cost intelligent sensors that are backed up by low-cost signal processing, enabling sophisticated detection, identification, and tracking functions

2. Long operational life of sensors housed in small and light packages

3. Wireless information flow between sensors in somehow reduced bandwidths that provide secure data flow and immunization to electromagnetic or deliberate interferences

4. Minimization of sensor network cost and maximization of sensor network lifetime

5. Provision of autonomous sensor management (self-management), such as automatic node selections and self-calibration

In recent years, IEEE-1451-compatible interfaces between the Internet/Ethernet serial port websensors have been developed. These sensors have direct Internet address. The interface is realized in IEEE-451, "Network Capable Application Processors (NCAPs)." The NCAP connects the Internet through Ethernet. NCAP is a communication board capable of receiving and sending information using standard the TCP/IP format. The sensor data are formatted to and from the serial port by one of the following: RS-232, RS-485, TII, Microlan/1-wire, Esbus, or I²C.

The future of sensors will probably be shaped by how they interface into the network to share information. In this respect, standards such as IEEE-1451 support network communications. Nevertheless, the importance of sensor networks has been realized by many vendors; hence, they promote proprietary solutions to connect smart sensors into TCP/IP-based networks. A typical arrangement of wireless websensor nodes is illustrated in Figure 4.24. Recent developments are taking place in two levels: low-level interfaces and high-level interfaces.

The low-level interface focuses on development of sensor clusters consisting of base stations and sensor nodes. It requires Web-enabled smart sensors that that can be directly coupled with the network. The high-level interface

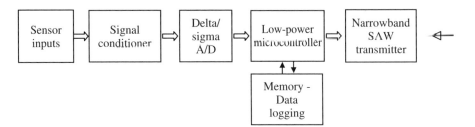

FIGURE 4.24
A wireless Web sensor.

focuses on various wireless network topologies (i.e., IEEE-802.11b) and information convergence in a central gateway. The high-level interface has been introduced to minimize the impact of the human operator on a smart sensor structure. There is a considerable amount of academic and commercial interest in both types of interface methods from hardware and software points of view.

In Table 4.10, a list of semiconductors and IC manufacturers and suppliers is provided. Only a few names are given; the list is not exhaustive. Many firms provide more than one or all types, but repetition is avoided whenever possible.

4.6   Operator Interface

The operator interface is the channel of communication between an instrument and the user involving both hardware and software. Through the interface, the instruments communicate with the user. Sets of interactions are fulfilled by encompassing the user view and understanding of the instrument. The rules of industrial design may be observed to develop a good man–machine dialogue and ultimately design the instrument in accordance with user needs, wants, and expectations. A good product should be easy to produce, learn, use, and maintain. The basic principles of interface must be worked out during the early stages of the design of an instrument, before the circuits are developed, software is written, and packages are built. Attractive and robust packages are equally important in the success of the instrument in the market.

The designers of a market-successful electronic portable instrument must have carefully identified the expectations of customers and considered user profiles, focus groups, and other influencing factors carefully. Such a product is usually a user-centered instrument that emphasizes the purpose of the device to serve the user. In this respect, a complete understanding of technology, programming, and the capabilities and knowledge of the user is essential. The user-centered approach focuses on analysis of the user, spec-

TABLE 4.10

Digital Components Manufacturers and Suppliers

Type	Manufacturers and Dealers (Phone No.)	
ASICs, Programmables, FPGAs, PLDs	Agilent Technologies (408-654-8675) Atmel Corp. (408-441-0311) Epson Electronics (800-228-3964) Fairchild Semiconductors (408-820-2200) Goal Semiconductors (800-943-4625)	Insight Electronics (800-677-7718) Micrel Semiconductors (408-944-0800) Motorola Semiconductors (480-413-4991) TLSI (631-755-7005) Xilinx Inc. (408-559-7778)
A/D, D/A converters	Advanced Linear Devices (408-747-1155) Analog Devices (800-262-5643) AVOX RIFA (847-848-9511) Cirrus Logic (512-445-7222) Linear Technology (408-432-1900)	Maxim Integrated Products (408-737-7600) Microsemi Corp. (949-221-7100) National Semiconductor (408-721-5000) SMSC Microsystems (631-435-6000) Texas Instruments (800-477-8924)
Data acquisition	AKM Semiconductors (408-436-8580) Intersil Corp. (888-468-3774) Micro Networks (508-852-5400)	Onset Computer Corp. (508-759-9500) Quatech Inc. (330-434-3154) Xicor Inc. (408-432-8000)
Discrete logic	Hitachi Semiconductors (408-433-1990) Hynix Semiconductors (408-501-6000) ON Semiconductors (602-244-6600)	Seiko Instruments USA (310-517-7771) TLSI (631-755-7005) Xilinx Inc. (408-559-7778)
DSPs	Blackhawk (877-983-4514) DY 4 Systems Inc. (613-599-9191) Infineon Technologies (408-501-6000)	Intersil Corp. (888-468-3774) Pentek (201-818-5900) Xilinx Inc. (408-559-7778)
LAN	Bill West Inc. (203-261-6027) Macronics (408-453-8088) RF Microdevices (336-664-1233)	Ubicorn Inc. (650-210-1500) Winbond Electronics (408-943-6666) Xicor Inc. (408-432-8000)
Memory storage (ICs, DRAM, EEPROM, flash, SRAM, CDs, discs)	Atmel Corp. (408-441-0311) Bill West Inc. (203-261-6027) Catalyst Semiconductors (408-542-1000) Chip Supplies Inc. (407-298-7100) Global American Inc. (603-886-3900) Legacy Electronics (949-498-9600) M-Systems Inc. (510-494-2090) Macronix (408-453-8088)	Mosel Vitelic Corp. (408-433-6000) Silicon Storage Technology (408-735-9110) SMART Modular Inc. (510-623-1231) Technoland Inc. (408-992-0888) Toshiba America (949-455-2000) Tri-M Systems (604-527-1100) Winbond Electronics (408-943-6666) WinSystems Inc. (817-274-7553)
Multiplexers	Chip Supply Inc. (407-298-7100) Micrel Semiconductors (408-944-0800) Mitsubishi Electronics (408-730-5900)	Silicon Laboratories Inc. (512-416-8500) Supertex Inc. (408-744-0100) Vishay Siliconix (408-988-8000)
Processors, controllers, coprocessors (4–32 bits)	Alacron Inc. (603-891-2750) Bill West Inc. (203-261-6027) Cypress Microsystems (425-939-1000) EBS Net Inc. (978-448-9340) Epson Electronics (800-228-3964) I-Logix Inc. (978-682-2100) Lineo Inc. (801-426-5001) Midwest Micro-Tek (605-697-8521)	National Semiconductor (408-721-5000) Sharp Microelectronics (360-834-2500) SMSC Microsystems (631-435-6000) Starcore (770-618-8522) Texas Instruments (800-477-8924) Ubicorn Inc. (650-210-1500) Winbond Electronics (408-943-6666) ZF Linux Devices (800-683-5943)
Sensors, transducers	Agilent Technologies (408-654-8675) Cherry Electrical Products (262-942-6500) Conventor Inc. (919-854-7500) Daido Corp. (732-805-1900)	Piezo Systems Inc. (617-542-1777) Rohm Electronics (858-625-3600) Saia-Burgess USA (847-549-9630) Schuster Electronics (513-489-1400) Sensor Scientific Inc. (973-227-7790)

TABLE 4.10 (Continued)

Digital Components Manufacturers and Suppliers

Type	Manufacturers and Dealers (Phone No.)	
Sensors, transducers (Continued)	Entran Devices Inc. (973-227-1002)	Sensotec Inc. (614-850-5000)
	Fujitsu Compound Semiconductor (408-232-9692)	Shaffner EMC Inc. (732-225-9533)
		Sharp Microelectronics (360-834-2500)
	Galco Industrial Electronics (248-542-9090)	Statek Corp. (714-639-7810)
	Grayhill Inc. (708-354-1040)	Sutron Corp. (703-406-2800)
	Hantronix Inc. (408-501-1100)	TLSI (631-755-7005)
	Insight Electronics (800-677-7718)	TT Electronics (714-447-2300)
	MTS Sensors (919-677-0100)	Vishay Siliconix (408-988-8000)
	National Semiconductor (408-721-5000)	
	Omron Electronics (800-556-6766)	
Transmitters, receivers, transceivers	Ace Technology (818-718-1534)	Omega-Vanzetti Inc. (781-784-4733)
	Aura Communications (978-527-4100)	Rohm Electronics (858-625-3600)
	Cironet Inc. (678-684-2000)	Silicon Wave (858-453-9100)
	Fujitsu Microelectronics (800-866-8608)	Vectron International (203-853-4433)
		Zarlink Semiconductor (613-592-0200)
	National Semiconductor (408-721-5000)	
CMOS transistor, FET	Central Semiconductors (631-435-1110)	Rohm Electronics (858-625-3600)
	Chip Supply Inc. (407-298-7100)	Supertex Inc. (408-744-0100)
	Mouser Electronics (800-346-6873)	TLSI (631-755-7005)
	RF Micro Devices (336-664-1233)	Vishay Siliconix (408-988-8000)

Source: Portable Design, March 2002, PennWell.

ification of performances, tasks, and methods to achieve those tasks. Standards such as DoD-2167 can provide some insights for the correct approach.

A well-designed electronic portable instrument is based on *cognition*, which involves the mental task and computations used in operating the instrument within the expectations of the user. Cognition includes learning about the instrument on its functionalities. The user's short-term and long-term memory about the sequences of operations constitute part of the cognition process. Ease of usability and utility of an instrument can differentiate the product in the market. Designers may have to prototype and test the instruments to assess the simplicity and minimal operation requirement of the device. Also, image and ownership of the instrument shapes the user perception, leading to his or her high level of commitment to the device. The image of an instrument may be shaped by physical appearance, perception, and operation. The operation must be logical, appropriate, and consistent, leading to smooth flow and pleasing to the ear, eye, and touch. Ownership is commitment to the product, which is mainly shaped by cost, good design, and good support for the product.

An electronic portable instrument may not need to have a full keyboard and a full feature display. Therefore, the user interface may be *menu driven*. A menu-driven instrument usually has a limited number of controls and is suitable for use by untrained operators. The user interface may also be *command driven*. Commands are entered at the full keyboard. These instruments may be complex, and they offer much greater flexibility. A good deal of hardware and software design aspects is involved in modern electronic portable instruments irrespective of whether they are menu or command driven.

Hardware involves arrangements of input components, switches, knobs, displays, and screens. They all assist in intuitive mapping, visual cue and utility, and control of the instrument. The input components can involve mouses or joysticks, or they can be as simple as a few simple push-button keyboards or as complex as full keyboards and complex touch-panel displays.

Keyboards are essential parts of the user interface. They are basically a special application of switches. In electronic portable instrument applications, robust and easy-to-use keyboards should be selected to be able to withstand harsh environmental conditions and abusive use by untrained operators. Keyboards are most suitable for rapid data entry. In most electronic portable instruments, the keyboards are simplified due to the small sizes of the devices. Touch screens require minimum training for effective use and require small space, but they need extensive software and a good understanding of user capabilities.

The software for human interface is often difficult to develop. It involves visibility, feedback, reduction of errors, and error recovery. Programs are usually developed in modular forms and integrated as a whole in the final product.

4.6.1 Displays

Suitable displays are very important in electronic portable instruments since they are one of the main means of man–machine interface. The most common displays in portable instruments are the light-emitting diodes (LEDs) and liquid crystal displays (LCDs).

4.6.1.1 Light-Emitting Diodes

Light-emitting diodes are small, low-voltage, low-power components that provide visual indication. They are simple to implement on circuit boards or at the front panels. The forward voltage drop across LEDs can be between 1 and 2 V, and it may require current-limiting resistors (100 to 2000 Ω) in series with the LED to drive it from digital circuits. However, some LEDs already have current-limiting resistors built inside, thus reducing additional circuit requirements. There are many LED arrays to display alphanumeric characters, but they can consume considerable amounts of power and cannot normally be driven by CMOS circuits.

4.6.1.2 Liquid Crystal Displays

Liquid crystal displays are low-power devices and therefore are most suitable in portable instruments. LCDs can provide large panels of characters

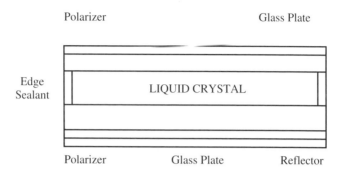

FIGURE 4.25
A typical LCD with basic components.

and custom-made shapes and symbols. LCDs require AC drives, but these drives generally are built into the LCD module to make them easy to interface with digital components. LCDs are less expensive and consume much less power than equivalent LED displays. They can be driven digitally to give flexibility in the man–machine interface. Many electronic portable instruments come with some intelligent LCD units that contain their own control logic circuits, character displays, and data storage inside. Since LCDs are an important part of electronic portable instruments, detailed information is provided below.

Basically, the LCD is a sandwich of an organic compound (can be nemadic, smectic, dichroic, dynamic scattering, or cholesteric) between clear plastic or glass plates equipped with suitable electrodes such as indium oxide and a back-glass plate electrode. When a voltage field activates the electrodes, the long cylindrical organic molecules align themselves with the segment electrodes. The sandwich has a front vertical light polarizer and a back horizontal light polarizer followed by a reflector. The reflected light cannot pass through the aligned molecules, and the activated segment appears dark. Figure 4.25 illustrates a typical complete LCD unit.

Some of the common features in the construction of LCDs are as follows:

- *Substrate glass*: In the majority of LCDs, use SiO_2 barrier-coated soda lime glasses having compositions similar to that of normal window glass. The typical thickness of the substrate is 0.3 to 0.5 mm.

- *Electrodes and patterns*: Indium oxide- and tin oxide-transparent electrically conducting electrodes are used to generate patterns. These thin transparent films are deposited to adhere on the glass in a variety of ways, such as thermal evaporation, reactive sputtering, and chemical vapor deposition. The electrode patterns on front and rear glass plates are designed in such a way that in the assembled cell, they overlap each other only in the desired information format. In most of LCDs, electrode patterns are formed by screen printing or photolithographically covering the desired portions of the coating on glass and chemically etching the unwanted portions.

- *Alignment and aligning materials*: Alignment of liquid crystals is achieved in a number of methods, for example, twisted nemadic, homogeneous, homeotropic, or grandjean. Some of the agents can be long-chain polymers like polyvinyl alcohol (PVA), proper silanes, and so on.

- *Sealing and spacer materials*: Liquid crystals between the glass plates are spaced and sealed by suitable sealants and spacers of uniform thickness. Glass frit sealing and epoxy sealing are the most popular ones that are effective against moisture.

- *Polarizers*: Polarizers appear to be the weakest link in LCDs, subject to bleaching, delamination, scratches, cosmetic blemishes, and air bubbling at high temperatures and humidity. In fact, an ideal polarizer with 100% efficiency cannot transmit more than 50% of light. Most polarizers are iodine-based dyes deposited as thin films over the glass. Diffusion types are also available.

- *Low-level AC voltage power LCDs*: A segment driver consists of a CMOS-exclusive OR gate whose output goes to the segment electrodes. Basically, the whole device functions as a capacitor; therefore, it is important not to cycle the signal at too high frequencies. If the frequency is too low, the display will flicker. A typical clock frequency ranges from 40 to 60 Hz. Each segment driver is commonly built into an LCD driver chip that comes with the display. For example, a typical four-digit LCD driver such as National Semiconductor's MM74C945 has a counter/latch/decoder driver built in.

Many LCDs contain glass plates with the desired electrode patterns, alignment coating, liquid crystal material, spacer, sealant, and reflectors if necessary. In addition, some LCDs require polarizers. The presentation of visual information by a display requires a method or methods of activating multiple positions in a display medium. The act of transmitting signal information through the display medium is termed *addressing*. Addressing may be achieved in a number of ways, such as beam, grid, direct, shift, and matrix. Suitable addressing techniques are determined by the nature of application requirements. For example, direct wiring is used in watches and industrial displays by using time division multiplexing methods.

A commonly used addressing method is *active matrix addressing*. In this method, local integrated circuits are used for the addressing of individual cells. The cell control devices are placed in series with the liquid crystal element acting as a capacitive load. Typical devices are zinc oxide varistors; they can be manufactured by MOSFET, CdSe thin film technology or TFT, and silicon TFT methods.

In Table 4.11, a list of display and user interface manufacturers and suppliers with telephone numbers is provided. Only a few names are provided;

TABLE 4.11

User Interface and Display Manufacturers and Suppliers

Type	Manufacturers and Dealers (Phone No.)	
Components (audio, buzzers, speakers)	AKM Semiconductors (408-436-8580)	Neutric USA Inc. (732-901-9488)
	Challenge Electronics (631-595-2217)	Rohm Electronics (858-625-3600)
	E-A-R Speciality (877-372-4332)	Shokai Far East Ltd. (914-736-3500)
	Golden Pacific Electronics (714-993-6970)	Technoland Inc. (408-992-0888)
Backlights and drivers	Buhite (973-423-2800)	Monolithic Power Systems (408-243-0088)
	Chip Supply Inc. (407-298-7100)	National Semiconductor (408-721-5000)
	DCI Inc. (888-824-9412)	
	Metromark (952-912-1700)	Omron Electronics (800-556-6766)
		Schuster Electronics (513-489-1400)
LEDs	Agilent Technologies (408-654-8675)	Rohm Electronics (858-625-3600)
	Densitron Corp. (562-941-5000)	Sharp Microelectronics (360-834-2500)
	Mouser Electronics (800-346-6873)	Toshiba America (949-455-2000)
LCDs	Advanced Display Systems (972-442-4586)	Mosaic Industries (510-790-1255)
		Omron Electronics (800-556-6766)
	Display Technologies Ltd. (760-918-6722)	Seiko Instruments (310-517-7771)
	Emerging Display Technologies (949-206-0255)	Sharp Microelectronics (360-834-2500)
		Toshiba America (949-455-2000)
	Fujitsu Microelectronics (800-866-8608)	
	Kopin Corp. (508-824-6696)	
Touch screens	Computer Dynamics (864-627-8800)	Laube Technology (805-388-1050)
	Global American Inc. (603-886-3900)	Mosaic Industries (510-790-1255)
	Grayhill Inc. (708-354-1040)	3M Touch Systems (978-659-9000)
	L-3 Communications (714-758-0500)	Tri-M Systems (604-527-1100)
Keyboards	ACT Components Inc. (805-987-2960)	Grayhill Inc. (708-354-1040)
	Cherry Industrial Products (262-942-6500)	ICS Advent (858-677-0877)
	Chirque Corp. (801-467-1100)	PI Engineering Inc. (517-655-5523)
	Fujitsu Microelectronics (800-866-8608)	Viziflex Seels Inc. (800-307-3357)
Keypads	Bill West Inc. (203-261-6027)	Metromark (952-912-1700)
	DataMetrics Corp. (407-251-4577)	Noble USA Inc. (847-364-1700)
	Fujitsu Microelectronics (800-866-8608)	PI Engineering Inc. (517-655-5523)
	Laird Technologies (570-424-8510)	Xycom Automation (734-429-4971)

Source: Portable Design, March 2002, PennWell.

the list is by no means exhaustive. Many firms provide more than one or all types, but repetition is avoided whenever possible.

4.7   Construction and Assembly

The physical construction, packaging, and enclosure of electronic portable instruments are governed by cost, performance, size, weight, power consumption, serviceability, reliability, user interface options, and ruggedness and versatility. Packaging is the mechanical structure, support, and orientation of an electronic portable instrument. The environments in which the portable instruments are likely to operate may frequently change and can be very harsh. Therefore, these instruments must have good internal and external packaging with strong enclosures and well-deliberated design features that balance performance against power consumption, weight against ruggedness, and simple user interface characteristics against the convenience of operations. All the main parts of the instruments — the power supply, sensors, electronic components, processors, packaging, and enclosures — must be designed and tested carefully.

4.7.1 Enclosures and Casing

Physical environment is an influencing factor in the construction and enclosure of electronic portable instruments. Many industrial, military, and medical products have to comply with regulations and standards that specify factors such as temperature, vibration and shock, humidity and exposure to water, corrosion, pressure, EMI, radiation, etc. Examples of such regulations and standards are MIL-STD-461D for enclosures, MIL-STD-901C for shock testing, and National Electrical Manufacturers Association (NEMA) rules.

For special environmental conditions, various requirements on casing and enclosures are imposed by various authorities such as NEMA. The following NEMA rules can be taken as typical guidelines to develop enclosures for industrial and other types of electronic portable instruments:

- TYPE 1 — *General purpose*: A general-purpose enclosure is intended primarily to prevent accidental contact with the enclosed apparatus. It is suitable for general-purpose applications indoors where it is not exposed to unusual service conditions. A Type 1 enclosure serves as protection against dust, light, and indirect splashing, but is not necessarily dust-tight.
- TYPE 2 — *Drip-tight*: A drip-tight enclosure is intended to prevent accidental contact with the enclosed apparatus and, in addition, is so constructed as to exclude falling moisture or dirt.
- TYPE 3 — *Weather resistant*: A weather-resistant enclosure is intended to provide suitable protection against specified weather hazards. It is suitable for use outdoors.
- TYPE 4 — *Watertight*: A watertight enclosure is designed to meet the hose test conducted with set procedures. A Type 4 enclosure is suitable for application outdoors on ship docks and in dairies, breweries, etc.
- TYPE 5 — *Dust-tight*: A dust-tight enclosure is provided with gaskets or their equivalent to exclude dust. A Type 5 enclosure is suitable for application in steel mills, cement mills, and other locations where it is desirable to exclude dust.
- TYPE 6 — *Submersible*: A Type 6 enclosure is suitable for application where the equipment may be subject to submersion, as in quarries,

mines, and manholes. The design of the enclosure will depend on the specified conditions of pressure and time.

- TYPE 7 (A, B, C, or D) — *Hazardous locations, Class I*: These enclosures are designed to meet the application requirements of the National Electrical Code for Class I hazardous locations.

- TYPE 8 (A, B, C, or D) — *Hazardous locations, Class I (oil immersed)*: These enclosures are designed to meet the application requirements of the National Electrical Code for Class I hazardous locations where the apparatus may be immersed in oil.

- TYPE 9 (E, F, or G) — *Hazardous locations, Class II*: These enclosures are designed to meet the application requirements of the National Electrical Code for Class II hazardous locations.

- TYPE 10 — *Bureau of mines (explosion-proof)*: A Type 10 enclosure is designed to meet the explosion-proof requirements of the U.S. Bureau of Mines. It is suitable for use in gassy coal mines.

- TYPE 11 — *Acid and fume resistant (oil immersed)*: This enclosure provides for the immersion of the apparatus in oil such that it is suitable for application where the equipment is subject to acid or other corrosive fumes.

- TYPE 12 — *Industrial use*: A Type 12 enclosure is designed for use in those industries where it is desired to exclude such materials as dust, lint, fibers and flyings, oil seepage, or coolant seepage.

- TYPE 13 — *Oil-tight and dust-tight, indoor*: A Type 13 enclosure is intended for using indoors to protect against lint, dust, seepage, external condensation, and spraying of water, oil, or coolant.

Most consumers prefer electronic portable devices with enclosures complying with ingress progress (IP) tests, which originated from Europe. In IP tests, according to IEC-60529 (2001–2002), the enclosure can be defined as a combination of two numerals: the first shows the protection against solid foreign objects, and the second defines the protection against water. The numerals indicate the following:

First numeral:

     0   Nonprotected

     1   Protected against solid foreign objects of 50-mm diameter and greater

     2   Protected against solid foreign objects of 12.5-mm diameter and greater

     3   Protected against solid foreign objects of 2.5-mm diameter and greater

     4   Protected against solid foreign objects of 1.0-mm diameter and greater

  5 Dust protected

  6 Dust-tight

Second numeral:

  0 Nonprotected

  1 Protected against vertically falling water drops

  2 Protected against vertically falling water drops when enclosure tilted up to 15°

  3 Protected against spraying water

  4 Protected against splashing water

  5 Protected against water jets

  6 Protected against powerful water jets

  7 Protected against the effects of temporary immersion in water

  8 Protected against the effects of continuous immersion in water

Two additional letters can be used optionally, the first one to indicate the degree of protection against access to hazardous parts, the second one to indicate the degree of protection in special conditions of operation. In instrumentation, these additional letters are not used. The typical value for indoor cabinets in an instrument room is IP20 or IP21; the typical value for field-mounted transmitters is IP65. An instrument complying with IP65 is dust-tight (6) and protected against water jets (5).

Electronic portable instruments must be strong and versatile. They are susceptible to damage caused by the users or the environment. For example, they may be exposed to a harsh environment while being operated (under water, high temperature, reactive gases), or may simply be abused or dropped by the user. Therefore, to reduce the chance of damaging or losing an expensive and very sophisticated tool, a proper enclosure is needed.

Also, the enclosure itself has a very important role since all the input and output devices are planted in it. Designing a user-friendly enclosure can help the durability and enhance the man–machine interface. User-friendly enclosure means that it helps the user with providing clear information about the feature of the detector such as the on–off button, clear and readable output display, and noticeable alarms. Moreover, the enclosure itself is an important part of the device since it carries the power supply. The position of the power supply within the instrument is also important for efficiency purposes, considering the need to replace or charge the battery. An important reason for selecting good casing is due to the need of avoiding any interference from the environment. Those interferences are varied. It can be radio or signal interference, light interference, radioactivity, etc.

In addition to weight and size, reliability, system voltage, power distribution, and failure mode analysis of electronic portable instruments, the maintenance and disassembly require special attention for serviceability purposes. For the serviceability human factors, reliability, qualified personnel, and necessary tools should be considered.

4.7.1.1 Weight and Size

Normally, electronic portable instruments will be expected to have features such as a good operating life, easy repairability, and ruggedness in construction so that damage against mistreatment can be minimized. All these features have weight and size implications. The weight and size of an instrument are usually determined and finalized in the final design stages. For desirable weight and size, the following factors may be observed:

1. Batteries and power supplies usually tend to be the heaviest and bulkiest components of the instrument; therefore, they need to be selected carefully to meet the requirements. Whenever possible, smaller batteries must be selected since they allow convenience in locations inside the instrument and ease in the weight distribution. Also, as a rule of thumb, the selection of commercially easily available standard type batteries has some advantages. Examples of common standard type batteries are the D size nickel and cadmium types. These batteries are easily available; hence, the users of the instruments can have ready access for purchasing replacements.

2. The case of the instrument is another major contributor to the weight and size. Although a smaller case may be desirable, it must be strong enough to protect the components inside. There are many manufacturers producing cases particularly for portable instruments. Many cases for portable instruments are made from aluminum or fiberglass with rounded corners and ribbed sides. A conductive metal coating inside the fiberglass provides EMI protection.

3. Carefully designed printed circuit boards are necessary. These boards should be arranged relative to each other carefully and located conveniently such that back-planes and connectors are minimized.

4. Other necessary parts for thermal, acoustic, and electromagnetic insulations may add extra weight and cost to the instrument; therefore, they need to be considered carefully in the early stages of the design.

4.7.2 On-Board Circuit Layout

Packaging, component layout, and the routing of signal traces greatly affect the operation of electronic portable instruments. Poorly selected placement, packages, and routing increase the noise, susceptibility, and EMI of the device. This can influence the operations of components that are already optimized and used sparingly.

4.7.2.1   Wiring of Connectors and Components

Wiring and cabling are used for interconnecting components of electronic portable instruments. Many portable instruments fail due to the mechanical failure of connectors, solder joints, and cabling. Generally wiring is decided by current-carrying capacity, requirement for mechanical strength, insulation properties, and shielding requirements.

In electronic portable instruments, mostly copper wires are used. Weight and complexity of cabling are important. Careful routing of cables and securely tying them down reduces mechanical problems. Routing is usually optimized for the shortest path, easy servicing, and EMI shielding.

Connectors are used for mechanical support for electrical connections. Connectors involve metal-to-metal contacts that raise several concerns, such as contact resistances and corrosion. Corrosion, for example, can take place due to galvanic action and electrochemical effects. Plating contacts reduces contact resistance and prevents corrosion and oxidation. Often, gold, tin, and nickel are used for plating. It is essential that portable instruments contain well-selected robust and good connectors.

4.7.2.2   Circuit Boards

Circuit boards combine electronic components through electrical connections and provide mechanical support into functional systems. There are many different configurations: wire wrap, chip-on-board, multichip modules (MCMs), and printed circuit boards. Wire wrap is suitable for prototype development because connections can be changed, thus allowing modification and corrections of circuits. PCBs have etched and plated connections that are suitable for mass production. MCMs achieve higher levels of circuit density by bonding ICs onto a substrate. The compact packaging improves signal speeds and reduces capacitive effects.

As in the case of nonportable instruments, portable instruments are developed by breadboarding and prototyping first. The developments take place on the solderless and wire-wrap boards. Once a successful prototype is constructed, in the operational instruments, either single-sided or double-sided PCBs are used.

PCBs are made of layers of glass cloth impregnated with epoxy or polymer resin. A copper-clad foil is laminated on one or both sides. The material in a PCB affects circuit performance and mechanical reliability; therefore, laminates with low dielectric constants result in faster signal operation. The deposited layer of copper coats vies, forms signal traces, and provides conducting paths. Unwanted copper foils are etched by electrochemical reaction to remove unwanted copper foils. The board is covered by photoresist to support the photographic patterns of the signal traces. The entire board, except solder pads, is covered by a resist to eliminate moisture and scratches.

PCBs can be single sided, double sided, or multilayered. Signal-sided PCBs are easy to produce. The circuit board does not contain plated-through holes since signal traces are on only one side. Double-sided PCBs have signal traces on both sides of the circuit board. They contain higher circuit connection densities than single-sided PCBs. They also support higher frequencies. Multilayer PCBs contain a stack of boards and support very dense circuit connections. They support high-frequency circuits. In general, all types of PCBs are less bulky, easy to manufacture, rugged, reliable, and electrically less noisy. Cost depends on the volume of production, process, and quality of components used.

Soldering and placement of components affect the circuit operation, manufacturing ease, and probability of design errors. Soldering can be done by trained human operators or by automatic machines. There are two forms of soldering: wave soldering and vapor-phase reflow soldering. Wave soldering coats the bottom of a PCB in a bath of molten solder attaching all components to the board simultaneously. The vapor-phase reflow method melts a solder paste on each pad and solders surface mount components to PCBs.

There are many different types of computer-aided design (CAD) software for PCB design, such as Protel; much of it requires extensive artwork. However, in order to reduce contacts and extra wiring, PCBs must be carefully designed, thus requiring minimum external routing.

While on the subject of routing, in the PCB design, circuits should be grouped according to their characteristics to maintain the correct operation of each circuit. For example, high-frequency circuits should be grouped near connectors to reduce the path length, cross talk, and noise; low-frequency and low-power circuits should be placed away from high-current and high-frequency circuits; also, analog circuits should be separated from digital circuits so that current pulses from the digital circuits cannot corrupt sensitive analog parts.

4.7.3 Interference and Noise Protection

Noise is undesired electrical and magnetic activity that couples one circuit to another. It always involves a source, a coupling mechanism, and a receiver. The noise sources can be external or internal. Examples of external noise sources are power lines, motors, high-voltage equipment such as televisions, radio transmitters, mobile telephones, etc. Internal noise is generated by any current-carrying conductors, buses, clocks, and so on. The coupling mechanism can be conductive, thus requiring a closed-circuit loop; inductive, such as mutual inductance; capacitive, associated with high voltages and electric fields; or electromagnetic.

External electromagnetic interference and external fields can affect the performance of the instruments seriously. It is known that moderate magnetic fields are found near power transformers, electrical motors, and power lines. These small fields produce currents and voltages in the electronic circuits. One way of eliminating external effects is accomplished by proper grounding of signals inside the instrument and by effective means of shielding. Grounds and shields improve safety, and reduce internal and external interference from noise. There are various standards regulating noise and interference issues, such as CNELEC, IEC-801, VDE, and VG NORM of Europe; FCC and MIL-STD-461D of the U.S.; and VCCI of Japan. In this section, discussions will mainly concentrate on grounding and shielding.

In electronic portable instruments, conductive noise is the most common form of noise. Conductive coupling occurs at lower frequencies and is commonly caused by improper grounding. At high frequencies, conductive circuit paths can have significant capacitive and inductive properties that dominate the coupling.

Due to power and space availability, electronic portable instruments have a tendency to use the minimum number of components; thus, the noise may be expected to be minimum. If noise is a problem, identification of dominant sources of noise and elimination of noise are important during the design and service.

Almost all sensors produce some form of noise in addition to useful signals. As discussed in Chapter 1, they are very likely to generate either systematic or stochastic distortions of the output signals. The noise and distortion of sensor signals can be related to dynamic characteristics, transfer functions and linearities of sensors, and manufacturing tolerances and quality of materials used during manufacturing.

The noise in electronic portable instruments can be inherent or interference noise. The inherent noise, mostly Gaussian, may be due to thermal noise, shot noise, excess noise, burst noise, spot noise, etc., as explained in Chapter 1. Other possible sources of noise and distortions can be due to not well taken care of input offset voltages and bias currents and drift of these two in time. The inherent noise can be minimized by selection of good components with low-noise properties and the use of filters.

Transmitted noise is generally known as the electromagnetic interference. The EMI can affect the performance of electronic portable instruments severely. In many cases, electronic portable instruments are designed to operate in intense electrical and magnetic fields such as electric motors, high-voltage power supplies, high-voltage switching, etc. They may be used in environments where strong electromagnetic fields exist, such as radio or television stations and microwave transmitters. The transmitted noise can be counteracted by various techniques, for example:

1. By filtering, decoupling, shielding of leads and components.
2. By using differential and ratiometric techniques. In differential techniques multiple sensors are used, some of which are exposed to transmitted noise, while others are kept as references.
3. By use of guarding potentials and elimination of ground loops.
4. By eliminating parasitic capacitances.
5. By eliminating reflections in cables using suitable termination techniques.

6. By isolating the inputs and outputs using optical or magnetic isolation circuits.

7. By isolating the switching components from other sensitive circuits and the physical reorientation of leads that carry high currents and high-frequency signals.

8. By the deliberate choice of low impedances wherever possible.

9. By designing PCBs to minimize transmitted noise. One method of PCB design is the *ground-plane* concept, which covers a large portion of the board with grounded metal foils.

10. By selecting ground and power supply traces that are wide and direct.

11. By guarding the inputs of high-impedance and noise-sensitive parts of the circuits.

12. By surrounding the input terminals of highly sensitive semiconductors with guard rings etched into suitable patterns.

13. By placing low-power, high-impedance, and sensitive sensors closer to the preconditioning circuits.

14. By selecting the serial communication methods in digital instruments.

15. By installing the complete device inside a suitable metal case.

16. By using bypass capacitors.

Some of these elimination techniques are self-explanatory, but others such as bypass capacitors, grounding, and shielding require explanations.

4.7.3.1   Bypass Capacitors

Bypass capacitors are often used to maintain low-power supply impedance at the point of load. Parasitic resistance and inductance in supply lines can cause high-power supply impedance. At high frequencies, the inductive parasitic can become troublesome, resulting in circuit oscillations and ringing. The bypass capacitors are important to eliminate these high-frequency oscillations that may be transmitted to other parts of the device. Bypass capacitors must be selected carefully to eliminate oscillations, ringing, and cross talk, particularly in digital systems.

4.7.3.2   Grounding

Grounding provides a signal reference and is often the most important concern in laying out a circuit board. The objective in the design is to reduce the voltage difference between the components with respect to a reference point. It is always good practice to separate grounds for analog and digital circuits. Grounding can be arranged to be single-point grounding if the instrument has a high-impedance ground structure, or multipoint grounding with low-impedance ground structures.

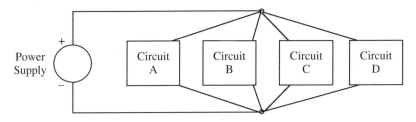

FIGURE 4.26
Single-point grounding.

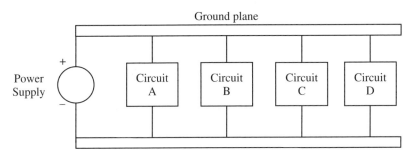

FIGURE 4.27
Multipoint grounding.

In single-point grounding, as shown in Figure 4.26, the separate ground conductors isolate the noise in the return paths of separate circuits and eliminate the ground loops. Single-point grounding is appropriate low-current and low-frequency (typically less than 1 MHz) circuits.

Multipoint grounding is usually applied in digital devices in which ground planes or grids are used. It is preferred at high frequencies. A ground loop is a complete circuit that comprises a signal path and a ground structure, as shown in Figure 4.27. It is most suitable for circuits above 10 MHz. Ground planes dimensionally need to be designed carefully and located strategically to minimize noise in the instrument.

4.7.3.3 Shielding

Shielding suppresses noise energy or prevents it from coupling between circuits. The noise energy can be through magnetic flux, electric field, or electromagnetic wave propagation. Shielding can be provided by the careful design of circuit boards. For example, in a PCB design, a return (or ground) plane proves to be the most effective shield for any circuit. Power and return planes provide circuit paths with the lowest impedance, which reduces electromagnetic radiation, noise, and cross talk. Shielding can be provided by suitable mumetal and steel enclosures. It should be remembered that prevention of coupling is more effective than suppression.

4.7.3.3.1 Electromagnetic Shielding

Electromagnetic shielding reduces emission and reception. Emission sources are usually radio and microwave transmitters and other high-frequency circuits. Electromagnetic interference begins as a conductive noise in wires and becomes radiative again. Good circuit layout and routing, shielding of enclosures, and limiting the bandwidth can significantly reduce EMI.

4.7.3.3.2 Electric Field (Capacitive) Coupling

Electric field (capacitive) coupling can effectively be blocked by proper electrostatic shielding using suitably shaped metals. This is because an electric charge cannot exist on the interior of a closed conductive surface. Also, shielding reduces noise coupling by rerouting the electric charge in an electric field. The effectiveness of shielding can be improved by reducing the noise voltage and frequency and signal impedance at frequencies less than 1 MHz.

The electrostatic shields serve two purposes. First, they confine noise to small regions, and by providing adequate grounding, they prevent it from propagating. Second, if noise exists, shielding prevents the noise from getting into sensitive electronic circuits. In both cases, while designing suitable means of shielding, it is important to identify the noise sources, the nature of the noise, and how it is coupled with the sensitive circuits.

There are several useful rules in designing electrostatic shields:

- The shield should be connected to the reference potential of the circuit that it is shielding.
- In the case of cables, the shield must be connected to the signal reference terminal at the source side.
- If there are multiple shields, they all must be connected at a single point at the signal reference node.
- If there are a number of independent signals, separate and independent shields must be used for each signal. If they share a common reference potential, they must be connected to a single point using separate wires with short lengths.
- The shield must be grounded at one point only. Multiple grounding of the shield can cause ground currents to flow.

4.7.3.3.3 Magnetic Fields (Inductive)

Magnetic fields (inductive) generated by the external sources may interfere with the instruments. Also, they can be produced by currents in wires and more strongly by the coils inside the instrument itself, propagating as mutual inductance between circuits. Proper shielding may significantly reduce noise resulting from electrostatic and electric fields. However, it is much more difficult to shield against a magnetic field, since it can penetrate through conducting materials, although its amplitude may decrease exponentially.

Magnetic noise coupling largely depends on the loop area and current within the emitting and receiving circuits. Therefore, effective inductive shielding minimizes the loop area and separates the circuits. Also, twisting of the signal-and-return conductors in cables reduces the mutual inductance and improves the common mode coupling. Coaxial cables at high frequencies (above 1 MHz) provide a minimal loop area.

To provide magnetic shielding, one or more shells of high-permeability magnetic materials can be used as enclosures or metal components that surround the parts to be shielded. The ends of each shell are separated by insulations so that the shells do not act as single shorted turns, thus accommodating the flow of high currents. Similarly, in the case of multiple shielding, shells are isolated from each other by proper insulation. Metal enclosures may be heavy, expensive, and frequency dependent, but they may be the only solution in some electronic portable instruments.

Magnetic coupling may be a significant problem if there are two or more coils in the instrument. Magnetic coupling may be controlled by large spacing between coils, the relative orientation of coils, the shape of the coils, and shielding each coil separately. Time-varying magnetic fields can produce a significant current in stationary, closed conductors. Problem coupling must be attended carefully if there are inductive sensors in the instrument.

At high frequencies, alternating magnetic fields need to be screened by interposing highly conductive metal sheets such as copper or aluminum on the path of the magnetic flux. Eddy currents induced in the shield give a counter magnetomotive force (mmf) that tends to cancel the interfering magnetic field.

Practical guidelines to minimize magnetic interference are as follows:

- The receiving circuit must be placed as far as possible from the offending magnetic source.
- Wires should run perpendicular to the magnetic fields.
- Appropriate shielding materials must be selected that are suitable for the strength and frequency of the magnetic field. At high frequencies, aluminum, copper, or steel may be used as the shielding material. But at low frequencies, high-permeability magnetic materials such as mumetal or steel must be used.
- Twisted pairs of conductors must be used to eliminate magnetic effects of high currents carried by the conductors.
- Shielded cables must be used wherever possible.

In Table 4.12, a list of packaging, materials and other hardware, manufacturers, suppliers, and service providers with telephone numbers is supplied. Only a few names are provided; the list is not exhaustive. Many firms provide more than one or all types, but repetition is avoided whenever possible.

TABLE 4.12

Packaging, Materials and Hardware Manufacturers, Suppliers, and Service Providers

Type	Manufacturers and Dealers (Phone No.)	
Connectors (coaxial, filtered, PCB, USB, micro)	AVIEL Electronics (702-739-8155) AVX Corp. (843-448-9411) Hirose Electric USA (805-522-7958) JAE Electronics (949-753-2600) JST Corp. (800-947-1110)	Meritec (888-637-4832) Mouser Electronics (800-346-6873) RF Industries (800-233-1728) Schuster Electronics (513-489-1400) Yamaichi Electronics (408-456-0797)
Enclosures	Babcock Inc. (714-994-6500) Chip Supply Inc. (407-298-7100) Galco Industrial Electronics (248-542-9090) OKW Enclosures Inc. (800-965-9872)	Phillips Plastic Corp. (715-381-3344) SBST Technologies (760-438-6900) TENBA Quality Cases (718-222-9870) UFP Technologies Inc. (978-352-2200)
Shielding materials	Amuneal Manufacturing (215-535-3000) Cuming Microwave Corp. (508-580-2660) Delafoil Inc. (610-327-2312)	Orion Industries (978-772-6000) Tecknit Inc. (908-272-5500) Vanguard Products (203-744-7265)
Services (assembly, design, calibration)	ARMA Design (858-373-1320) Bit-7 Inc. (608-224-0377) DCI Inc. (888-824-9412) e-Digital Corp. (858-679-1504) Hybricon Corp. (978-772-5422)	KLN Klein Product (604-530-1491) Logic Product Development (612-672-0726) Midwest Micro-Tek (605-697-8521) Mosaic Industries (510-790-1255) Toronto Microelectronics (905-625-3203)
Testing (compliance, EMI/RFI, environment)	Analab LLC (579-689-3919) Elite Electronics (630-495-9770) GM Nameplate (800-366-7668) KLN Klein Product (604-530-1491)	Lineo Inc. (801-426-5001) Motorola Energy Systems (770-338-3742) Phoenix Co. of Chicago (630-595-2300) Schaffner EMC Inc. (732-225-9533)

Note: RFI = Radio frequency interface.
Source: Portable Design, March 2002, PennWell.

4.8 Software Aspects

In the past, most instruments were hardware driven with very little customized software support. Today, software is pervasive in the majority of electronic products, including in electronic portable instruments. Software accounts for up to 75% of projects involving microprocessors and microcontrollers. Modern portable instruments are largely software driven, and therefore can be programmed for the implementation of various forms of measurement techniques. Their front- and back-end analogy hardware too can be standardized and designed to cover large classes of similar applications.

Software is developed by writing appropriate programs from which various instructions and actions can be implemented. The instructions and actions are made from algorithms, which describe the general actions to be taken. Algorithms can be independent of each other and also independent of specific programming languages used. These algorithms can be complex or very simple, depending on the task at hand. In developing the algorithms and final software, the requirements of what they should do must be identified clearly in the initial stages. Once the software is written to address the prespecified requirements, the performance of the final product must be tested and modified if necessary. Also, software must be accountable, traceable, and flexible enough to accommodate changing requirements. There are various guidelines and standards for software development. Some of these are ISO 900 for quality; DoD-STD 2168 for software quality evaluation; and ANSI/IEEE STD 731, 828, and 830 for software quality assurance plans, software configuration management plans, and software requirement specifications.

In the design and construction of electronic portable instruments, three kinds of programs are necessary: (1) a host computer assisting in the design of hardware and software to develop the portable instrument, (2) programs developed during the prototyping to test operations of components, and (3) final operating software. In all cases, the programs can be developed using a variety of low-level and high-level languages, such as machine and assembly languages on the low-level side and Basic, Visual Basic, Ada, Fortran, C, VC, and C++ on the high-level side. The selection of a particular language depends on the knowledge and expertise of the development team, the hardware, and supporting tools.

Programs for electronic portable instruments are written for the management of firmware, peripheral interface and drives, operating systems, user interface, application programs, and so on. Tools available for the programming tasks include compilers, disassemblers, debuggers, emulators, monitors, and logic analyzers, as well as extensive software libraries. The software elements in a design may be listed as:

- Operating system and procedures
- Communication and I/O
- Monitoring process
- Fault recovery and other special procedures
- Diagnostic features
- Synchronization between tasks
- User interface and displays
- Built-in testing and verification procedures
- Fault tolerances
- Factors against misuse, etc.

Machine and assembly languages constitute instructions for a particular processor architecture, and hence closely reflect the physical operation of the processor. Especially, the machine codes can control individual bits and can set registers directly. Both machine and assembly languages are suitable for electronic portable instruments that have minimum memory and need a high execution speed. They allow the precise control of the peripheral devices, but writing programs in these languages is tedious and requires attention to exacting details.

High-level languages make programming much easier. They can handle details and provide good structure and readability. They are most suitable for large and complex systems, since they are capable of handling low-level details by simple instructions. High-level languages require more memory and their execution speeds are much slower. Beyond high-level languages, object-oriented programming (OOP, or simply OP) can be implemented to ease complexity in some applications.

Independent of the language selected, the design level involves several architectures: structured, object oriented, and computer-aided software engineering (CASE). Structured design involves planning, identifying logical parts, and interaction between the parts before starting to write the necessary codes and algorithms. Object-oriented architecture involves incorporation of data abstraction, modularity, and information hiding. CASE architecture involves blending of developed tools, improvement of structures, and managing the design of the software as a complete system.

Another type of software is the virtual instrument, such as LabView. Extensive discussions on this topic can be found in Chapter 3.

4.8.1 Software for User Interface

The user interface is a major concern in electronic portable instruments. The software plays a particular role in the interface and can make or break an instrument. The user interface includes issues such as help facilities, response time, and error handling. In portable devices, the interface should be correct, easy to follow, and understandable. A good layout technique should be used, only the relevant information should be displayed, and the help facilities should be on-line and context sensitive. Reversal of actions should be permitted, and critical actions should be verified. The command sequences should be useful, self-explanatory, and consistent. Error handling should be clear and give remedial action. Response time should have a reasonable interval and be consistent.

In Table 4.13, a list of software producers and suppliers with telephone numbers is supplied. Only a few names are provided; the list is by no means exhaustive. Many firms provide more than one or all types, but repetition is avoided whenever possible.

TABLE 4.13

Software Developers

Type	Manufacturers and Dealers (Phone No.)	
CAE/CAD	Conventor Inc. (919-854-7500)	Sonnet Software Inc. (315-453-3096)
	Fluent Inc. (603-643-2600)	Tanner EDA (626-792-3000)
	I-Logix Inc. (978-682-2100)	VAMP Inc./McCAD (323-466-5533)
Communication	BSQUARE Corp. (888-820-4500)	PCTEL Inc. (408-965-2100)
	EBS Net Inc. (978-448-9340)	Texas Instruments (www.ti.com)
	OSE Systems Inc. (866-844-7867)	Ubicorn Inc. (650-210-1500)
Compilers,	Byte Craft Ltd. (519-888-6911)	Microchip Technology (480-792-7200)
languages	Cypress Microsystems (425-939-1000)	Signum Systems (805-523-9774)
	Green Hills Software (805-965-6044)	SynaptiCAD Inc. (540-953-3390)
Digital signal	Blackhawk (877-983-4514)	International Data Sciences Inc.
processing	DSP Research Inc. (408-773-1042)	(401-737-9900)
	Forte Design Systems (408-487-9340)	Pentek (201-818-5900)
		Starcore (770-618-8522)
Emulation	Atmel Corp. (408-441-0311)	Hitachi Semiconductors (408-433-1990)
simulation	Elanix Inc. (818-597-1414)	OSE Systems Inc. (866-844-7867)
	Fujitsu Microelectronics (800-866-8608)	SoftTools Technology (408-973-7828)
Graphics image	Accelerated Technology (251-661-5770)	Mosaic Industries Inc. (510-790-1255)
processing	Alacron Inc. (803-891-2750)	Technoland Inc. (408-992-0888)
	KADAK Products Ltd. (604-734-2796)	VAMP Inc./McCAD (323-466-5533)
Operating	Annasoft Systems (800-690-3870)	I-Logix Inc. (978-682-2100)
systems	Bill West Inc. (203-261-6027)	Micro Digital Inc. (714-473-7333)
	CMX Systems (904-880-1840)	Signum Systems (805-523-9774)
	Eyring Corp. (801-561-1111)	Texas Instruments (800-477-8924)
Real-time	Accelerated Technology (251-661-5770)	Express Logic (888-847-3239)
OS/kernels	BSQUARE Corp. (888-820-4500)	Lineo Inc. (801-426-5001)
	EBS Net Inc. (978-448-9340)	Signum Systems (805-523-9774)

Source: Portable Design, March 2002, PennWell.

5

Examples and Applications of Portable Instruments

Introduction

In the previous chapters, we dealt with the necessary information on the design and construction of electronic portable instruments. We pointed out that today's portable instruments are largely digital, using microprocessors and microcontrollers. They use intelligent sensors and have wireless communication capabilities, which allows them to be networked with other digital devices. In this chapter, we bring all the information given in the previous chapters and provide some examples of portable instruments that are currently available in the marketplace. It is worth noting that electronic portable instruments and instrumentation is a fast developing field and there are many other devices available.

Recent progress in digital electronics, sensors, and communication technology has resulted in significant progress in the development and application of portable instruments. More and more novel portable instruments are appearing in the marketplace every year. They are becoming available in many diverse ranges and capabilities with improved or added features. In some areas, they are replacing conventional nonportable instruments. They can be manufactured small enough to fit in wristwatches, as altimeters and pulse measuring devices for personal use, and large enough to be mobile geological surveying and mineral prospecting instruments and instrumentation system in environmental applications.

Even within the same class of instrument that performs similar functionalities, there are many different types of portable instruments offered by a diverse range of manufacturers. It is not possible to cover all ranges of portable instruments, but in this chapter some typical examples will be given. In this chapter, nine different major classes of electronic portable instruments are discussed. These classes are:

Laboratory instruments and oscilloscopes
Environmental instruments

Chemical instruments

Biomedical instruments

Mechanical, sound, and vibration instruments

Radiation instruments

GPS and telemetry-based instruments

Computer-based and intelligent instruments

Domestic, personal, hobby, and leisure portable instruments

Some details of electronic portable instruments in these classes are given in the ensuing sections and subsections.

5.1   Laboratory Instruments

Laboratory instruments were one of the first and most commonly available portable instruments. They were used in the measurements of currents, voltages, and resistances, such as the well-known analog AVO meters. Today, there are many different types of electronic portable instruments suitable for laboratory applications, ranging from simple multimeters to measure currents and voltages to highly advanced digital storage oscilloscopes. Some of these portable laboratory instruments are discussed next.

5.1.1   Multimeters

Multimeters are one of the simplest, oldest, and most useful of the multifunctional electronic portable instruments. They can be found throughout the electronic and electrical industries, in laboratories and maintenance shops, in repair facilities, and in private homes. They are usually the first instruments to resort to for troubleshooting any electrical and electronic problems.

A classic multimeter performs three basic measurements: voltage, current, and resistance. Some microprocessor-based multimeters include frequency measurements; component evaluation functions, such as transistor and diode tests; and inductance and capacitance measurements, and they may have data-logging functions and serial ports to interface with computers. In some cases, multimeters are equipped with special sensors to enable measurement of high currents, temperatures, insulation resistances, and magnetic fields.

Digital multimeters (DMMs) nowadays are widely used, despite, historically, the fact that almost all the multimeters were analog. A standard DMM is designated to measure voltage, current, and resistance; however, it can contain additional features depending on the price and functionalities required.

The heart of most of the digital multimeter is the direct current (DC) voltage measurement circuit that is augmented by a current-to-voltage converter (typically, low-resistance shunts), a current source for measuring resistance, and a rectifier or a root mean square (rms) converter to perform the alternating current (AC)-to-DC conversion. Signal conditioning circuits include attenuation/amplification to extend the dynamic range of the instrument. The voltage measurement circuit contains an analog-to-digital (A/D) converter, which may be critical when designing high-accuracy digital multimeters. Existing A/D converters have up to 22-bit resolution at low conversion frequencies of up to 100 Hz, which allows implementation of 6 ½ digit readout DMMs. The sizes, features, and accuracies of multimeters vary widely, but some handheld ones can now rival the range and sensitivity of bench-top instruments comfortably. Most DMMs also can measure low-frequency AC signals. This is accomplished by rectifying the AC input signals and measuring their DC average values. The "true rms" DMMs calculate and display the rms value of the input, as long as the input signal is within the specified crest factor limits. The DMM performance is frequently specified in the number of displayed digits, which is the number of 9's the meter can measure plus the partial digit display, which cannot reach full 9, but still extends measurement range beyond the integer number of digits.

Many DMMs are capable of making very accurate measurements, but care must be taken to avoid common measurement errors. One of the most common sources of error is the loading, which is caused by insufficient meter impedance, especially at high-frequency signals. This type of error results from the meter drawing enough energy from the circuit under test to cause excessive voltage drops, and hence appreciable changes occur on the actual measurement values. Other possible errors include thermal electromotive force (emf), which is produced by connecting dissimilar metals. Errors such as settling time errors during high-resistance measurements and errors due to the change in internal resistor divider values because of the excessive power dissipations need to be considered carefully.

A typical commercial digital multimeter is depicted in Figure 5.1(a) and a range of multimeters is given in Figure 5.1(b). Typical characteristics of such devices are:

- Digital display with analog bar graph
- Backlight display and beeper
- Frequency, capacitance, resistance and diode measurements
- AC/DC measurements with .07% basic accuracy
- DC voltage range: 1 mV to 600 V, ±0.07%
- AC voltage range: 1 mV to 600 V, ±1.0%
- DC current range: 1 mA to 10 A, ±1.0%
- AC current range: 10 mA to 10 A, ±1.5%
- Resistance range: 0.1 Ω to 40 MΩ, ±1.9%

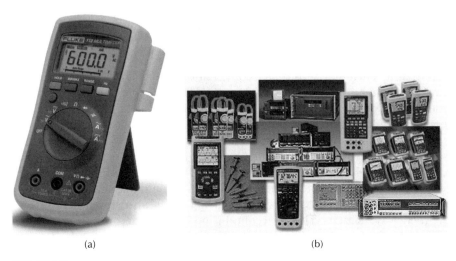

(a)   (b)

FIGURE 5.1
(See color insert following page 320.) (a) A digital multimeter. (b) A range of multimeters. (Courtesy of Fluke Electronics, http://www.flukecanada.ca/)

- Capacitance range: 1 nF to 10,000 μF, ±1.9%
- Frequency range: 0.01 Hz to 100 kHz, ±0.1%
- Operating temperature: –10°C to 50°C
- Weight: 350 g
- Battery life: Alkaline 300 hr

5.1.2 Power and Energy Meters

Historically, power measurements were done by portable wattmeters or by means of the complex arrangement of a number of instruments such as ammeters and voltmeters. Nowadays, many forms of digital single-unit handheld power meters are available, such as the power meters of Fluke Electronics of Canada, NanoVip by Elcontrol of Italy, and Valhalla of the United States. Many of these devices are suitable to use in single- and three-phase voltage systems. They can measure and display the values of currents, power in watts, energy in the kilowatt-hour, cost, power factor (pf), total harmonic distortions, transients, spikes, frequency, duty cycles, and power cycles. They can store information in nonvolatile memory for further analysis.

Some of these instruments can operate as harmonic energy analyzers. They track percent of total harmonic distortion (%THD), which indicates the degree of nonsinusoidal waveshapes that voltages or currents exhibit. As line disturbance analyzers, they track voltage or current disturbances. In this case, the amplitude, duration, and rise time are monitored; the largest transient data are logged; and swells and sags are detected and logged. As

FIGURE 5.2
(See color insert following page 320.) A range of digital powermeters. (Courtesy of Fluke Electronics, http://www.flukecanada.ca/)

power–energy data loggers, they track power consumption, energy, cost, harmonic distortion, true watts, kilowatt-hours, true RMS volts, amps, and true power factor. In some instruments, the graphs and individual waveforms can be uploaded to a personal computer (PC) for analysis.

A range of typical digital powermeters are illustrated in Figure 5.2. A typical portable power energy meter has the following features:

- Three voltage inputs and four current inputs — A, B, C and neutral
- Data logged may be uploaded to a computer for display and analysis
- Internal rechargeable battery pack can operate unit for 10 h between charges
- PC control/analysis — includes the serial communications interface hardware and input/output (I/O) cabling with the software
- Frequency and duty cycle display
- Accuracy: volts/amps/frequency, 0.5% nominal
- Power, energy, cost: 1.0% nominal
- Inputs: 0 V to 600 V_{RMS}, three phase; 0.01 to 3000 A, three phase; 45 to 66 Hz
- Transients: 32 μsec minimum, 2000 V_{pk}

5.1.3   Oscilloscopes

Standard oscilloscopes are used to measure signals in voltage forms. On the simplest level, an oscilloscope has a pair of terminals for connecting the voltage to be measured and a display, which produces a graph of the voltage as a function of time. Controls are supplied for adjusting the scale of both the voltage (vertical) and time (horizontal) axes. Before making any measurement, the vertical, horizontal, and time scales should be adjusted so that the signal trace appears as large and as clear as possible on the display without going beyond the boundaries.

The portable oscilloscope is a useful and powerful tool for the measurement of electrical variables and for troubleshooting and testing of most kinds of electrical and electronic systems. Existing oscilloscopes are mostly luggable; that is, they are externally powered instruments that use a cathode ray tube (CRT) for their primary display. Further miniaturization of the CRT-based scopes is difficult because of the size, fragility, and power consumption of the CRTs; therefore, they find very little application in portable oscilloscopes. Modern portable oscilloscopes use flat-panel computer displays such as liquid crystal displays (LCDs) backed up by digital technology.

There are two different types of oscilloscopes: analog and digital. Although analog oscilloscopes are still available, because of the additional features and many other advantages, digital scopes are widely used. Typical characteristics of these scope meters are:

- Dual input — 200, 100, or 60 MHz bandwidth with up to 2.5 BD/s real-time sampling
- Color or black and white display
- Fast display update for dynamic behavior
- Automatic capture and replay of 100 screens
- 1000-V CAT II and 600-V CAT III safety
- Four-hour rechargeable Ni-MH battery operation
- RS-232 interface

When selecting and using digital oscilloscopes and digital storage oscilloscopes (DSOs), the user must be aware that the signal on the screen is sampled and digitized prior to being displayed. The theory of sampling has been given in detail in Chapter 3. To recreate a continuous signal waveform from the discrete signal, the samples must be interpolated. Efficient and accurate interpolation is usually best achieved if the digitization rate is at least four to five times higher than the signal bandwidth. This simple high-speed digitization technique is called real-time sampling (RTS). RTS imposes no limitations on the nature of the signal to be acquired, except for its bandwidth. Although RTS is simple and does not impose limitations on the signal to be acquired, it requires high-performance hardware. For example, a 200-MHz bandwidth device should sample and store at a 1-GHz rate, which may be a demanding performance even for a low-resolution (8 to 10 bits) A/D converter, typically used in DSOs.

If the input signal is periodic, which is true for many practical signals, the equivalent-time-sampling technique (ETS) is often used to lower the acquisition rate. The user must be aware that ETS-based oscilloscopes emulate the effective acquisition rate of the RTS-based scopes using much lower actual sampling rates. A waveform that may appear to consist of a finely

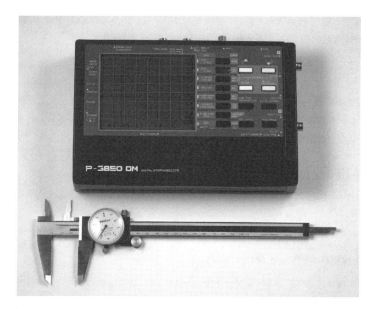

FIGURE 5.3
(See color insert following page 320.) A portable digital oscilloscope. (Courtesy of Protek T&M, http://www.hcprotek.com/)

spaced sequence of samples in fact has been assembled from the samples acquired over many signal periods. As a result, ETS cannot be used to digitize short and fast signal events and cannot reliably acquire transient features of periodic signals. ETS-based oscilloscopes specify sampling rates that are equal or lower than their bandwidth. This means that the front-end circuitry of the oscilloscope can accurately sample signals within the specified bandwidth, but digitize them at much lower rates.

The main issue in the design of battery-powered DSOs is their performance at low-power consumption. Higher-signal acquisition rates demand higher-power consumption and usually can only be realized with low-density semiconductor devices. This is true of all main DSO components — the memories, the A/D converters, the operational amplifiers, and the digital signal processors.

Modern commercial digital oscilloscopes can be manufactured having many additional features such as digital multimeter capabilities and logic analyzer functions. A typical example is depicted in Figure 5.3. It has the following features:

- Handheld, battery-operated DSO, DMM, and logic analyzer
- 50 MS/s sampling rate
- RS-232 interface and supporting software for data transfer
- Data storage for display, printing, and analysis

- 16-bit logic analyzer, frequency counter, capacitance measurement
- Sensitivity: 5 mV/Div to 20 V/Div in 12 speeds
- Accuracy: ±3%, ±1 pixel
- Bandwidth DC to 10 MHz
- Triggering 25%, 50%, 75% of trace
- Display modes: CH 1 and CH 2, dual, add, subtract, XY
- Power requirements: DC six 1.5 V "AA" and two 1.5 V "AAA" batteries
- Maximum input voltage for logic analysis: 40 V
- Threshold voltage for logic analysis: +1.4 V TTL, +2.4 V CMOS, –2.5 V to 5.5 V

5.1.4 Spectrum Analyzers

Spectrum analyzers (SAs) are state-of-the-art instruments in electronic portable devices and involve much of the information provided in this book. Therefore, more discussion is given on this topic in this section. SAs find applications in the analysis of many different types of signals generated by physical phenomena.

The spectrum analyzer is an instrument that performs spectral analysis of electrical signals and displays them as an amplitude vs. frequency plot. Because many complex systems can be easily evaluated in the frequency domain, the spectrum analyzer serves as an instrument of great value, used extensively to measure the performance of communication and other signals. Existing SAs achieve a high degree of accuracy, sensitivity, and resolution. They are often used for high-precision signal noise and distortion measurements, evaluating system frequency responses, tuning electronic circuits, performing incoming inspections and production tests, and evaluating electromagnetic emission and spectrum use compliance. There are two basic types of spectrum analyzers that use two distinct measurement techniques: the analog spectrum analyzers and the digital spectrum analyzers.

5.1.4.1 Analog Spectrum Analyzers

There are many types of analog spectrum analyzers. One type uses superheterodyne receivers. These receivers are designed as swept-tuners to scan selected frequency ranges. They measure signal powers at each of the selected frequencies and display them as a graph of power vs. frequency. Because super-heterodyne-based SAs are sequentially tuned to scan the band of interest, they cannot be used to measure fast changing signals. Analog spectrum analyzers can span a wide range of measurement frequencies, from 10 Hz up to 20 to 30 GHz, and they are uniquely suitable for high-frequency applications.

The second type of analog spectrum analyzer is the real-time spectrum analyzer, which uses a large number of bandpass filters to separate spectral components and provide simultaneous or real-time measurements. These spectrum analyzers perform basic measurements of signal amplitude and power at selected frequencies and the frequency measurement of the unknown spectral components. Spectrum analyzers also can be used to perform other, more complex measurements, such as noise measurements over the selected frequency band, harmonics analysis and total harmonic distortion measurements, system frequency response measurements, and so on.

The basic parameters of all spectrum analyzers are:

- Frequency resolution — the ability to distinguish two closely spaced signals
- Frequency range — the operational extent of the SA in the frequency domain
- Dynamic range — the range of the largest and smallest signals that can be measured and displayed simultaneously
- Sensitivity — the smallest signals that can be measured; sensitivity is limited by the SA internal noise
- Amplitude and frequency measurement accuracy

Another type of analog SA is the swept-tuned spectrum analyzer, which is widely used for measurements in communications, cable TV and video systems, and antenna and microwave equipment. These devices generally have the following major components:

- Input attenuators that scale the power of input signal to the dynamic range of the internal circuitry
- RF amplifier and filter, which are necessary to amplify input signal and reject mirror image of the input signal before the mixer
- Mixer that generates sum and difference frequencies of the heterodyne process
- Bandpass filters/amplifiers that select and amplify the spectral component of interest and reject all others
- Log amplifiers that amplify the incoming signal logarithmically to compress its dynamic range to the limited scale of the visual display
- Detector and video filters that form the amplitude measurement circuit

Some of the swept-tuned spectrum analyzers are manufactured as digital instruments so that the output of video filters can be displayed right away on the screen or further digitized and processed digitally. In order to maintain accuracy of the measurement, digitization is performed by a high-resolution A/D converter (at least 16 bits) operating at speeds greater than the largest-resolution bandwidth.

5.1.4.2   Digital Spectrum Analyzers

Digital spectrum analyzers mainly use the fast Fourier transform (FFT), which is an algorithm to calculate the set of spectral components of the digitally acquired signal record. FFT analyzers calculate all spectral components at the same time and, if fast enough, can perform real-time measurements. Because FFT analyzers do not miss any of the dynamically changing data, they are frequently called dynamic signal analyzers. FFT analyzers are powerful and flexible instruments, but because precise digitization of very high frequency signals is still a difficult task, they operate at the lower end of the spectrum, typically 0 Hz to 100 kHz, and up to tens of megahertz for some expensive high-end devices. At present, FFT analyzers largely replaced swept-tuned instruments in low-frequency applications.

FFT analyzers are good examples to illustrate the common trend in modern electronic instrumentation that shifts instrument functions and complexity from the analog domain to the digital domain, also from benchtop instruments to portable ones. A typical FFT analyzer is a conceptually simple instrument that can perform a variety of highly sensitive and complex measurements and can be easily implemented as a handheld device. The operation of a typical FFT analyzer is very simple. Input signals are conditioned, digitized, and stored in the processor memory. Each new sample is acquired at a rate equal to the sampling frequency, f_s. After an N-long block of signals is acquired, the processor begins calculating the N-point FFT algorithm. Meanwhile, as the processor is calculating, a new block of data is being acquired. As long as the processor can finish its calculations before new data become available, it is said that the FFT analyzer performs real-time calculations, without omitting any input data. The frequency resolution of FFT analyzers is determined by the sampling rate of the A/D converter and length N of the FFT data block.

Although FFT analyzers use digital techniques to perform spectral analysis, they critically depend on the accuracy of their analog front-end and digitizer circuitry to perform accurate measurements. Each acquisition channel has a preamplifier, anti-aliasing filter, and A/D converter. Inexpensive delta-sigma (Δ/Σ) A/D converters reach 16-bit resolution at a 200-kHz sampling rate and are often used as the main converters. These converters require very simple anti-aliasing filtering, so a high-accuracy data acquisition channel can be built with just a few inexpensive components. Similarly, an optional output channel can be implemented with Δ/Σ digital-to-analog (D/A) converters, which require the backup of very simple analog interpolation filters. Frequently, audio frequency (20 Hz to 20 kHz) Δ/Σ, A/D and D/A converters are combined into a single audio codec that is available in 16- to 20-bit resolutions. Low-cost high-performance signal processors are readily available, so a high-accuracy FFT analyzer can be a relatively inexpensive

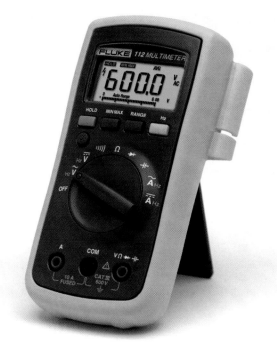

(a)

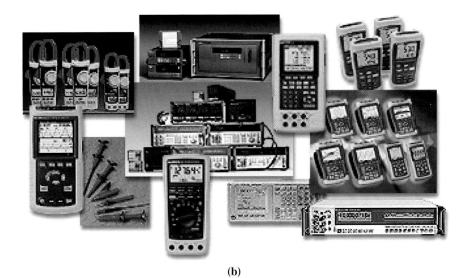

(b)

COLOR FIGURE 5.1 (a) A digital multimeter. (b) A range of multimeters. (Courtesy of Fluke Electronics, http://www.flukecanada.ca/)

COLOR FIGURE 5.2 A range of digital powermeters. (Courtesy of Fluke Electronics, http://www.flukecanada.ca/)

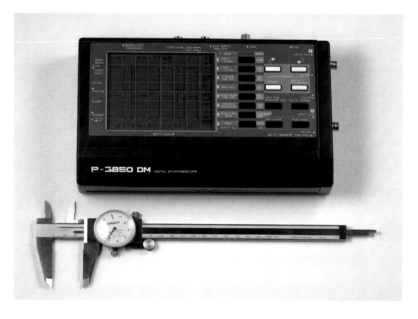

COLOR FIGURE 5.3 A portable digital oscilloscope. (Courtesy of Protek T&M, http://www.hcprotek.com/)

PSA-33B
Portable Spectrum Analyzer

Satellite and CATV Industry

- 1 to 2100 MHz in 2 Bands
- 3 MHz & 300 KHz Resolution Bandwidth
- LCD Center Frequency Readout
- 22 KHz Signal for Dual LNB
- Power for LNA & LNB's
- Internal Battery & Charger
- Rugged Reliable & Field Proven

Ideal CATV Analyzer
The PSA-33B spectrum analyzer is the perfect analyzer for CATV industry or private cable operator. The analyzer covers the whole CATV band from 1 MHz to 1100 MHz in one range and the L band from 950 to 2100 MHz in the other range. With these features you can view the entire CATV spectrum for precise measurements and signal analysis with a dynamic range of -90 dBm you have the ability to evaluate the signals anywhere in your system. You can also check your satellite antennas and feeds without investing in additional test equipment. Low band coverage from 1-1100 MHz highband from 950-2100 MHz, for CATV Test & Measurement downlink installation, service, and maintenance. With two resolution BW settings and 4-digit center frequency display, you can align dishes and identify and resolve terrestrial interference problems quickly and precisely.

Dish Alignment a Snap!
If you are doing DBS installs, you need to be on and off the job fast! DBS signals such as DISH NETWORK, PRIMESTAR, and DIRECT-TV can be seen with AVCOM'S PSA-33B. Normal DBS antenna alignments, such as roof, wall, or backyard, can be accomplished in seconds because the DBS receiver and TV monitor do not have to be at the antenna. Think of the benefit of doing just one more install per day!

Do the Impossible!
The PSA-33B can really help with more creative DBS dish installations (inside buildings, looking through sparse foliage or windows, smaller antennas, etc). When the signal obtained is not optimum or there is uncertainty where to point the dish, it is difficult to use the receiver's built-in alignment program. Using the PSA-33B, the signal can be quickly located and measured even though it is below the receiver's signal strength requirement.

Technical Specifications

Frequency Coverage:	Low Band: 1 to 1100 MHz High Band: 950 to 2100 MHz Extendable with MFC's
Frequency Display:	4-Digit LCD
Span:	Over 1000 MHz to Zero
Resolution Bandwidth:	3 MHz & 300 KHz Selectable
RF Sensitivity:	> -90 dBm Typical
Reference Levels:	0, -20, & -40 dBm /+49, +29, & +9 dBmv
Scale:	2 dB/DIV or 10 dB/Div., Selectable
Amplitude Accuracy:	± 2 dB Typical
Dynamic Range:	65 dB on Screen
Input Connector:	BNC, DC Blocked, 50 Ohm Impedance
LNB Power:	+13/18 VDC Switchable 22 KHz Signal for Dual LNB
Size/ Weight:	11.5"w x 5.5"h x13.5"d/ 17 lbs 29cm x 14cm x 34cm/7.72kg
Power Requirements:	115 VAC, 60Hz (220 VAC Avail.) 12 VDC Cigar Lighter Cord Included **Internal Batteries & Charger**
Display:	5" Diagonal CRT Green Phosphorous

Accessories
- AVSAC Cordura Nylon Carrying Case
- ADA-10A Computer Display Adapter
- QRM-35 Quick Release Rack Mount
- TISH-40 Terrestrial Interference Survey Horn
- MFC-Frequency Extenders Available

500 Southlake Blvd.
Richmond, VA 23236
Phone: 804-794-2500

Of Virginia Inc.
www.avcomofva.com

Available from
AVCOM
Stocking Distributors

COLOR FIGURE 5.4 A portable spectrum analyzer. (Courtesy of AVCOM of Virginia, Inc., http://www.avcomofva.com/)

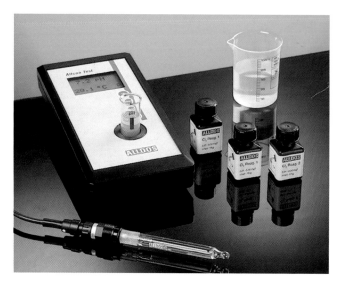

COLOR FIGURE 5.5 A water quality analyzer. (Courtesy of ALLDOS Eichler GmbH, http://www.alldos.com/)

Nilsson Model 400
Cat. No. 44106

Leads
Cat. No. 37009

MCM Soil Box
Cat. No. 37008

COLOR FIGURE 5.6 A typical soil resistivity-measuring instrument and ancillaries. (Courtesy of MC Miller Co., http://www.mcmiller.com/)

COLOR FIGURE 5.7 An automatic portable air analyzer. (Courtesy of Mine Safety Appliances Co., http://www.MSAnet.com/)

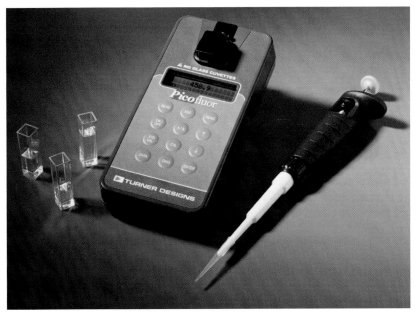

COLOR FIGURE 5.8 An automatic chemical analyzer. (Courtesy of Turner BioSystems, http://www.turnerbiosystems.com/)

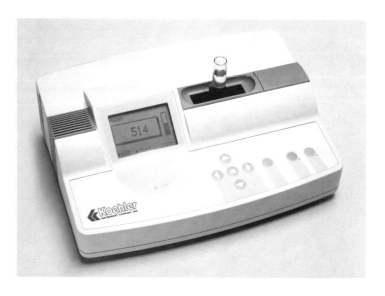

COLOR FIGURE 5.9 A typical colorimeter. (Courtesy of Koehler instruments Company, Inc., http://www.koehlerinstrument.com/)

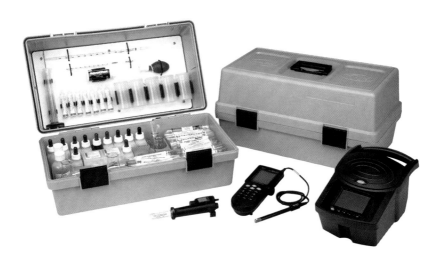

COLOR FIGURE 5.10 A portable chemical analyzer laboratory. (Courtesy of Hach Co., http://www.hach.com/)

COLOR FIGURE 5.11 A typical portable gas [detect]or. (Courtesy of Draeger Safety, www.draeger.com/)

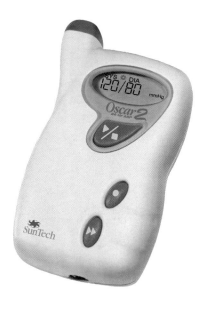

COLOR FIGURE 5.12 A blood pressure monitoring instrument. (Courtesy of SunTech Medical Instruments, http://www.suntechmed.com/)

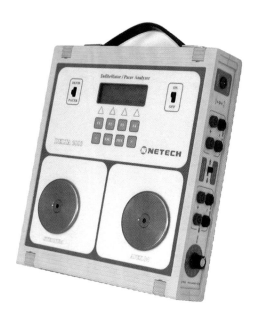

COLOR FIGURE 5.13 A biomedical kit. (Courtesy of NeTech Corporation, http://www.gonetech.com/)

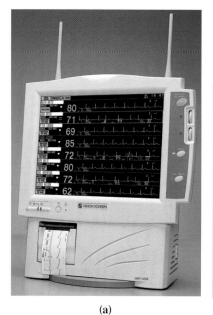

(a)

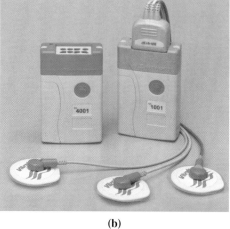

(b)

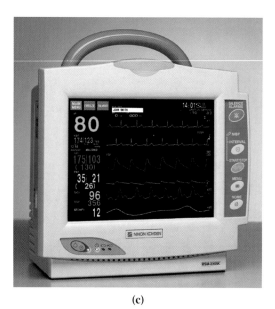

(c)

COLOR FIGURE 5.14 A telemetric patient monitoring system. (a) A bedside monitor. (b) A wireless transmitter. (c) A base station. (Courtesy of Nihon Kohden Corporation, http://www.nihonkohden.com/)

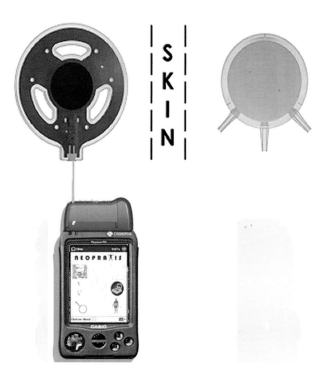

COLOR FIGURE 5.15 An underskin sensory system equipped with telemetric features. (Courtesy of Neopraxis Pty. Ltd., http://www.neopraxis.com.au/)

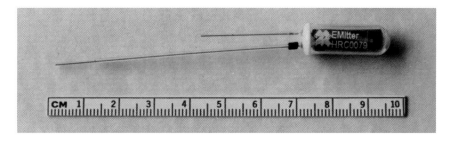

COLOR FIGURE 5.16 An implantable sensor/telemetry transmitter. (Courtesy of Mini Mitter Company, Inc., http://www.minimitter.com/)

COLOR FIGURE 5.17 An RF telemetric measurement of parameters of rotating shafts. (Courtesy of Advanced Telemetrics International (ATI), http://www.atitelemetry.com/)

COLOR FIGURE 5.18 A typical electronic scale. (Courtesy of Ohaus Corporation, http://www.ohaus.com/)

COLOR FIGURE 5.19 A handheld laser speed detector. (Courtesy of Lion Laboratories Ltd., and MPH Industries, Inc., http://www.lion-breath.com/; http://www.mphindustries.com/)

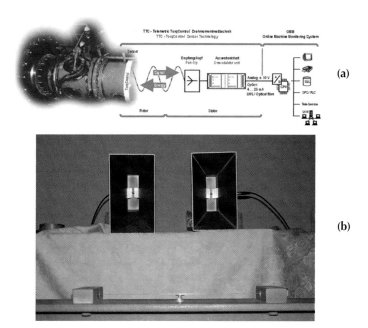

(a)

(b)

COLOR FIGURE 5.20 (a) A telemetric vibration/torque sensing arrangement. (Courtesy of ACIDA TorqControl GmbH, http://www.acida-torqcontrol.de/). (b) A microwave telemetric vibration/stress sensing arrangement. (Courtesy of VIBSTRING, http://www.vib-string.com/)

COLOR FIGURE 5.21 A sound level meter. (Courtesy of NorSonic AS, Norway, http://www.nor-sonic.com/)

(a)

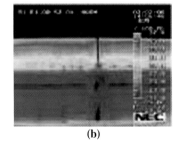

(b)

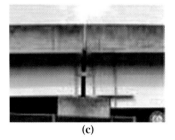

(c)

COLOR FIGURE 5.22 A short-range infrared detector. (a) The detector. (b) The infrared image. (c) The visual image. (Courtesy of NEC San-ei Instruments, Ltd., http://www.necsan-ei.co.jp/)

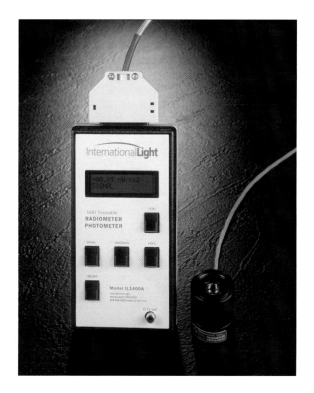

COLOR FIGURE 5.23 A photometer. (Courtesy of International Light, Inc., http://www.intl-light.com/)

COLOR FIGURE 5.24 A typical digital dosimeter. (Courtesy of S.E. International, Inc., http://www.seintl.com/)

COLOR FIGURE 5.25 A solar power digital thermometer. (Courtesy of Winters Instruments, Inc., http://www.winters.ca/)

COLOR FIGURE 5.26 A typical GPS receiver suitable for navigations. (Courtesy of Garmin, Ltd., http://www.garmin.com/)

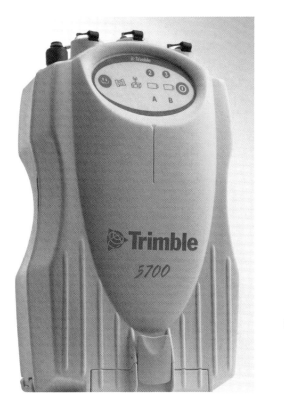

COLOR FIGURE 5.27 A surveying GPS receiver equipped with RF transmitters. (Courtesy of Trimble Navigation Ltd., http://www.trimble.com/)

COLOR FIGURE 5.28 A solar powered RF networkable weather station. (Courtesy of Oregon Scientific, Inc., http://www.oregonscientific.com/)

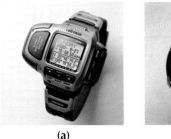 

(a)   **(b)**

COLOR FIGURE 5.29 (a) A GPS altimeter wristwatch. (Courtesy of Shriro Australia Pty., Ltd., http://www.shriro.com/). (b) A barometric altimeter wristwatch. (Courtesy of Oregon Scientific, Inc., http://www.oregonscientific.com/)

COLOR FIGURE 5.30 A personal alarm system. (Courtesy of D&D Security Products, Inc., http://www.ddsp.com/)

COLOR FIGURE 5.31 A microprocessor-controlled metal detector. (Courtesy of A&S Company, Inc., http://www.detection.com/)

device, compared to the analog spectrum analyzers of similar performance. The dynamic range of existing FFT analyzers can reach 90 to 100 dB, which is enough for most demanding measurement tasks. In comparison to the scanning SA, FFT analyzers can provide phase information, if triggered data acquisition can be implemented.

FFT analyzers frequently add a rich set of related information to the FFT signal processing functions. Through their software-defined instrument modes, they can provide the measurement functionality of a network analyzer, acoustic analyzer, vibration analyzer, audio distortion analyzer, or even waveform analyzer in a single portable device. While the simplest instruments may have just one input channel, two-channel instruments support more sophisticated measurements, such as calculation of correlation and coherence measurements and a wider range of applications, such as sound intensity and system transfer function determinations.

Typical applications of FFT analyzers include spectral measurements of mechanical, acoustic, and audio systems. Specialized variations of the FFT analyzer are widely used for troubleshooting mechanical equipment, where they can detect and diagnose excessive vibrations, faulty bearings, rotational unbalances, and other dynamic mechanical problems. A typical existing hand-held FFT analyzer might be a one- or two-channel battery-powered device with a 12- or 16-bit A/D converter with a preamplifier sensitive enough to interface with accelerometers and microphone sensors, and they also are suitable for vibration and acoustical applications. The typical frequency range could be 2 Hz to 20 kHz, with 100 to 400 lines of frequency resolution, a dynamic range of 60 to 80 dB, an amplitude accuracy of 0.3 dB, and a frequency accuracy of 0.01%. Larger battery-powered notebook-size devices usually are two-channel instruments. Their dynamic range may extend to 80 to 90 dB with 1000 to 4000 lines of frequency resolution and built-in auxiliary software-based measurement functions, such as sound octave analysis, order analysis, a three-dimensional waterfall spectrum display, and so on.

Design of a high-accuracy analog SA is a difficult task. Although many swept-tuned spectrum analyzers are portable, only a few are battery-powered handheld or compact devices. High-frequency operation still requires small- to medium-scale integration electronic components with relatively high power consumption. Because small devices inevitably have high-density packaging, close proximity of high-frequency switching digital components to the highly sensitive analog radio frequency (RF) components produces interference and makes it difficult to design sensitive and accurate RF front-end circuitry in small packages. However, development of high-integration RF semiconductor technology, mostly for applications in wireless communications, is now making these limitations less relevant. Most of the battery-powered and handheld spectrum analysis devices now are FFT analyzers. These digital instruments greatly benefit from the general miniaturization in portable computing.

There are many different types of spectrum analyzers, depending on the frequency range of the electromagnetic spectrum and frequency range of

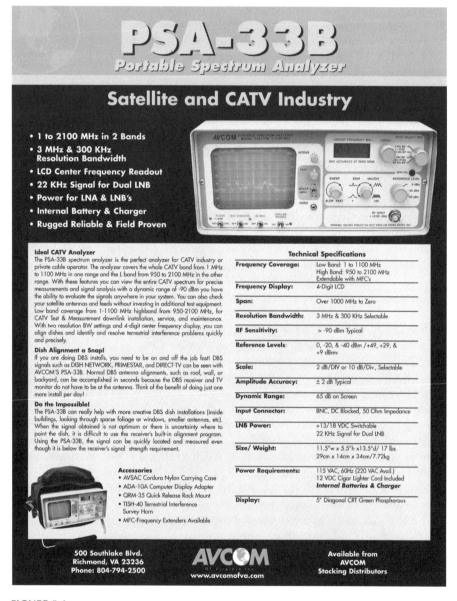

FIGURE 5.4
(See color insert following page 320.) A portable spectrum analyzer. (Courtesy of AVCOM of Virginia, Inc., http://www.avcomofva.com/)

interest. An example of a modern portable spectrum analyzer is illustrated in Figure 5.4. This particular device has the following characteristics:

- Frequency coverage from 1 to 2100 MHz in two bands
- Four-digit LCD frequency readout

- Accurately measures wideband signals commonly used in the satellite communications and CATV industries
- Lightweight (17 lb) and portable
- Battery- or line-operated
- Suitable for field test measurements, Ku and L band compatible
- 220- to 240-V models available
- Internal battery charger
- Selectable vertical sensitivity of either 2 or 10 dB/div

5.2 Environmental Instruments

5.2.1 Water Analyzers

Water is a resource fundamental to all life. Both its quality and quantity are important, particularly for humans. Fresh, clean, unpolluted water is necessary to human health. Different water qualities are acceptable for different uses. Water for human consumption should be free of disease-causing microorganisms, harmful chemicals, and objectionable taste and odor, and should have an acceptable level of color and suspended materials. As a general principle, water quality problems can fall into two categories: biological and chemical.

- Biological agents such as bacteria, viruses, and some higher organisms can exist naturally or can be human induced. They can cause infections and outbreaks of acute diseases.
- Chemical agents such as suspended sediments, toxins, and nutrients are generated by various forms of land use, industrial and agricultural activities, waste, and air pollution.

Many pollutants, through terrestrial runoff, direct discharge, or atmospheric deposition, mix with surface waters. However, water quality varies from one location to the next, depending on local geology, climate, biological activity, and human impact. One of the major causes of water pollution is due to cycles of nitrogen and phosphorus. The main affecting pollutants in rivers are in the form of chemical and biological variables such as soil erosion, salt, nutrients, wastewater with high organic contents, metals, acids, and other chemical pollutants. As far as lakes are concerned, an important environmental concern is the problem of eutrophication, which signifies the enrichment of lakes with nutrients. Underground water, on the other hand, suffers from damping wastes and agricultural activities. Underground water is an important source of drinking water in both developing and developed

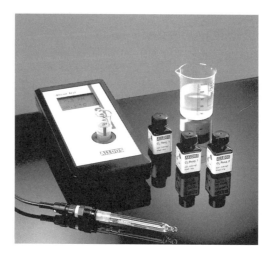

FIGURE 5.5
(See color insert following page 320.) A water quality analyzer. (Courtesy of ALLDOS Eichler GmbH, http://www.alldos.com/)

countries. Water quality in seas is particularly important for risk of contamination of fisheries.

With an increasing number of chemicals being released into the environment by man, the possible number of variables that can be monitored by both fresh- and marine water is growing all the time, currently exceeding hundreds of variables. Consequently, there are many different types of instruments to determine water quality. Some of these instruments are nonportable; others are portable. Portable instruments have the advantage of determining the water quality on-site, which may be far away from laboratories.

Some portable water analyzers are based on optical spectrometry, while others are based on ultrasonic principles. Many portable water analyzers include different sensing mechanisms and need human intervention. Another class of water analyzers operates in remote stations monitoring the water quality of the concerned reservoirs, rivers, and lakes and transmitting the information to a base station. All of these instruments can have many other additional features, such as built-in calculators, self-diagnostics, real-time data, alarms for pollution detection, computer interface, and so on.

Opto-electronic instruments measure the absorption of light by a sample as a function of wavelength. It is used for the qualitative identification and quantitative determination of a variety of colored substances. Some of these instruments, illustrated in Figure 5.5, operate on the principle of four-beam measurement, which allows the compensation of disturbing influences by turbidity of the sample, external light, and aging of the light source. They use microprocessors so that parameters of the measurements can be selected for different applications. They run on 9-V batteries and can interface a printer or PC via RS-232. The device in Figure 5.5 has the following specifications:

- Error detection, automatic zero-point adjustment, real-time clock
- Measuring up to 17 parameters
- RS-232 interface
- Suitable for drinking water, ground and surface water, industrial process and wastewater applications
- Power requirement: 9-V alkaline battery

Opto-electronic water analyzers are capable of identifying chemicals and metals such as aluminum, chlorine, chlorine dioxide, cyanuric acid, fluoride, ammonium, iron, nitrate, nitrite, ozone, pH, manganese, nickel, and phosphate.

5.2.2 Soil Analyzers

Soils are prone to degradation due to human influences in a number of ways: (1) crops remove nutrients from soils, and hence chemical deterioration takes place; (2) management practices influence soil quality such as waste damping, silting, and salinization; and (3) erosion removes soil. Soil contamination refers to quantitative changes of soil constituents, due to domestic, industrial, and agricultural activities, by additional compounds that were originally absent in the system. Soil contamination has two different meanings. One is the slow but steady degradation of soil quality (e.g., organic matter, nutrients, water holding capacity, porosity, lack of contamination) due to practices in land usage such as domestic and industrial wastes, and chemical inputs from agriculture. The other is the concentrated pollution of smaller areas, mainly damping or leakage of waste materials.

A number of metal and chemical concentrations are commonly regarded as contaminants of soil. Several are referred to as heavy metals; the list may contain metalloids as well as nonmetals. The main contaminants are arsenic, cadmium, chromium, copper, fluorine, lead, mercury, nickel, and zinc, while beryllium, bismuth, selenium, and vanadium may also be considered contaminants. Waste products are one of the major factors in degradation of soil. Waste products can be classified as municipal wastes, wastewaters, wastes dumped at sea, oil and oil products, hazardous waste, and radioactive effluents.

Electronic portable instruments are developed to determine a number of properties of soils: the resistivity, the moisture, and chemical contents.

5.2.2.1 Soil Resistivity Instruments

Soil resistivity instruments utilize a number of electrodes, known as the Wenner four-pin method. The resultant value of the readings is the average resistivity of the soil to a depth equal to the spacing between adjacent

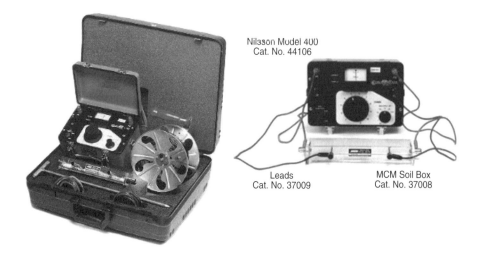

FIGURE 5.6
(See color insert following page 320.) A typical soil resistivity-measuring instrument and ancillaries. (Courtesy of MC Miller Co., http://www.mcmiller.com/)

electrodes (soil pins). A standard measurement requires about 5- to 6-m pin spacing.

A complete soil resistivity-measuring instrument consists of a resistivity meter, soil pins, leads, and plastic carrying case. The soil resistance values are read either in analog or digital form. The reading range of a typical meter varies from 0.1 Ω to 1.1 MΩ. A typical soil resistivity meter is illustrated in Figure 5.6. This particular soil resistivity test kit has 4× pin resistivity reels and maximum spacing (depth) between the electrodes is 7 m.

5.2.2.2   Soil Moisture Instruments

Soil moisture instruments can be divided into two main groups: those suitable for outdoor field applications and those used on samples. Field application devices are similar in operation to soil resistivity measurement instruments. However, sample-based instruments have additional features.

In sample type moisture measuring devices, fixed amounts of samples with maximum grain sizes are used. They can be programmed to measure the moisture contents of different types of materials, such as sand, aggregates, ores, coal, soils, ceramics, abrasives, and other powders and chemicals. They offer the unique advantage that the samples can be tested on-site, eliminating risk of moisture loss during transport.

In the implementation of the instruments, the weighed samples are placed in the tester with a measured quantity reagent (e.g., calcium carbide) and sealed with a cap. Upon agitation, free moisture in the test sample reacts with the reagent to form acetylene gas. The pressure gauge reads directly in percent moisture within a few minutes.

5.2.2.3 Soil Chemistry Analysis Instruments

Soil chemistry analysis instruments are suitable to measure organic or inorganic compounds, such as pH, nitrate, phosphorus, potassium, humus, calcium, magnesium, ammonia nitrogen, manganese, aluminum, nitrite, nitrogen, sulfate, chloride, and ferric iron, as well as tests for nitrogen, phosphorus, and potassium.

5.2.3 Air Analyzers

Air analyzers are mainly used for air pollution detection. Air pollution may be defined as the unwanted change in the quality of Earth's atmosphere caused by the emission of gases and solid or liquid particulate. It is considered to be one of the major causes of climatic change, greenhouse effect, and ozone depletion, and may have serious consequences for all living organisms in the world. Polluted air is carried everywhere by winds and air currents and is not confined by national boundaries. Therefore, air pollution is a concern for everybody, irrespective of what the sources are and where it is generated.

Air pollutants can be classified according to their physical and chemical composition as:

Inorganic gases — sulfur dioxide, hydrogen sulfide, nitrogen oxides, hydrochloric acid, silicon tetrafluorade, carbon monoxide and carbon dioxide, ammonia, and ozone

Organic gases — hydrocarbons, terpenes, mercaptans, formaldehyde, dioxin, fluorocarbons, etc.

Inorganic particulates — lime, metal oxides, silica, antimony, zinc, radioactive isotopes, etc.

Organic particulates — pollen, smuts, fly ash, etc.

There are many different types of air analyzers, ranging from large fixed stations to laboratories carried by airplanes. A typical portable air/gas analyzer is illustrated in Figure 5.7. This portable air analyzer contains an internal pump to draw air in and senses four different gases, O_2, H_2S, CO, and combustible gas. It operates on NI-MH or alkaline batteries for a 16- to 20-h run time in pumped mode. It has an environmental protection rating of IP54. The performance of the device can be improved by changing the sensors.

Air analyzers developed in recent years need to satisfy the monitoring requirements of National Ambient Air Quality Standards (40 CFR 58) and the U.S. Environmental Protection Agency (EPA) requirements, which lay out the particulate air monitoring rules. Some of these rules are:

Aldehydes/ketones in air by U.S. EPA TO-11, IP-6A; ASTM DS197

Formaldehyde by OSHA Method 52

Inorganic acids by NIOSH Method 7903

FIGURE 5.7
(See color insert following page 320.) An automatic portable air analyzer. (Courtesy of Mine Safety Appliances Co., http://www.MSAnet.com/)

Isocyanates by OSHA Methods 42 and 47, ASTM D5836

Pesticides by U.S. EPA TO-10A, IP-8; ASTM D4861

Volatile compounds by U.S. EPA TO-17

There are private companies that have developed methods of sampling. They have worked out guidelines to handle air samples that contain different analyte compounds, and descriptions of sample parameters suitable for a diverse range of sampling media. A typical analyzer has the following features:

- Should meet all requirements of EPA Compendium Method TO-14/ TO-15 and protocols for collection of ambient volatile organic compounds (VOCs)
- Remote starting of sample channels from the site data logger
- Sampler operates on battery, car lighter, and solar panels
- Microprocessor-based, menu-driven control, and data recording system
- Data transfer by a small data link retrieval system or modem
- Programmability allows the operator to set the days of the week on which the sampling will take place; device keeps the total elapsed time of each channel and pump run; each channel page shows the site time and run day of the week

Real-time measurement of indoor airborne particulate concentrations in some industries bears particular importance because of health and safety regulations, and there are some automatic portable devices to perform these tasks. These instruments measure the mass concentrations of airborne dust, smoke, mists, haze, and fumes and provide continuous real-time readouts and data storage for future analysis. The real-time aerosol monitor ranges from 0.0001 mg/m^3 (0.1 µg/m^3) to 400 mg/m^3. The technical specification of such a device is 0.1 to 999.9 µg/m^3 (resolution); particle size range of maximum response, 0.1 to 10 µm; total number of data points in memory, 10,000; real-time and date data, seconds, minutes, hours, day, month, and year; LCD readout; internal battery rechargeable sealed lead–acid, 6.5 A-h, 6 V nominal; operating time, >24 h; operating environment, 0 to 40°C, 0 to 95% RH, noncondensing; weight, 5.3 kg (11.7 lb).

5.3 Chemical Instruments

Portable chemical analyzers are usually used for safety reasons in industry, analysis of combustion in machines, environmental safety, pharmaceutical testing, geological and scientific explorations, and emergency situations where accurate information is required by the officials on the chemical content of air, liquids, or solids in the day-to-day operations of manufacturing and processing plants. The main purpose of industrial chemical sensors is to detect levels of chemicals that might endanger employees or traces of the harmful chemicals. The substances being detected in a safety environment are the chemicals that may cause burns, cause explosions, or carry health hazards when inhaled or contacted. These include hydrochloric acid and various hydrocarbon compounds, such as natural gas and kerosene. Such chemicals can exist in the following industrial operations:

- Printing, laminating, and coating processes
- Solvent vapor carbon beds and incinerators
- Process ovens and dryers
- Coil coating and painting
- Oil and gas pipelines and storage facilities
- Chemical processing
- Pulp and paper mills
- Semiconductor manufacturing
- Pharmaceuticals
- Wastewater treatment

Some chemicals require very small sensors, whereas others require larger, bulkier sensors. Therefore, the sizes and shapes of these analyzers depend on the chemicals under investigation.

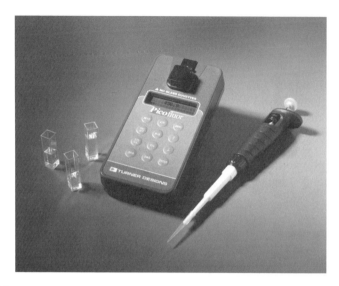

FIGURE 5.8
(See color insert following page 320.) An automatic chemical analyzer. (Courtesy of Turner BioSystems, http://www.turnerbiosystems.com/)

There are many electronic portable chemical analyzers, such as flame ionization, catalytic, electrochemical, diffusion sensors, flame photometers, fluorimeters, colorimeters, and various forms of semiconductor detectors, as discussed in detail in Chapter 2. Brief technical details of some of these electronic portable instruments are given below.

Portable *flame photometers* are designed mainly for determination of sodium (Na), potassium (K), lithium (Li), barium (Ba), and calcium (Ca). They are fitted with automatic flame failure detection for user safety, and they are ideal for use in clinical, industrial, and educational applications. They display the amount of chemical substances in millimoles per liter. These devices are extensively supported by software-driven monitoring and control functions and the results of analysis can be transmitted via RS232.

Portable *fluorimeters* are equipped with microprocessors, thus providing good sensitivity, low drift, and high reliability. Figure 5.8 illustrates a typical example of fluorimeter, which has the following features:

- Two separate optical channels to measure two different fluorophores in the same essay
- Channel A: UV excitation 365 to 395 nm, emission 430 nm
- Channel B: Blue excitation 425 ± 15 nm, emission 515 ± 10 nm
- Internal data logging package with serial interfacing features
- 1000 data points can be logged
- Alarm and auto shutoff

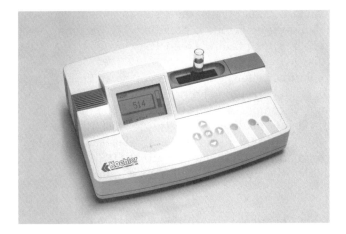

FIGURE 5.9
(See color insert following page 320.) A typical colorimeter. (Courtesy of Koehler Instruments Company, Inc., http://www.koehlerinstrument.com/)

- Detector: photodiode
- Power requirements: Four AAA batteries for over 1000 measurements

One of the key determinants of fluorimetric sensitivity is the degree of excitation energy available from the source lamp. A typical fluorimeter utilizes a high-intensity xenon source to provide energy output across the spectrum. The lamp is fitted into an ozone-free lamp house, thus eliminating the need to install additional extraction equipment. Excitation and emission wavelengths are selected by means of interchangeable filters over the operating range of 200 to 600 nm. The emission detector is a high-sensitivity photomultiplier. The emission signals are processed by an on-board microprocessor and displayed digitally.

The fluorimeter is used for quality control in areas such as the food industry and clinical, environmental, and public health laboratories, where stability and sensitivity are of utmost importance.

Colorimeters, shown in Figure 5.9, are used extensively, as they are easy to construct as portable instruments. This particular device is a single-beam filter colorimeter that utilizes reference beam path technology to measure samples over eight spectral lengths ranging from 400 to 700 nm. It uses a krypton lamp light source and is equipped with RS-232 for communication with other devices.

In the majority of colorimeters, hermetically sealed filters are provided for use over a range of wavelengths (typically 400 to 710 nm). This arrangement enables the user to select the appropriate measurement wavelength for the sample under test. Samples are presented to the unit in either glass or disposable cuvettes. Usually, the measurements and calibrations are initiated via a sealed membrane keypad. Calibration is automatic from zero to 100%

values. In standard digital types, the results are displayed via a 3¼-digit LCD, and a data-logging facility is provided. The data may be downloaded via the RS-232 link for printing or analysis, or may be recalled onto the instrument display.

A typical colorimeter for petroleum products applies to products with a color of 0.5 or darker, including lubricating oils, heating oils, and diesel fuel oils. The sample is compared against standard color discs in the colorimeter. The comparisons are made according to ASTM D1500 specifications. The color discs situated on either side of the sample contain standards conforming to the chromaticity coordinates of ASTM D1500. The discs are rotated until the sample color matches the relevant standards.

5.3.1 Portable Chemical Laboratories

In industrial, agricultural, municipal, mining, and environmental applications, comprehensive information on the chemical contents of substances may be necessary. Analyses of the chemical substances can be made on-site by portable instruments that can loosely be termed as *electronic portable laboratories*. These laboratories represent an advanced level of electronic portable instruments that contain a main electronic unit and many add-on type ancillaries that are designed to measure different chemical contents of the substance. In many instances, computers are used together with the electronic unit for information loading and further data analysis. In many cases, the computers are programmed as virtual instruments.

An example of portable laboratories is the portable chemical laboratory used as a water analysis instrument and is shown in Figure 5.10. This analyzer operates on the nephelometric principle of measurement that is the monitoring light scattered by the sample at 90° to the incident beam to be picked up by a scattered light detector. This instrument operates on four

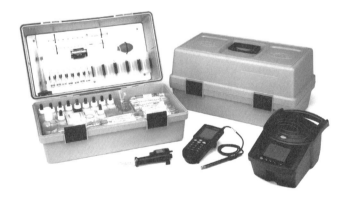

FIGURE 5.10
(See color insert following page 320.) A portable chemical analyzer laboratory. (Courtesy of Hach Co., http://www.hach.com/)

AAA batteries and its accuracy is better than 5%. The light source is an infrared emitter with a wavelength of 880 ± 10 nm. Many chemical analyzers are largely based on spectrophotometers. The complete chemical analysis system also includes auxiliary instruments such as a pH meter, a conductivity meter, a digital titrator, reagents, and other equipment and apparatus to measure a wide range of parameters. There are many different versions of such instruments, for example, the soil and irrigation laboratory, the aquaculture laboratory, and the water conditioning laboratory. A typical computerized laboratory can be used to determine the following: acidity, alkalinity, bromine, calcium, carbon dioxide, chloride, chromium, copper, fluoride, hardness, iodine, iron, manganese, nitrogen, ammonia, pH, phosphorus, salinity, silica, sulfate, and dissolved solids.

5.3.2 Gas Detectors

Apart from the portable air pollution detectors, discussed in Section 5.2.3, there are gas detectors designed for specific applications. An example of such a gas detector is illustrated in Figure 5.11. This gas detector has the following features:

- A broad range of gases, e.g., toxic gases and vapors, oxygen, carbon monoxide, hydrogen sulfide, hydrocarbons, etc.
- Full graphics with point matrix display, and audible alarm signal
- Menu-guided operation
- 8000 data points can be logged

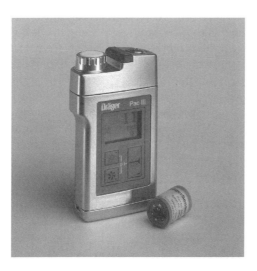

FIGURE 5.11
(See color insert following page 320.) A typical portable gas detector. (Courtesy of Draeger Safety, http://www.draeger.com/)

- Gas intake: diffusion or by external pumps
- Power requirement: 9-V alkaline, zinc charcoal, or lithium batteries
- Operating temperature from –20 to 55°C
- Housing: interface immune

Gas detectors analyze the sample and identify the types of gases in the sample. The corresponding output is usually displayed on an LCD readout or a warning sound is generated. Suitable sensors for the detection of specific gases are necessary. Typically, combustible gases require catalytic sensors, oxygen and most toxic gases require electrochemical sensors, and most hydrocarbons require metal oxide semiconductors (MOSs). The data produced by the sensor is an analog signal. The signal is converted into a digital signal. Further, the signal is compared with the user-defined reference level of concentration of gases so that the processing unit can determine when to respond once the gas concentration reaches the preset level. The processing unit analyzes the digital signal and hence determines what type of gas was detected, as well as the level of the gas.

5.4 Biomedical Instruments

Portable biomedical instruments are extensively used in clinical applications and home health care. They are frequently used as patient monitors, pacemakers, and defibrillators, and also for ambulatory electrocardiogram (ECG) and blood pressure recording. Most of the patient monitors are now computer-based instruments that contain many elements of the basic digital electronics. The portable patient monitors vary in size from compact to palm-size instruments. Depending on the their sophistication, portable patient monitors may measure many vital parameters, such as heart rate, respiration rate, arterial blood oxygen saturation, and noninvasive blood pressure measurement. Some of these measurements are performed continuously and others intermittently. Because of the need to provide noninterrupted measurements, the instrument power supply is usually battery-backed. Some patient monitors can be part of a larger monitoring system to which they can be integrated over a wireless connection. Most patient monitors include alarms, and more sophisticated units may also include some diagnostic functions, such as arrhythmia detection. Some monitors are equipped with graphic displays.

The safety of biomedical instruments is a matter of great concern. Nerves, muscles, and especially the heart are highly sensitive to electrical signals. Because many electrophysiological instruments are designed to use low-contact resistance electrodes, these instruments can present a real danger of accidental application interfering with the natural electric signals of the human body. One simple safety precaution could be to insert high-value

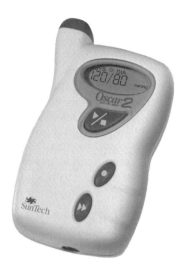

FIGURE 5.12
(See color insert following page 320.) A blood pressure monitoring instrument. (Courtesy of SunTech Medical Instruments, http://www.suntechmed.com/)

current-limiting resistors in series between the electrodes and instrument inputs. Another common solution is to use isolation amplifiers in the instrument front. Typically, a transformer or optical coupling between input and output of such amplifiers prevents DCs from flowing between the electrodes and the rest of the circuitry. Instruments that are powered by small-capacity batteries may provide an additional degree of safety, because of their inherently limited capacity to deliver large voltages and currents. In the U.S., the Federal Food and Drug Administration (FDA) requires an in-depth review of the safety, efficacy, and clinical utility of medical instruments before they can be used clinically. Examples of biomedical instruments will be given next.

The earliest and one of the most popular portable biomedical instruments are the blood pressure monitors. They can be used without medical supervision. A typical example of a blood pressure monitor is given in Figure 5.12, which has the following features.

- Works on step deflation oscillometry technique
- Accuracy: –3 mmHg
- LCD display
- 52 hours recording duration and 250 samples of systolic, diastolic, and heart rate
- Power requirement: 2 × 1.5 V AA alkaline or Ni-MH batteries
- RS-232 interface capability

There are ambulatory instruments for recording transient arrhythmia events for 24 hours. In these instruments, two types of monitors are used:

cassette recorders and solid-state recorders. The solid-state recorders use flash memory technology and provide relatively reliable and maintenance-free operation. Because the large memory needs to support full 24-hour recording, most solid-state monitors rely on data compression at the expense of the decreased fidelity of recording. In the future, data compression may become unnecessary as flash memory densities continue to increase. Another type of ambulatory recorder is the event recorder. This device requires the patient to manually start recording on the onset of symptoms and continues recording for a preset length of time. To extract sporadic diagnostic information from the large amount of recorded data and to isolate artifacts, recordings are post-processed by a trained operator, using PC-based scanning analysis software.

There are many different types of portable *blood testing* instruments that measure glucose levels and other parameters in the blood. Many of these devices use advanced biosensors to determine glucose levels from 20 to 600 mg/dL across a 20 to 70% hematocrit range. Some of these devices can operate on very small amounts of blood (3 to 5 µL) to provide accurate results within 20 s. The digital versions of these devices typically have the following features:

- An audible signal that tells the operator that an adequate sample has been provided.
- Automatic initiation of the test procedure once a sample has been detected.
- A laser bar code scanning for data entry of information, such as patient ID, operator ID, and test strip information.
- Patient results are displayed on a backlit LCD screen.
- Results for up to 4000 patient tests and 1000 control tests can be stored.
- A data port and an optional docking station, which can be used to automatically downloads results to a point-of-care data management system, laboratory information system (LIS), or hospital information system (HIS).
- Data ports that can be used for bidirectional communication with the laboratory and nursing PCs.
- A bar code label mechanism that holds the calibration information about the test strip, including the lot number, the expiration date, and expected control solution ranges.

Pulse oximeters are typical examples of widely used portable biomedical instruments. These are devices that provide measurement of arterial blood oxygen saturation. The oxygen saturation level of blood is defined as the percentage of oxygenated hemoglobin divided by the total amount of hemoglobin (for a healthy person, this value is 96 to 98%). These noninvasive

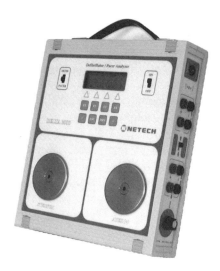

FIGURE 5.13
(See color insert following page 320.) A biomedical kit. (Courtesy of NeTech Corporation, http://www.gonetech.com/)

instruments do not need calibration or site preparation; the measurement is quick and does not require skilled operators. Pulse oximeters range in size from handheld, single-measurement devices to compacts with built-in measurement memory and printers. This device works by sending infrared light, through a finger or earlobe, to a photodetector on the other side. The amount of red light absorbed by the blood depends on the amount oxygen in the blood, while absorption of the infrared light is constant. During arterial pulse, the fractional change in signal is due only to the arterial blood. This method leads to the reduction of most uncertainties commonly present during transmission of light in tissue and eliminates the need for calibration. These devices are commonly available, and some manufacturers offer portable biomedical instruments in kit forms. Also, manufacturers offer additional devices that combine the functions of a number of instruments in one unit, as illustrated in Figure 5.13. This particular device operates as a defibrillator, pacemaker analyzer, and ECG arrhythmia simulator and has the following features:

- Menu driven operation and LCD display.
- RS-232 interface.
- Two 9-V alkaline batteries or AC operation.
- Defibrillator tester operates on automatic or semiautomatic manner. It measures energy, peak voltage, current and cardio sync.
- Pacemaker analyzer tests all external pacemakers.
- ECG simulator can provide 12 arrthymia waveforms.

A portable *apnea monitor* is a cardiorespiratory monitoring device that detects apnea, a life-threatening occurrence of respiratory arrest in infants and adults. These devices are used in hospitals and in homes to record the apnea events and generate real-time alarms. Two physiological parameters are monitored: the heart rate and respiration rate. The heart rate is usually measured by analyzing the signals. The respiratory measurement typically is based on impedance pneumography, which is the measurement of the chest impedance.

A portable *stethoscope* produces powerful, crystal clear amplification of heart sounds, respiratory sounds, and faint murmurs. They have fingertip volume control and a filter switch to differentiate sounds of variable intensity and pitch. They are equipped with special noise reduction circuitry to minimize the external noise. The auxiliary output is used for connection to external devices such as printers and computers. In some advanced versions of stethoscopes, the application of software comes with the device to record sounds and phonocardiograms. The PC-based software with the analyzer captures sounds as Microsoft wav files. These files can be stored, called up at a later date, and e-mailed. The sound files can also be viewed as charts and graphs, allowing the practitioner to visually compare the patient's condition and progress over time.

In recent years, telemetry applications have resulted in substantial progress in biomedical instruments. At least two areas of telemetric biomedical instrumentation are worth discussing in detail: human biotelemetry and animal biotelemetry.

5.4.1 Human Biotelemetry

For humans working in hazardous environments, telemetry is a way to ensure their safety. In these cases, instruments equipped with telemetry are used to detect the potential onset of distress to take the necessary precautions. For example, in manned space flights, the heart rates of the astronauts are one of the signals sent to ground stations. The astronauts' ECGs are monitored by using surface electrodes that at the same time can be used to monitor respiratory rhythm. During the first manned landing on the moon, the voices of the astronauts were combined with 900 other signals, some of them physiological parameters from the astronauts.

The use of medical telemetry systems is on the rise worldwide despite a growing problem of overcrowding on the airwaves. Because this technology is used to monitor at-risk patients who may experience cardiac problems during ambulation, a clear interrupted transmission from the patient to the monitoring station is critical. Different countries are approaching the congestion of the airwaves in different manners, ranging from assigning less crowded regions of the electromagnetic spectrum to telemetry services to developing modulation and compression algorithms that minimize the amount of bandwidth used by these systems. In the U.S., the Federal Communications Commission (FCC) is the agency responsible for assigning portions of the electromagnetic spectrum to users other than the federal government.

Until recently, telemetry services, for example, in a hospital were assigned a band of the electromagnetic spectrum shared with other services. This meant that the telemetry services had to accept the interference from the services that shared the bandwidth. In 1998, experimental digital high-definition television signals strayed into a hospital's airspace and obliterated vital signs monitors for at least 12 heart transplant and open-heart surgery patients monitored by telemetry inside the hospital. Once the problem was pinpointed to the experimental signals for the station that shared the band with the telemetry services, the station agreed to cease the experimental broadcastings until the hospital could adopt solutions to prevent further interference and possible hazards to patients. This was a wakeup call to all the telemetry services in the country, as they became aware of the higher degree of possible interference coming from stations experimenting with digital signals. As a result of this incident, the FCC issued new rules for medical telemetry in which it assigned portions of the spectrum to telemetry services to operate in an interference-protected basis. The new rule will also assign new portions of the spectrum to clinical telemetry on a primary basis. With this approach, hospitals will have reasonable assurance that the next generation of medical telemetry devices will operate in this expanded spectrum with minimal interference over their useful lifetime. This will, in turn, protect the patients' safety.

A telemetric nurse-based biomedical instrumentation system is illustrated in Figure 5.14. Figure 5.14(a) shows telemetry bedside monitors. It has a 7-in. CRT screen with two waveforms. The vital signs can be graphically or tabularly displayed. The vital signs of patients, such as ECG waveforms, are transmitted, normally in a wireless manner, to a remote station for observation and recording. Figure 5.14(b) shows patient transmitters for wireless bedside monitors. The base station, in Figure 5.14(c), can monitor waveforms, data, trends, and alarms of up to eight telemetry patients. It has dual-antenna diversity reception, 100-arrhythmia event memory, and a vital signs list. Other features of this system are:

- Operates on touch screen monitors with screen menus
- Uses special protocols to support up to 300 beds using TCP/IP
- Transmitter operates up to 10 days on single set of batteries

Wireless telemetric methods free patients from being bedridden, allowing them to move in a limited area since mild exercise is an important facet of recovery from cardiac illnesses. Yet even this limited exertion can sometimes lead to another heart attack, which fully justifies the careful monitoring of such patients. ECG telemetry devices monitor patients for cardiorespiratory dysfunction while allowing them to be ambulatory in the hospital before

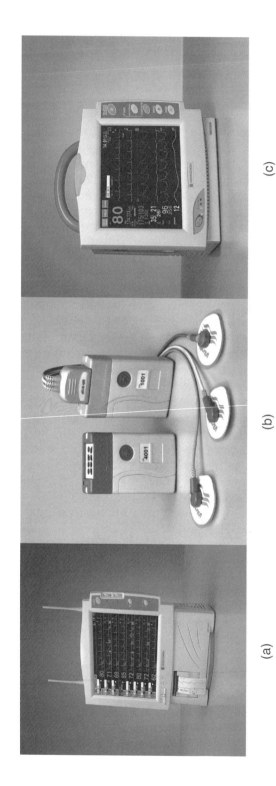

(a)   (b)   (c)

FIGURE 5.14
(See color insert following page 320.) A telemetric patient monitoring system. (a) A bedside monitor. (b) A wireless transmitter. (c) A base station. (Courtesy of Nihon Kohden Corporation, http://www.nihonkohden.com/)

they are discharged. This freedom to move about during recovery shortens overall recovery time and improves clinical outcomes for cardiac patients.

Today's telemetry systems typically consist of a small digital transmitter worn by the patient, a set of electrical leads for gathering patient vital signs, a distributed receiving antenna system, and a modular receiver. Displays, controls, and recordings are often controlled from a central computer located at the nursing station. In addition, entire units dedicated to the diagnosis and treatment of chest pain are rapidly emerging. These clinics triage patients presenting chest pain in the emergency department and often use telemetry systems during the standard 3- to 12-h observation period. Over the past several years, manufacturers have made tremendous strides in improving the design and functionality of telemetry systems, allowing more acute patients to become ambulatory during recovery. These systems are now capable of monitoring several different parameters, from ECG waveforms, arrhythmias, and heart rate to pulse oximetry and other vital signs, thus increasing the concept of cardiac telemetry to a whole range of clinical parameters.

Figure 5.15 illustrates an underskin sensory system that provides real-time information on the position of the lower extremities and the trunk of a patient. The sensor pack is microprocessor controlled and contains a miniature rate gyroscope and a pair of two-dimensional accelerometers. A navigator, which is worn on the body, acts as the controller. The navigator runs software customized for each patient, collecting body position data from the sensors, allowing it to implement closed-loop control strategies. Menu-driven operation is displayed on a touch-sensitive color display. The communication takes place via USB/RS-232 protocol. The navigator powers and controls the implanted sensors via a transmit coil that is held in place on the skin surface, thus forming RF data.

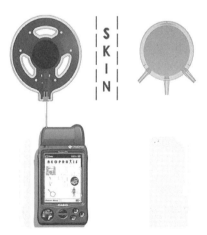

FIGURE 5.15
(See color insert following page 320.) An underskin sensory system equipped with telemetric features. (Courtesy of Neopraxis Pty. Ltd., http://www.neopraxis.com.au/)

Modern cardiac pacemakers, for example, rely on telemetry to improve the quality of life of the patients who wear them. The telemetry process is initiated by the cardiologist by sending a special signal to the pacemaker, normally by placing the transducer on the patient's chest. This telemetry is normally bidirectional, and it is used from the physician to the patient to program a different code or pacing algorithms in the implanted pacemaker, and from the pacemaker to transmit performance and clinical data back to the cardiologist. The typical data that pacemakers normally telemeter include the voltage, impedance, and current consumption of the battery; the indication of low battery if necessary; the impedance of the lead contact, voltage, current, width, charge, and energy of the delivered pulses; the count of sensed and paced events; and the ECG. This last signal has a very important value for the cardiologist, as it records the electrical activity of the heart without the distortion that the signal suffers when it is recorded superficially. All the performance parameters of telemetry can be programmed to be read back or interrogated, which allows confirmation of the programming of adequate parameters in the pacemaker.

For patients who are located in rural or isolated areas and need routine monitoring of their physiological parameters, wireless transmission is not possible. In these cases, it is necessary to use the public switched telephone network for telemetry purposes. In some cases, simultaneous speech and data communication are useful to aid patients in the self-use of a defibrillation device and also for patients who need to self-administer drugs, while the physicians need to monitor their vital signs to prevent undesirable effects. This increased bandwidth that needs to be transmitted demands the use of some kind of preprocessing of the data in order to accommodate all the signals in the telephone bandwidth.

5.4.2 Animal Biotelemetry

Life sciences have benefited a great deal from the use of telemetry techniques. Although in some cases telemetry does not add anything else (rather than unnecessary complexity), in other situations it either makes an otherwise impossible experiment feasible or yields in more productive results being obtained in a given time. These methods prove important, particularly in those situations where it is desirable to leave the subjects in a relatively normal physiological and psychological state by interfering with their normal pattern of activities as little as possible. With uncooperative animals or wildlife, this brings additional value. For example, the fact that a very small radio transmitter can be swallowed opens the possibilities for study of body regions that would otherwise be difficult to explore, while leaving the subject totally unaware of its presence. Also, the absence of electrical leads connecting the instrumentation to the subject's body is especially important in long-term studies, as they can greatly interfere with the subject's normal activities.

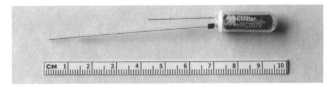

FIGURE 5.16
(See color insert following page 320.) An implantable sensor/telemetry transmitter. (Courtesy of Mini Mitter Co., Inc., http://www.minimitter.com/)

Implanted transmitters have been used to study habits and patterns of animals. In these cases, the transmitter is surgically implanted in the body of the animal. In these types of applications, the life of the battery and the stability of the recording transducer become critical to ensure the success of the project. This type of approach has been used, for example, to study peristalsis on cold-blooded animals by using a simple transmitter of pressure and temperature implanted in a mouse that was subsequently swallowed by a snake. The researchers recorded temperature and pressure data for 22 days until the transmitter was passed in the usual way.

Figure 5.16 illustrates an implantable sensor/telemetry transmitter for measuring the temperature, gross motor activity, and heart rate of an animal. This sensor/transmitter arrangement does not require a battery.

The range of the transmitter also becomes an important parameter to consider while using telemetry to study animals. For example, it can be from a few centimeters in laboratory experiments to several miles while tracking wildlife. Another critical parameter is the selection of the transducer, as it is necessary to accurately sense the desired range of physiological change, while at the same time, it needs to be resistant against corrosion from body fluids. A third important parameter to consider is the range of temperature that the telemetry transmitter will have to withstand. This range can be from subfreezing temperatures to the strong heat of the desert. The lives of the batteries that power the transmitter will be seriously affected by the operating temperatures.

5.5 Mechanical, Sound, and Vibration Measurement Instruments

5.5.1 Mechanical Measurement Instruments

Mechanical parameter measurements can be divided into:

- Acceleration and velocity measurements
- Mass, volume, density, and weight measurements
- Power and torque measurements

- Force, stress, and strain measurements
- Displacement measurements
- Rotation and distance measurements
- Pressure measurements

There is a diverse range of electronic portable instruments to measure all the parameters mentioned above. Only a few will be discussed here.

At first sight, it appears that in most of the mechanical measurements, it is necessary to attach sensors to the subject under investigation. This is true in many applications; however, in recent years, advances in telemetry technology have enabled detachment of sensors from the main processing unit. That is, an instrument can now be built having different components separated from each other. Figure 5.17 illustrates such an arrangement. This system transmits data from rotating shafts or machinery to a stationary receiver. It has the following features:

- Power to transmitter is supplied by a 9-V battery or an induction power supply
- The receiver is powered by a 12-V DC supply
- Transmitters can be configured for strain, torque, pressure, voltage, and temperature measurements
- The system is immune to electromagnetic interference, and sealed appropriately for dust, oil, moisture, etc.
- Remote calibrations and automatic operations are possible

FIGURE 5.17
(See color insert following page 320.) An RF telemetric measurement of the parameters of rotating shafts. (Courtesy of Advanced Telemetrics International (ATI), http://www.atitelemetry.com/)

A typical telemetry transmitter allows sensors to be wireless, thus bridging the gap between moving sensors and stationary data recorders. The data obtained from a rotating shaft can be monitored at all times within a distance of up to 3 to 4 km. The transmitters can be coupled to measure torque, temperature, vibration, strain, displacement, thrust, deformation, and so on.

The noncontact, portable electronic displacement and gap measurement devices are based on capacitive sensors. There are many different types of probes (button, threaded, cylindrical, etc.) that can be changed to suit the application requirements. Many of these instruments are microprocessor based, thus giving them all the advantages of digital systems.

Electronic scales and balances constitute an important part of portable instruments. There are many different types available suitable to use in a diverse range of applications from domestic scales to ultraprecise laboratory balances. A typical example of balance measures microweights with 0.1-µg readability. In these instruments, the temperature compensation and the elimination of the effects of static electricity are extremely important. Some instruments contain digital outputs via the RS-232 to meet stringent traceability requirements. The software takes care of density determination, tare memory, check weighing, diameter determination, counting, and back-weighing. As an example, an electronic scale is depicted in Figure 5.18. This particular device is termed a handheld scale with a 320-g capacity. It operates on two 1.5 V AAA batteries. TS readibility is 0.1 g with a linearity of –0.2 g. Its stabilization time is 3 s.

Another example of electronic portable instruments in mechanical applications is in velocity and speed measurements. Velocity and speed measure-

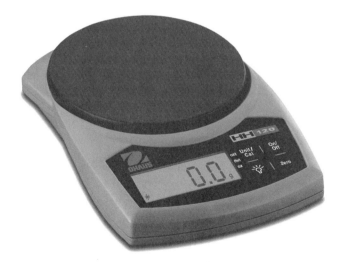

FIGURE 5.18
(See color insert following page 320.) A typical electronic scale. (Courtesy of Ohaus Corporation, http://www.ohaus.com/)

FIGURE 5.19
(See color insert following page 320.) A handheld laser speed detector. (Courtesy of Lion Laboratories Ltd., and MPH Industries, Inc., http://www.lion-breath.com/; http://mphindustries.com/)

ments can be done by using many different methods, depending on the application. A typical speed-measuring instrument is depicted in Figure 5.19. This device is used for 16- to 270-kp speed detection of vehicles by the police. Its accuracy is −1.6 kph detecting vehicle speeds at distances of 1.6 kilometers. It uses a two-stage trigger at K-band frequencies for positive target identification and track-through lock. The batteries of the device allow 8 h of continuous use.

Other types of speed detectors operate on laser beams and they use time of flight or Doppler shift principles. A typical laser speed gun emits a very narrow and intense beam, with a frequency of 904 nm, which reflects back from a vehicle. The speed gun calculates distance by measuring the length of time it takes for the beam to travel to the target and back to the gun. Many such readings are taken over a brief period of time to determine how fast a vehicle is traveling.

5.5.2 Vibration Measurement Instruments

Vibration, jerk, and shock measurements are mostly conducted largely by using accelerometers together with suitable signal processors. The applications can vary from the chemical industry to airplanes. Figure 5.20(a) illustrates a typical arrangement of a telemetric vibration sensor used in heavy industry. It operates on the basis of strain gauges. The measured signals are amplified and sent to stator inductively. In other types of vibration sensors, accelerometers are bolted on the shaft together with the RF transmitter. In these applications, the transmitter needs to be located close to the test piece, 3 or 4 m.

Another type of telemetric vibration/stress mechanism, depicted in Figure 5.20(b), can be regarded as a portable instrument that operates on microwave

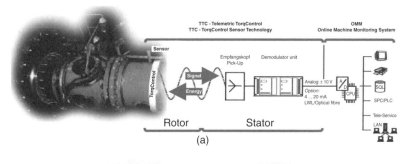

(a)

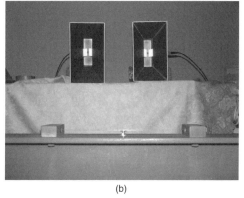

(b)

FIGURE 5.20
(See color insert following page 320.) (a) A telemetric vibration/torque sensing arrangement. (Courtesy of ACIDA TorqControl, GmbH, http://www.acida-torqcontrol.de/). (b) A microwave telemetric vibration/stress sensing arrangement. (Courtesy of VIBSTRING, http://www.vib-string.com/)

principles. The energy is transmitted from a stationary microwave generator to be reflected back from an oscillating electrical conductor (musical string) that is located on the vibrating member. The musical string is firmly attached between two points on the measurement object and pretensioned to a selected resonance frequency. When the beam is strained along the string direction, the string tension level changes thereby altering its resonance/ filter frequency. In effect, the suspended string becomes a tunable mechanical filter operating as a strain transducer. The reflected microwave signals are essentially amplitude modulated to be picked up by the microwave receiver. Minute variations in the pretensioned string oscillations can be picked up easily. This allows the detection of the vibrating string frequency at distances of many meters. This system can be configured to sense other parameters such as torque, and stress. For example, in a torque sensor configuration, 4 vibrating strings are used to sense 2 main stress direction in 2 different planes. The strings are set to 400, 700, 1000, and 1300 Hz resonances at zero shaft loads. Each of these string frequencies would then have a freedom of moving ±200 Hz with varying shaft strain loading, without interrupting each others' frequency bands.

5.5.3 Sound Measurement Instruments

Sound measurements are frequently performed to evaluate the level of noise generated by a product or process for regulatory acoustic noise compliances inside or outside the workplace. Sound level meters (SLMs) are instruments that measure the amplitude of a sound pressure wave or sound pressure level (SPL) at a selected measurement point according to the following formula:

$$SPL = 10 \log P^2 / P_r^2 \qquad (5.1)$$

where P is the sound pressure in the audible frequency range (20 Hz to 20 kHz) and P_r is the threshold of hearing (20 µPa).

SPL is the simplest form of sound measurement. It produces a scalar value that does not provide information on the direction of the sound wave, and it cannot be used to identify the source of the sound. The sound intensity is a vector value that includes the velocity of the sound. The sound velocity is difficult to measure, but it can be approximated as a sound pressure gradient between two points. Modern sound level meters perform most of the required processing in the digital domain. A typical SLM picks up the acoustic signal by microphone, to be amplified and filtered before being digitized by an A/D converter. The SLM measurement algorithm typically performs the A-weighting filtering as an approximation to the spectral sensitivity of the human ear, or octave filtering per standard regulatory specifications. In the latter case, the signal power is determined by squaring and averaging the derived signal, and a log conversion is made for the display. Because most SLM instruments are digital, they offer many convenient and measurement features, such as maximum and minimum and average levels of sound over a period of time.

Advanced SLM instruments measure only the overall intensity of the sound in the frequency band of interest and do not provide spectral analysis of the sound or its temporal characteristics. In these instruments, apart from the input amplifier, all signal processing is done digitally. An extensive memory capability permits large amount of data to be stored and transferred to a PC via an RS-232 port. Instrument capabilities can be expanded to filters and automated or manual frequency analysis, reverberation time calculations, logging of sound levels vs. time, auto-sequencing, event recording, and various statistical calculations. They can be configured for vibration measurements. Other typical features of those instruments are:

- 240×64 pixels LCD display with backlight switch that can be turned on or off.
- Control of an instrument is divided between hard and soft controls. Hard controls are implemented by push buttons, while soft controls are accessed via the instrument delay.

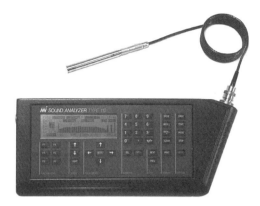

FIGURE 5.21
(See color insert following page 320.) A sound level meter. (Courtesy of NorSonic AS, Norway, http://www.norsonic.com/)

- Dynamic range: 90 dB.
- 22 to 135 dB measuring range with 50 mV/Pa microphone sensitivity.
- Measuring time varies between 2 ms to 99 h.
- Two independent RS-232 interfaces.
- Converter type is sigma delta and sampling rate is 50 kHz.
- Frequency range: 1 Hz to 20 kHz.
- Digital filter is 8-pole Butterworth IIR filter.

A typical sound level meter is illustrated in Figure 5.21. A simple SLM instrument measures only the overall intensity of the sound in the frequency band of interest and does not provide spectral analysis of the sound or its temporal characteristics. A typical SLM is a small, one-channel handheld device with a built-in microphone and is powered by two AA batteries that can last for about 2 days of continuous operation. The frequency range of the meter is from 20 Hz to 8 kHz. These devices are useful for quick field measurements with a measurement range from 30 to 130 dB. Further, the improvement of performance SLM can be achieved by the implementation of several types of weighting filters (e.g., A, C, and D), enhancing the frequency range to 20 Hz to 20 kHz.

Another portable sound measurement device is the sound analyzer. Opposed to SLM, these devices are larger in size, battery powered, perform real-time FFT analysis, offer 1/1 and 1/3 octave digital filtering, graphically display sound spectrum and have some other features normally present in computer-based instruments (e.g., data storage, built-in printer, etc.). Sound analyzers usually do not have built-in microphones and require separate high-performance microphones for precision measurements. Also, many existing FFT analyzer instruments are designed for use in sound measurements and offer some of the same features as the sound analyzers.

Many commercial sound analyzers provide a comprehensive range of functions required for sound and vibration. They typically operate 1/3-octave and full-octave bands, giving power, maximum, minimum, and percentile spectra. Sequential measurement mode also includes 1/12 and 1/16-octave bands with freely selectable center frequencies. The frequency range covers 1 to 20 kHz with a 0.1 dB resolution. Other acoustic functions include reverberation time measurements and spectrum averaging. Direct connection of accelerometers allows to the subject under investigation vibration levels in decibels or engineering units. They have event modes that allow the detection of threshold-based events to trigger a standard data recorder to capture the event. They can be programmed to carry out complex sequences of measurements switching between time and frequency mode automatically to give a complete picture of an acoustic environment. Serial ports allow direct connection to printers or personal computers. Internal nonvolatile memory can hold up to 100,000 results. The display gives high-resolution results in numeric and graphic formats. They come with data transfer and postprocessing software for specialist applications.

5.6   Radiation, Light, and Temperature Instruments

Radiation is the emission and propagation of energy through space. There are two types: (1) particles or subatomic particles (e.g., protons, neutrons, and electrons) moving at high speeds, and (2) electromagnetic radiation, including x-rays, gamma rays, ultraviolet rays, visible light, infrared rays, microwaves, and radio waves. Electronic portable instruments are available to measure both types of radiation; some examples are given here.

5.6.1   Long-Wave Radiation Instruments

There are different types of long-wave radiation. Thermal radiation measurements are based on the exchange of energy between hotter and cooler objects. The infrared (IR) devices are suited for long-wave radiant energy measurements. All objects (with temperatures above $-273°C$) radiate infrared energy with an intensity relative to the temperature of the object. As an object becomes hotter, the peak spectral response moves from the infrared to the visible spectrum. IR applications range from spectroscopy, process control, gas detection, security systems, and military instruments to medical instruments.

Figure 5.22(a) illustrates a handheld infrared thermal detector. It can be used as a digital camera for visible as well as thermal images. The visible image makes it easy to pinpoint the position with thermal images. Typical

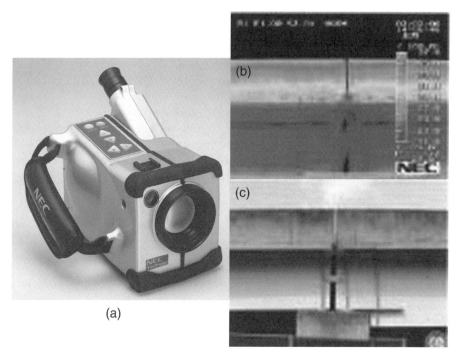

FIGURE 5.22
(See color insert following page 320.) A short-range infrared detector. (a) The detector. (b) The infrared image. (c) The visual image. (Courtesy of NEC San-ei Instruments, Ltd., http://www.necsan-ei.co.jp/)

thermal and visible images are shown in Figure 5.22(b) and (c). Other features are: Temperature measuring range: −20 to 250°C with 0.08°C resolution. It uses uncooled Focal Plane Arrays, and UFPA detectors. One hundred thermal images can be stored on a 16-MB memory card and the memory can be extended to 832 images. It operates 110 minutes on a single battery. Data can be downloaded to a PC via IEEE-1394 interface. Focusing range is from 50 cm to infinity. Environmental protection is IP-54 (dustproof and splash proof). Many of the infrared detectors find application in hostile environments such as coke ovens, metal processing, glass and glass fibers, borosilicates, refractories, ceramic kilns, etc. These devices have built-in eye protection filters for high temperatures. The internal display shows temperature, while the external display shows temperature, emissivity, out of range and battery status. Adjustable focus capabilities, through-the-lens viewing, allow sighting the target while reading temperatures. Temperature ranges can vary from a few degrees to 3000°C with an accuracy of about ±0.5% of reading. In some instruments, the narrow spectral range of 0.8 to 1.1 m allows measurements through glass ports and minimizes errors due to emissivity and atmospheric absorption. A standard RS232C digital output sends data to a PC or a printer.

FIGURE 5.23
(See color insert following page 320.) A photometer. (Courtesy of International Light, Inc., http://www.intl-light.com/)

5.6.2 Short-Wave Radiation Sensing

Solar light is a typical example of short-wave radiation. The light is radiation primarily in the wavelength range of 0.25 to 4 μm (99% of the total incoming solar radiation is in this range). In practice, short-wave instruments generally measure in the range of 0.3 to 3 μm. Solar radiation flux densities vary significantly among regions due to season, time of day, surrounding terrain elevation, and obstructions. Short-wave solar radiation can be separated into two components: the direct beam and the diffuse beam. The direct beam is the portion of radiation that reaches the Earth's surface in relatively parallel beams. The diffuse beam is the portion of radiation that has been scattered by gas molecules and suspended particles in the atmosphere and reaches the Earth's surface from multiple directions.

Pyranometers are typical portable instruments used to measure direct and diffuse short-wave radiation arriving from the whole hemisphere. Pyranometers are largely used in climate prediction applications. Some pyranometers can be used to measure reflected short-wave radiation from a surface. In addition, with the aid of shading devices that can obstruct the direct solar beam, pyranometers can measure diffuse radiation as well.

Photometers are another type of instrument that is suitable for short-wave radiation measurements. They are commonly used often in photographic applications and in cameras. Figure 5.23 illustrates a typical microprocessor-controlled photometer. It has the following features:

- 10-decade dynamic range
- 10 calibration factors in memory

- Dynamic range: 0.2 pA to 2.0 mA
- Linearity: ±0.1%
- Input power: 6 C-cells or 8 to 15 V external DC
- RS-232 and TTL outputs

The calibration of these devices may be made for any nonstandard units: W/cm^2, W, lm/ft^2, lx, $E/sec/m^2$, cd/cm^2, cd/m^2, A (ampere), $W/cm^2/sr$, and $W/cm^2/nm$. In the integrate mode: J/cm^2, J, $lm \times sec/ft^2$, $lx \times sec$, E/m^2, $cd \times sec/cm^2$, C (coulomb), $J/cm^2/sr$, and $J/cm^2/nm$.

5.6.3   Dosimeters

Dosimeters are radiation measuring devices that are used in medical instruments, pharmaceuticals, nuclear weapons research and manufacturing, genetic engineering, nuclear reactor plants, accelerators, biomedical and materials research, space radiation studies, food processing, environmental monitoring, and so on. They need to be reliable and consistent for administrative, legal, technical, medical, educational, and psychological reasons. During the use of dosimeters, the following factors need to be taken into consideration: (1) type of radiation, (2) dose range, (3) accuracy of readings, (4) convenience of the readout, (5) cost factors, and (6) location of the dosimeter.

Dosimeters make use of principles of interaction of radiation with matter. Different types of radiation react with matter in different ways. The interactions between radiation and matter can be divided into heavy positively charged particles, beta particles, gamma and x-rays, and neutrons. As radiation interacts with matter and penetrates into the material, it can cause ionization. Radiation ionization takes place when one or more of the electrons in an outer orbit are knocked loose and the atom becomes a positive ion. In nonionizing radiation, the atoms are intact but the radiation energy is absorbed. These include ultraviolet radiation, visible light, infrared radiation, and radio waves.

There are many different types of dosimeters since one instrument cannot detect all types of radiation. A typical handheld microprocessor-based instrument is illustrated in Figure 5.24. This instrument can measure low levels of alpha, beta, gamma radiation, and x-rays. It has a red light flasher and audible beeper sound for warning. The typical specifications of such instruments are:

- Sensors are suitable for alpha, beta, gamma radiation, and x-ray detection
- Internal tubing is a halogen-quenched uncompensated GM tube with mica window
- Operating range: 1 R to 100 mR/h, 0.010 Sv to 1000 Sv/h, 0 to 300,000 counts per minute (CPM)
- Sensitivity: 15% from 0 to 30 mR/h, and 20% from 30 to 100 mR/h

FIGURE 5.24
(See color insert following page 320.) A typical digital dosimeter. (Courtesy of S.E. International, Inc., http://www.seintl.com/)

- Gamma sensitivity: 3500 CPM/mR/h referenced to Cs-137
- Outputs: dual miniature jack provides counts to a computer or data logger
- Power: one 9-V alkaline battery; battery life of 2000 h at normal background
- Software: IBM PC-compatible on 3.5-in. floppy disk

5.6.4 Temperature Measurements

Electronic temperature measuring devices are the most commonly used portable instruments. Consequently, they come in many different shapes and sizes to suit different application requirements. They find a wide variety of applications from heavy industry and the health industry to domestic applications. Consequently, there are many different types manufactured in all shapes and sizes. A typical thermometer operating on solar power is illustrated in Figure 5.25. This solar-powered digital thermometer has a temperature range from –50 to 150°C with 1% accuracy. It has a backup lithium battery and generally finds applications requiring fast and accurate temperature readings.

5.7 GPS and Telemetry-Based Instruments

Global positioning system (GPS) technology is used in many applications, from personal use for hiking, hunting, camping, boating, flying, and driving

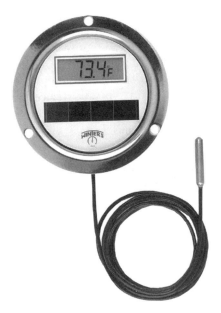

FIGURE 5.25
(See color insert following page 320.) A solar power digital thermometer. (Courtesy of Winters Instruments, Inc., http://www.winters.ca/)

FIGURE 5.26
(See color insert following page 320.) A typical GPS receiver suitable for navigations. (Courtesy of Garmin Ltd., http://www.garmin.com/)

to industrial applications for fleet tracking and geological surveying. GPS devices find extensive military applications from aircraft and missiles to individual soldiers. Consequently, there are many competitors in the marketplace offering a diverse range of GPS products. These products vary from simple handheld receivers to mountable systems backed up with accessories to perform specific tasks, such as ship and aircraft navigation. Prices range from a few hundred dollars to thousands of dollars per unit.

A typical GPS receiver, illustrated in Figure 5.26, offers cartographic capabilities in automotive navigation and marine chart plotting units by plugging it into a PC. Such a device is used for navigation purposes assisted by highly sophisticated maps. Typical features of this device are:

- It offers access to shortest and fastest routes, directions and estimated arrival times to a destination
- With a touch button, locations of the nearest restaurants, highway exits, gas stations, ATMs, hospitals, etc., can be viewed
- Built-in base map with oceans, lakes, rivers, cities, airports, interstate highways, borders, etc.
- Voice navigation instructions and warnings
- PC interface: RS-232
- It has a differential-ready 12-parallel channel GPS receiver that continuously tracks and uses up to 12 satellites
- Update rate is 1 s, continuous
- GPS accuracy: 15 m with a velocity 0.05 m/s steady state
- Differential GPS accuracy: 3 to 5 m, 95% typical
- Power: Six AA batteries with a battery life 2 to 20 h

Many of these GPS devices have multichannel receivers with submersible waterproof constructions. Some of these devices have large databases that include highways, major roads, parks, waterways, airports, cities, etc. Their capabilities can be enhanced to show rural roads, waterway information, and other maps in 8-, 16-, 32-, and 64-MB sizes. Other features include backtrack, built-in help, North finder with sun/moon positions, and sunrise/sunset calculator. They can support several languages.

Another example of GPS receivers, shown in Figure 5.27, is used for surveying applications. This particular receiver combines in the same unit a dual-frequency GPS receiver, UHF radio and power source. Bluetooth technology is used for short-range data communication. Other features include the following:

- Code differential GPS positioning: ±0.25 m + 1 ppm RMS horizontal and ±0.50 m + 1 ppm RMS vertical
- Surveying: ±5 mm + 0.5 ppm RMS horizontal and ±5 mm + 1 ppm RMS vertical
- Uses Advanced Maxwell 4 Custom Survey GPS chip
- PC interface: RS-232, 450 MHz or 900 MHz UHF radio modem, and 2.4 GHz fully integrated Bluetooth communication port

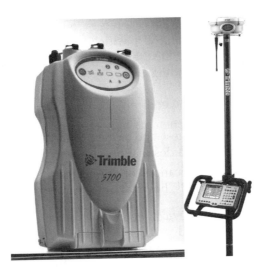

FIGURE 5.27
(See color insert following page 320.) A surveying GPS receiver equipped with RF transmitters. (Courtesy of Trimble Navigation Ltd., http://www.trimble.com/)

- Data storage: 128-Mb memory for over 3400 raw data
- Construction: IPX7, 100% humidity condensation

Many similar devices integrate GPS receivers, RF communications, survey controllers, and software. They are capable of collecting real-time dynamic data of mobile objects. They use RF transmission for communications while the object is on the move.

5.8 Computer-Based and Intelligent Instruments

There are a number of different types of computer-based electronic portable instruments. In the first type, laptop and Palm computers are used as portable instruments. In the second type, computers are used as base stations receiving information from portable instruments and intelligent sensors via telemetry. In the third type, the instrument itself is powerful enough that it contains a computer as part of its normal operation.

Computer-based instruments are equipped with data acquisition cards (DAQs) together with appropriate software such as LabView. They provide the measurement of physical variables in a digital environment as well as enable the control of instruments if they are equipped with telemetric communication channels. Ethernet data acquisition systems (EDAS) are also finding applications in portable instrumentation systems. They are based on open architecture standards allowing for development and deployment into existing and future plant-wide data capture systems. EDAS units allow users to remotely monitor and control a wide range of sensors, instruments, machines, and processes over the Ethernet network. EDAS units are used in a range of applications, including:

- Machine monitoring and control
- Networked or stand-alone data logging
- Environmental monitoring and control
- Remote data acquisition
- Security and access control

Manufacturers of such devices generally produce wireless broadband equipment, including extended-range wireless local area network (LAN) systems and wireless Internet solutions for service providers. In a typical assembly, wire-free networking can provide an 11 Mb/sec over-the-air data rate to connect remote stations as far as 50 km away. Multiple security mechanisms work to prevent unauthorized access to both wired and wireless network resources. These devices support point-to-point or point-to-multi-point bridging operations. They also support roaming and, with a wide range of third-party opens, air devices. They generally have a frequency-hopping spread spectrum modulation suitable particularly for mobile computing, and wire-free networking is all about the freedom to roam.

A typical example of equipment for wireless computer networks suitable for portable instruments is given in Figure 5.28. This wireless outdoor system operates on solar power. It can monitor temperature, humidity, barometric

FIGURE 5.28
(See color insert following page 320.) A solar powered RF networkable weather station. (Courtesy of Oregon Scientific, Inc., http://www.oregonscientific.com/)

pressure, wind speed and direction, and rainfall. A wireless link (433 MHz) allows networking of up to seven similar devices on to PCs within a 100-m range. With the use of repeaters, the range can be extended up to a 48-km radius. Other features of the device are:

- Temperature range: −50 to 70°C with a 0.1°C resolution
- Relative humidity: 2 to 98% RH with a 1% RH resolution
- Barometric pressure: 795 to 1050 mb with a 1-mb resolution
- Wind speed: 0 to 56 m/s with a 0.2 m/s-resolution
- Rainfall: 0 to 999 mm/h with a 1 mm/h-resolution
- Power requirement: solar power or four AA and AAA batteries

Handheld computers or Palm computers are also used as portable instruments. Most of these computers are powerful enough to be able to act as any portable instrument. They can send and receive data files in a wireless manner with other digital devices via infrared or RF ports. These computers are used, for example, in education, as real-time monitoring instruments. Some of the applications include environmental monitoring, such as temperature changes over time, atmospheric and water pollutions, and so on.

5.9 Domestic, Personal, Hobby, and Leisure Portable Instruments

Domestic, personal, hobby, and leisure instruments constitute a large industry with many competitors in the marketplace. Because of the market size, a diverse range of instruments is offered, from simple disposable types to highly sophisticated and robust instruments. In some of these instruments, the power efficiency is of paramount importance, as in the case of biomedical applications, while in others, power efficiency is not considered at all, as in the case of many disposable instruments such as some toys. Some examples of domestic, personal, and hobby instruments are discussed below.

A typical example of an electronic portable instrument is illustrated in Figure 5.29(a). This device is a wristwatch with a built-in GPS receiver capability. It provides current position latitude and longitude in degrees, minutes, and seconds. Position determination is carried out continuously in one-second intervals. Other features are as follows:

- 200 landmark memory with a 16-character name
- Memory for 400 measurements for latitude, longitude, and altitude
- Navigation and track capabilities

(a)

(b)

FIGURE 5.29
(See color insert following page 320.) (a) A GPS altimeter wristwatch. (Courtesy of Shriro Australia Pty., Ltd., http://www.shriro.com/). (b) A barometric altimeter wristwatch. (Courtesy of Oregon Scientific, Inc., http://www.oregonscientific.com/)

- A 16-channel parallel receiver
- Position accuracy 10 m RMS
- Windows PC data compatible
- Power requirement: rechargeable lithium iron battery

There are many other similar watches in the marketplace. They provide data access to other PCs and come with a built-in battery that can be recharged from an AC outlet. With the auto time correction activated, current time settings are set automatically based on the data received from satellites. The position measurements are displayed as latitude (degrees E or W, minutes, seconds) and longitude (degrees N or S, minutes, seconds). The navigation screen on the watch depicts the relationship of current position to the target destination, specified checkpoints, or landmarks. They also show the current heading, current speed, and maximum attained speed. The altimeter uses measurement data from satellites to produce three-dimensional data covering altitudes as well as latitudes and longitudes. The altitude data may be stored automatically for recall in a graphical form displaying altitude changes.

In Figure 5.29(b), an altimeter wristwatch is illustrated, which works on the principles of barometric measurements. This watch can measure altitudes up to 9000 m displaying ascending and descending speeds m/s. It also has other functions such as: heart rate monitor, barometer, thermometer, clock, and stopwatch. The presentation can be in bar graphs.

Another example of a portable instrument for personal use is the electronic motion detection alarm designed to alert the bearer if a door, window, or other personal property is touched or moved. Figure 5.30 illustrates a per-

FIGURE 5.30
(See color insert following page 320.) A personal alarm system. (Courtesy of D&D Security Products, Inc., http://www.ddsp.com/)

sonal motion detection system. Such devices can be attached to bicycles, motorcycles, golf clubs, luggage, bags, and other items containing valuables. They have a number of modes. Mode 1 is as a personal alarm, which has a strap attached to a pin that, when pulled, activates a sound. Mode 2 is a doorbell, attached to a door with the plate sounding musical notes when someone knocks. Mode 3 is the motion detection mode; it is deactivated with a programmable password.

The final example of electronic portable instrument given in this book is the metal detector. There has been a sizable range of metal detectors in the marketplace for hobbyists and industrial users. Recently, they are manufactured in fairly sophisticated forms containing microprocessors and other digital components. An example of such a metal detector is illustrated in Figure 5.31. This device uses pulse induction (PI) with adjustable pulse-width. Pulse-width control allows tuning the unit to detect metals at preset depths. These types of detectors are suitable particularly for saltwater applications since specially designed single loop coils are not affected by conductivity of the media. Conductivity of media can severely influence the operation of very low frequency (VLF) type detectors. This detector has two distinct operating modes: VCO and normal. In VCO mode, as the target gets closer to the coil, the threshold generates a louder and higher pitched tone. In the normal mode, a single tone is used for beachcombing activities and other prolonged searches. The technical specifications of these metal detectors are as follows:

- Operating frequency: 10 kHz
- Audio output: 220 to 450 Hz with 600 pps

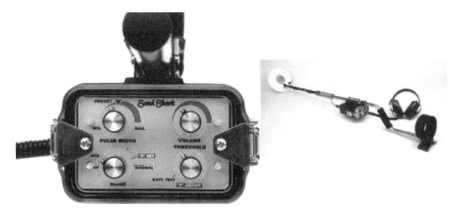

FIGURE 5.31
(See color insert following page 320.) A microprocessor-controlled metal detector. (Courtesy of A&S Company, Inc., http://www.detection.com/)

- 2 1/4″ speaker and headphone jack
- Battery requirement: Eight AA (alkaline)
- Battery life: 10 to 20 h
- Temperature range: 30 to 100°F
- Maximum depth: 70 m

There are many other types of low-frequency metal detectors, for hobbyists and industrial users, that utilize microprocessors to indicate the size and depth of the detected metals on LCD displays. The size of the metal is determined by the phase shift of the signal received, and the amplitude determines the possible depth.

A recent trend in portable metal detectors is by using Ground Penetrating Radar technology together with appropriate software that provides 3-D images. Ground penetrating radar detectors find many applications including noninvasive metal detection, archaeological, geological, construction, cave and tunnel locating, geotechnical, mining, prospecting, cable and pipe locating, geophysical, underground water explorations, etc. They use a range of antennas depending on the properties of targets to reach desired depths. Some of these devices use 100, 200, 300, 500, and 900 MHz, or 1.5 GHz antenna systems. The antennas can be surface coupled or air coupled, and shielded or unshielded.

6

Conclusions and Future Directions

Once you read this book, you can easily appreciate that electronic portable instruments are going through a revolutionary phase. With the availability and use of advanced technology on digital systems and communications, portable instruments are emerging as important tools that add value to our lives.

Portable instruments can communicate with other devices or digital systems while they are in service. They are used under all conditions, including remote environments where conventional instruments cannot operate. In parallel to the recent developments in communications, digital systems, intelligent sensors, microsensors, and virtual instruments are making significant progress in electronic portable instruments. The emergence of satellite communications and advances in other means of communication, such as optical, microwave, and radio frequency (RF), are greatly impacting the development and use of portable instruments. Also, the range of existing portable instrumentation now extends far beyond simple, low-accuracy portable devices to highly complex, network integrated, high-performance measurement systems.

The design of electronic portable instruments is centered on the trade-offs among the instrument cost, performance, limited size and weight, power consumption, user interface options, ruggedness, and ability to operate in a range of environments.

Today's portable instruments are largely software driven, and therefore can be programmed for the implementation of various measurement techniques. Their front-end and back-end hardware can be standardized and designed to address large classes of similar applications.

This book discusses three main types of portable instruments:

1. *Portable and handheld instruments* that are built for a specific application
2. *Intelligent sensor-based devices* that find wide applications in portable instruments
3. *Portable data systems* that contain fixed sensors and supporting mechanisms, but are capable of communicating with other digital devices and computers via communication networks

A general discussion in the initial stages is given on measurements, instruments and instrumentation systems, and their characteristics, such as errors and error control techniques, uncertainty, standards, design, testing, calibration, and their uses. Information on sensors and transducers that is highly relevant to portable instruments and digital techniques from theoretical, hardware, and software points of view have been elaborated. Details about design and construction of portable instruments and the associated techniques, together with subtleties and pitfalls, have been given. Some examples of commercial electronic portable instruments have been illustrated. Their technical specifications are provided in such a way that they can be related to the issues given in this book.

In my opinion, as technology progresses, electronic portable instruments will continue to change. Challenges will come from various directions, such as:

1. *Intelligent sensors* technology will progress considerably. Sensors will become smaller and more sophisticated, and, as a result, portable instruments will become smaller and more sophisticated.

2. *Integration of communication technology and improvement of networks* will improve the leading emergence of many different electronic portable instruments.

3. *Power supply* technology will shift from batteries to fuel cells and other alternative power supply methods. At the same time, battery technology will improve to better serve the needs of portable instruments.

4. *Research and design* efforts will concentrate on portable instruments, offering new added features.

5. *Growth in the manufacturing industry* will create competition, offering a diverse range of portables.

6. *Market acceptance* will increase, opening up new demands.

I hope you find this book informative and enjoyable.

Bibliography

Ackenhusen, J.G., (1999), *Real Time Signal Processing, Design, and Implementation of Signal Processing Systems*, Prentice Hall PTR, Upper Saddle River, NJ.

AT91 ARM Thumb, available at http://eu.atmel.com/atmel/products/prod35.htm (online).

AVR 8-bit RISC, available at http://eu.atmel.com/atmel/products/prod23.htm (online).

Baglio, S., Graziani, S., and Pitrone, N., (1996), An intelligent sensor for distance estimation based on fuzzy data fusion, *Trans. Inst. Meas. Control*, 18, 217–220.

Balarin, F., (1997), *Hardware-Software Co-Design of Embedded Systems: The POLIS Approach*, Kluwer Academic Publishers, Boston.

Bentley, J.P., (1988), *Principles of Measurement Systems*, 2nd ed., Longman Scientific and Technical, Burnt Mills, U.K.

Boashash, B., (1998), *Digital Signal Processing*, rev. ed., Queensland University of Technology, Signal Processing Research Centre, Brisbane, Australia.

Bollinger, J.G. and Duffie, N.A., (1989), *Computer Control of Machines and Processes*, Addison-Wesley, Reading, MA, 613 pp.

Brignell, J. and White, N., (1996), *Intelligent Sensor Systems*, rev. ed., Institute of Physics Publishers, Philadelphia.

Britton, S., (2002), Wireless sensor networks: battery operation of link and sensors, in *IEEE SI/Con Sensors for Industry Conference Proceedings*, Houston, TX, pp. 169–171.

Brooks, T., (2001), Wireless technology for industrial sensor and control networks, in *SIcon/01 Sensors for Industry Conference, Proceedings of the First ISA/IEEE*, Rosemont, IL, pp. 73–77.

Buchla, D. and McLachlan, W., (1992), *Applied Electronic Instrumentation and Measurement*, MacMillan Publishing Company, New York.

CALEX, (2002), Power Conversion: DC/DC Converters, Power Supplies, Application Notes, available at www.calex.com (online).

Caristi, A., (1989), *IEEE-488: General Purpose Instrumentation Bus Manual*, Academic Press, New York, 253 pp.

Cassioli, D., Detti, A., Loreti, P., Mazzenga, F., and Vatalaro, F., (2002), The Bluetooth technology: state of the art and networking aspects, *NETWORKING/2002, Networking Technologies, Services, and Protocols, Performance of Computer and Communication Networks*, Pisa, Italy, pp. 479–490.

Cavicchi, T.J., (2000), *Digital Signal Processing*, John Wiley, New York.

Chen, W.K., (1995), *The Circuits and Filters Handbook*, CRC Press, Boca Raton, FL.

Chugani, M.L., Samant, A.R., and Cerna, M., (1998), *LabVIEW Signal Processing*, Prentice Hall PTR, Upper Saddle River, NJ.

Connell, B., (1996), *Process Instrumentation Applications Manual*, McGraw-Hill, New York.

Considine, D.M. and Considine, G.D., (1984), *Standard Handbook of Industrial Automation*, Chapman & Hall, New York.

Coombs, C.F., Jr., Ed., (1995), *Electronic Instrument Handbook*, 2nd ed., McGraw-Hill, New York.

Coombs, C.F., Jr., (2000), *Electronic Instrument Handbook*, 3rd ed., McGraw-Hill, New York, pp. 47.1–47.26.

Cypress PSoC, available at http://www.cypressmicro.com/support/index.html (online).

Delaney, C.F.G. and Finch, E.C., (1992), *Radiation Detectors: Physical Principles and Applications*, Clarendon Press, Oxford.

Demler, M.J., (1991), *High Speed Analogue-to-Digital Conversion*, Academic Press, New York.

Doebelin, E.O., (1990), *Measuring Systems: Application and Design*, 4th ed., McGraw-Hill International Editions, New York, 992 pp.

Dorf, R.C., Ed., (1993), *The Electrical Engineering Handbook*, CRC Press, Boca Raton, FL.

Dunbar, M., (2001), Plug-and-play sensors in wireless networks, *IEEE Instrum. Meas.*, 4, 19–23.

Dyer, S.A., Ed., (2001), *Survey of Instrumentation and Measurement*, John Wiley & Sons, New York.

Eijkel, K.J.M. and Fluitman, J.H.J., (1990), Optimisation of the response of magnetoresistive elements, *IEEE Trans. Magn.*, 26, 311–321.

Eren, H., (1999), Inductive displacement sensors, in *The Measurements, Instrumentation and Sensors Handbook*, Webster, J.G., Ed., CRC Press, New York, p. 6:15.

Eren, H., (1999), Capacitive sensors, in *The Measurements, Instrumentation and Sensors Handbook*, Webster, J.G., Ed., CRC Press, New York, pp. 6:37–52.

Eren, H., (1999), Acceleration, vibration and shock, in *The Measurements, Instrumentation and Sensors Handbook*, Webster, J.G., Ed., CRC Press, New York, pp. 17:1–32.

Eren, H., (1999), Capacitance and capacitance measurement, in *The Measurements, Instrumentation and Sensors Handbook*, Webster, J.G., Ed., CRC Press, New York, pp. 45:1–27.

Eren, H., (2000), Density measurement, in *Mechanical Variables Measurement: Solid, Fluid and Thermal*, Webster, J.G., Ed., CRC Press, New York, pp. 2:1–16.

Eren, H., (2001), Instruments, in *Survey of Instrumentation and Measurement*, Dyer, S.A., Ed., Wiley, New York, pp. 1–14.

Eren, H., (2001), Magnetic sensors, in *Survey of Instrumentation and Measurement*, Dyer, S.A., Ed., Wiley, New York, pp. 46–60.

Eren, H., (2001), Displacement measurement, in *Survey of Instrumentation and Measurement*, Dyer, S.A., Ed., Wiley, New York, pp. 509–521.

Eren, H., (2001), Acceleration measurements, in *Mechatronics Engineers Handbook*, Bishop, R.H., Ed., CRC Press, Boca Raton, FL, pp. 19.12–19.33.

Eren, H., (2002), Analogue and discrete input/output, costs, and signal processing, in *Instrumentation Engineers Handbook*, 4th ed., Bela Liptak, Ed., CRC Press, Boca Raton, FL, pp. 123–141.

Eren, H., (2002), Recent technological progress in electronic portable instruments, in *IEEE SI/Con Sensors for Industry Conference Proceedings*, Houston, TX, pp. 78–83.

Erin, C. and Asada, H.H., (1999), Energy optimal codes for wireless communications, in *Proceedings of the 38th IEEE Conference on Decision and Control*, Phoenix, AZ, Vol. 5, pp. 4446–4453.

Everett, H.R., (1995), *Sensors for Mobile Robots: Theory and Application*, A.K. Peters, Wellesley, MA.

Ewing, R.L. and Zohdy, A.A., (2001), Design approach for biomedical smart sensors, in *Proceedings of the 44th Midwest Symposium on Circuits and Systems*, Dayton, OH, Vol. 1, pp. 474–477.

Ferrari, P., Flaminni, A., Marioli, D., and Taroni, A., (2002), A low cost smart sensor with Java interface, in *IEEE SI/Con Sensors for Industry Conference Proceedings*, Houston, TX, pp. 161–167.

Ferri, G., De Laurentiis, P., D'Amico, A., and Di Natale, C., (2000), A low voltage integrated CMOS analog lock-in amplifier for LAPS applications, in *Proceedings of Eurosensors XIV Conference*, Copenhagen, Denmark, pp. 783–786.

Ferri, G., Faccio, M., Stochino, G., D'Amico, A., Rossi, D., and Ricotti, G., (1999), A very low voltage bipolar op-amp for sensor applications, *Analogue Integrated Circuits Signal Process*, 20, 11–23.

Filipov, A., Srour, N., and Falco, M., (2002), Networked microsensor research at ARL and the ASCTA, in *IEEE SI/Con Sensors for Industry Conference Proceedings*, Houston, TX, pp. 212–218.

Fowler, K.R., (1996), *Electronic Instrument Design: Architecting for the Life Cycle*, Oxford University Press, New York.

Fraden, J., (1993), *AIP Handbook of Modern Sensors: Physics, Design and Application*, American Institute of Physics, New York.

Frank, R., (1996), *Understanding Smart Sensors*, Artech House, Norwood, MA.

Frenzel, L.E., (2002), After a slow start, Bluetooth shows its colors, *Electron. Design*, 50, 68–74.

Gardner, J.L., (1994), *Microsensors, Principle and Applications*, Wiley, New York, p. 344.

Garrett, P.H., (1987), *Computer Interface Engineering for Real-Time Systems: A Model Based Approach*, Prentice Hall, Englewood Cliffs, NJ.

Garrett, P.H., (1994), *Advanced Instrumentation and Computer I/O Design: Real-Time System Computer Interface Engineering*, IEEE Press, New York.

Georgopoulos, C.J., (1985), *Interface Fundamentals in Microprocessor-Controlled Systems*, D. Reidel Publishing Company, Dordrecht, Holland.

Gibson, J.D., Ed., (1997), *The Communications Handbook*, CRC Press, Boca Raton, FL.

Girson, A., (2002), Handheld devices, wireless communications, and smart sensors: what it all means for field service, *Sensors*, 19, 16–20.

Gopel, W., Hesse, J., and Zemel, J.N., (1989), *Sensors: A Comprehensive Survey*, WCH, Weinheim, Germany.

Gotz, A., Gracia, I., Plaza, J.A., Cane, C., Roetsch, P., Bottner, H., and Seibert, K., (2001), Novel methodology for the manufacturability of robust CMOS semiconductor gas sensor arrays, *Sensors Actuators*, B77, 395–400.

Grob, R.L. and Barry, F.E., (1995), *Modern Practice of Gas Chromatography*, 3rd ed., Wiley, New York.

Grover, D. and Deller, R.J., (1999), *Digital Signal Processing and the Microcontroller*, Prentice Hall PTR, Upper Saddle River, NJ.

Haris, C.M., *Handbook of Acoustical Measurements and Noise Control*, 3rd ed., McGraw-Hill, New York, 1997.

Haykin, S., (1994), *Communication Systems*, 3rd ed., John Wiley & Sons, New York.

Haykin, S. and Kosko, B., Eds., (2001), *Intelligent Signal Processing*, IEEE Press, New York.

Hitachi, available at http://www.hitachisemiconductor.com/sic/jsp/japan/eng/products/mpumcu/32bit/index.html (online).

Hoeschelle, D.F., (1994), *Analogue-to-Digital and Digital-to-Analogue Conversion Techniques*, John Wiley & Sons, New York.

Hofmann-Wellenhof, B. and Collins, L.J., (2001), *Global Positioning System: Theory and Practice*, 5th rev. ed., Springer-Verlag, Wein, Germany.

Holman, J.P., (1989), *Experimental Methods for Engineers*, 5th ed., McGraw-Hill, New York.

Horowitz, P. and Hill, W., (1991), *The Art of Electronics*, 2nd ed., Cambridge University Press, Cambridge, U.K.

Hyde, A., (2001), Medical net instruments: a new generation in telemedicine, *J. Telemed. Telecare*, 7, 183–185.

IEC 60584-1, (1995), *Thermocouples: Part 1: Reference Tables*.

IEC 60584-2, (1982), *Thermocouples: Part 2: Tolerances*.

IEEE, (1990–2002), *Spectrum*, IEEE, Piscataway, NJ.

IEEE, (1990–2002), *Transactions on Instrumentation and Measurement*, IEEE, Piscataway, NJ.

IEEE, (1998), *Special Issue: Integrated Sensors, Microactuators Microsystems* (MEMS), IEEE, Piscataway, NJ, pp. 1531–1811.

IEEE, IMTC, (1994–2001), *Instrumentation and Measurement Conference Proceedings*, IEEE, Piscataway, NJ.

Intel Corporation, (1992), *Embedded Microcontrollers and Processors*, Intel, Mt. Prospect, IL.

Jain, R., Werth, J., and Browne, J.C., (1996), *Input/Output in Parallel and Distributed Computer Systems*, Kluwer Academic Publishers, Boston.

Johnson, G., (1997), *LabVIEW Graphical Programming: Practical Applications in Instrumentation and Control*, McGraw-Hill, New York, p. 625.

Kaiser, V.A. and Liptak, B.G., (1995), DCS-I/O hardware and setpoint stations, in *Instrumentation Engineers' Handbook: Process Control*, 3rd ed., CRC Press, Boca Raton, FL.

Karim, M.A., Ed., (1992), *Electro-Optical Displays*, Marcel Decker, New York.

Klein, L.A., (1999), *Sensor and Data Fusion Concepts and Applications*, 2nd ed., SPIE Optical Engineering Press, Bellingham, WA.

Kohls, O. and Scheper, T., (2000), Setup of a fiber optical oxygen multisensor-system and its applications in biotechnology, *Sensors Actuators*, 70, 121–130 (www.elsevier.nl).

Lapsley, P., (1997), *DSP Processor Fundamentals: Architectures and Features*, IEEE Press, New York.

Lee, K., (2000), IEEE 1451: A standard in support of smart transducer networking, in *Proceedings of the 17th IEEE Instrumentation and Measurement Technology Conference*, Vol. 2, pp. 525–528.

Lee, K.B. and Schneeman, R.D., (2000), Distributed measurement and control based on the IEEE 1451 smart transducer interface standards, *IEEE Trans. Instrum. Meas.*, Baltimore, MD, 49, 621–627.

Liptak, B., (2002), *Instrumentation Engineers Handbook*, 4th ed., CRC Press, Boca Raton, FL.

Luo, R.C., (2001), Sensors and actuators for intelligent mechatronic systems, in *IECON'01, 27th Annual Conference of the IEEE Industrial Electronics Society*, Denver, CO, Vol. 3, pp. 2062–2065.

Mackay, R.S., (1993), *Biomedical Telemetry*, IEEE Press, Piscataway, NJ.

Madisetti, V.K. and Williams, D.B., (1998), *The Digital Signal Processing Handbook*, CRC Press, Boca Raton, FL.

Mariella, R.P., (2001), Development of a battery-powered hand-held, real-time PCR instrument, *Proc. SPIE Int. Soc. Optical Eng.*, 42, 58–64.

McDonald, K.D., (1993), *GPS Receivers: Survey of Equipment Characteristics and Performance*, Navtech Seminar Inc., Arlington, VA.

McKinlay, A.F., (1981), *Thermoluminescence Dosimetry*, Adam Hilger, Bristol, U.K.

Miliozzi, P., Kundert, K., Lampaert, K., Good, P., and Chian, M.A., (2000), Design system for RFIC: challenges and solutions, *Proc. IEEE*, 88, 1613–1632.

Mills, J.P., (1993), *Electromagnetic Interference Reduction in Electronic Systems*, Prentice Hall, Englewood Cliffs, NJ.

Moller, H., (1986), *Electro Acoustic Measurements*, Bruel & Kjaer, Naerum, Denmark.

Monkman, G.J., (1999), Advancements in infrared array detectors, *Sensor Rev.*, 19, 273–277.

Moore, J.O., Ross, T., and Johnson, R.N., (1999), Migrating legacy applications to the IEEE 1451 standards, in *Proceedings of the Sensors Exposition*, Baltimore, pp. 33–39.

Motorola, available at http://e-www.motorola.com/webapp/sps/site/homepage.jsp?nodeId = 03M0ym4t3ZG (online).

Mulgrew, B., Grant, P., and Thompson, J., (1999), *Digital Signal Processing: Concepts and Applications*, Macmillan, Basingstoke, U.K.

Nachtigal, C.L., (1990), *Instrumentation and Control, Fundamentals and Applications*, Wiley Series in Mechanical Engineering Practice, Wiley Interscience, Washington, D.C., p. 890.

Nagy, I., (1992), *Introduction to Chemical Process Instrumentation*, Elsevier, New York.

Nekoogar, F. and Moriarty, G., (1998), *Digital Control Using Digital Signal Processing*, Prentice Hall, Englewood Cliffs, NJ.

Neuricam, available at http://www.neuricam.com/Neuricam/Products/Totem/nc3002.htm (online).

Ogata, K., (1997), *Modern Control Engineering*, 3rd ed., Prentice Hall, Englewood Cliffs, NJ.

Pahlavan, K. and Levesque, A.H., (1995), *Wireless Information Networks*, Wiley, New York.

Pallás-Areny, R. and Webster, J.G., (1999), *Anologue Signal Processing*, John Wiley & Sons, New York.

Pallás-Areny, R. and Webster, J.G., (2001), *Sensors and Signal Conditioning*, 2nd ed., John Wiley & Sons, New York.

Paton, B.E., (1999), *Sensors, Transducers, and LabView*, Prentice Hall PTR, Upper Saddle River, NJ.

Penrose, W.R. and Penrose, S.E., (1987), *Designing Portable Computerised Instruments*, Tab Books, Inc., Blue Ridge Summit, PA, p. 262.

Peyton, A.J. and Walsh, V., (1993), *Analog Electronics with Op Amps*, Cambridge University Press, Cambridge, U.K.

Philips, available at http://www.semiconductors.com/pip/P87C51SBPN (online).

Pitt, M.J. and Preece, P.E., (1991), *Instrumentation and Automation in Process Control*, Ellis Harwood Ltd., Chichester, England.

Pollack, M.W., (1996), Communications-based signalling: advanced capability for mainline railroads, *IEEE Aerosp. Electron. Syst. Mag.*, 11, 13–18.

Pollock, D.S.G., (1999), *A Handbook of Time-Series Analysis, Signal Processing and Dynamics*, Academic Press, San Diego.

Portable Design, The Authority for Mobile Technologies and Products, PennWell, available at www.portabledesign.com (online).

Potter, D., (2002), Smart plug and play sensors, *IEEE Instrum. Meas.*, 5, 28–30.

Predko, M., (1999), *Handbook of Microcontrollers*, McGraw-Hill, New York.

Primdahl, F., (1979), The fluxgate magnetometer, *J. Phys. E Sci. Instrum.*, 1, 242–253.
Proakis, J.G. and Manolakis, D.G., (1992), *Digital Signal Processing — Principles, Algorithms, and Applications*, 2nd Edition, Maxwell-Mcmillan, New York.
Putten, A.F.P., (1996), *Electronic Measurement Systems: Theory and Practice*, 2nd ed., Institute of Physics Publishers, Philadelphia.
Randy, F., (2000), *Understanding Smart Sensors*, 2nd ed., Artech House, Boston.
Rathore, T.S., (1996), *Digital Measurement Techniques*, Narosa Publishing, London.
Rezazadeh, M. and Evans, N.E., (1990), Multichannel physiological monitor plus simultaneous full-duplex speech channel using a dial-up telephone line, *IEEE Trans. Biomed. Eng.*, 37, 428–432.
Reznik, L., (1997), *Fuzzy Controller, Newness*, Butterworth-Heinemann, Oxford.
Roden, M.S., (1996), *Analogue and Digital Communication Systems*, 4th ed., Prentice Hall, Upper Saddle River, NJ.
Scharf, L.L. and Demeure, C., (1991), *Statistical Signal Processing: Detection, Estimation, and Time Series Analysis*, Addison-Wesley, Reading, MA.
Schnell, L., (1993), *Technology of Electrical Measurements*, John Wiley & Sons, New York.
Schuler, C.A. and McNamee, W.L., (1986), *Industrial Electronics and Robotics*, McGraw-Hill International Editions, New York, p. 474.
Shen, G., (2000), The analyses and design of the mechanics sensor chip, *J. Transduction Technol.*, 13, 61–66.
Shults, M.C., Rhodes, R.K., Updike, S.J., Gilligan, B.J., and Reining, W.N., (1994), A telemetry-instrumentation system for monitoring multiple subcutaneously implanted glucose sensors, *IEEE Trans. Biomed. Eng.*, 41, 937–942.
Sinclair, I.R., (1992), *Sensors and Transducers*, 2nd ed., BH Newnes, Oxford.
Soclof, S., (1991), *Design and Application of Analog Integrated Circuits*, Prentice Hall, Englewood Cliffs, NJ.
Soloman, S., (1999), *Sensors Handbook*, McGraw-Hill, New York.
Sound Power Measurements, Hewlett-Packard Company Application Note AN1230.
Stanley, W.D., Dougherty, G.R., and Dougherty, R., (1984), *Ditial Signal Processing*, Prentice Hall, NJ.
Stoebe, T.G., Chen, T.C., Smith, K.R., Thibado, J., Gasiot, J., and Prevost, H., (1996), Solid state integrated radiation sensor development, *Radiat. Prot. Dosimetry*, 66, 427–429.
Sydenham, P.H., Hancock, N.H., and Thorn, R., (1989), *Introduction to Measurement Science and Engineering*, Wiley, New York.
Sze, S.M., (1994), *Semiconductor Sensors*, Wiley, New York.
Tang, A., Smith, B., Schild, J.H., and Peckham, P.H., (1995), Data transmission from an implantable biotelemeter by load-shift keying using circuit configuration modulator, *IEEE Trans. Biomed. Eng.*, 42, 524–527.
Taub, H. and Schilling, D.L., (1986), *Principles of Communication Systems*, 2nd ed., McGraw-Hill, New York.
Taylor, D.M., (1994), *Industrial Electrostatics: Fundamentals and Measurements*, Electronic and Electrical Engineering Research Studies, John Wiley, New York.
Togawa, T., Tamura, T., and Ake Oberg, P., (1997), *Biomedical Transducers and Instruments*, CRC Press, Boca Raton, FL.
Tompkins, W.J. and Webster, J.G., Eds., (1989), *Interfacing Sensors to the IBM PC*, Prentice Hall, Englewood Cliffs, NJ, p. 447.
Tooley, M.H., (1995), *PC-Based Instrumentation and Control*, 2nd ed., Newnes, Oxford.
Usher, M.J. and Keating, D.A., (1996), *Sensors and Transducers: Characteristics, Applications, Instrumentation, Interfacing*, 2nd ed., Macmillan Press, London.

Valkenburg, M.E., (1982), *Analog Filter Design,* HRW, New York.

Van de Plassche, R., (1994), *Integrated Analogue-to-Digital and Digital-to-Analogue Converters,* Kluwer Academic Publishers, Dordrecht, Netherlands.

Wanhammer, L., (1999), *DSP Integrated Circuits,* Academic Press, San Diego.

Watkinson, J., (1994), *The Art of Data Recording,* Focal Press, Oxford.

Webster, J.G., Ed., (1999a), *Wiley Encyclopedia of Electrical and Electronic Engineering,* John Wiley & Sons, New York.

Webster, J.G., Ed., (1999b), *The Measurement, Instrumentation, and Sensors Handbook,* CRC Press, Boca Raton, FL.

Webster, J.G., Ed., (2000), *Mechanical Variables Measurement: Solid, Fluid, and Thermal,* CRC Press, Boca Raton, FL.

Witzel, J., (2002), Inexpensive quality data acquisition, *IEEE Instrum. Meas.,* 5, 52–54.

Yen, J., Langari, R., and Zadeh, L.A., (1995), *Industrial Applications of Fuzzy Logic and Intelligent Systems,* IEEE Press Marketing, New York.

Yujin, L., Jesung, K., Sang, L.M., and Joong, S.M., (2001), Performance evaluation of the Bluetooth-based public Internet access point, in *Proceedings of the 15th International Conference on Information Networking,* IEEE Computer Society, CA, pp. 643–648.

Index

A

Absorption sensors, 103
AC, *see* Alternating current
Acceleration, 4, 124, 131, 145
    measurement, 57, 139, 343
    sensors, 64, 140
    signal, 142
Accelerometer(s), 34, 63, 67, 68, 124, 220, 273,
        283, 321
    differential-capacitance, 144
    piezoelectric, 141–142
    piezoresistive, 141
    seismic, 141
    two-dimensional, 341
    vibrating beam, 143
Accuracy, 3, 16, 19, 95
Acoustic measurements, 321
Acoustic sensors, 68
Actuators, 43, 151, 219, 266
A/D, *see* Analog-to-digital converter
Adapted synchronous model feedback
        systems (ASMFS), 216
Aerosol, 14, 104, 109, 237, 329
Air analyzers, 327
Air-core capacitors, 254
Air pollution, 109, 119, 120, 167, 323
Aliasing, 177, 184, 199, 201, 210
Alkaline batteries, 113, 242, 325, 327, 337, 354
Alkaline manganese dioxide batteries, 242
Alpha particles, 94, 95, 118
Alternating current (AC), 15, 47, 237, 313
ALU, *see* Arithmetic logic unit
Ambulatory monitoring, 17
American National Standards Institute
        (ANSI), 307
Amorphous sensing, 67
Amperimetric sensors, 100, 105, 106, 107
Amplifier(s), 267
    charge, 268
    instrumentation, 263
    inverting, 268
    lock-in, 126
Amplitude modulator, 196

AMR, *see* Anisotropic magnetoresistor
Analog components, 262
Analog-to-digital (A/D) converter, 9, 48, 152,
        232, 313
Analog instruments, 9
Analog spectrum analyzers, 318
Anisotropic magnetoresistor (AMR), 112
ANSI, *see* American National Standards
        Institute
Anti-aliasing, 16, 184, 210, 320
Application-specific chip (ASIC), 289
Arithmetic logic unit (ALU), 172
ARQ, *see* Automatic repeat request
ASIC, *see* Application-specific chip
ASMFS, *see* Adapted synchronous model
        feedback systems
Atmosphere, 87, 95, 104, 119
Autocorrelation, 198
Automatic repeat request (ARQ), 212

B

Back-plane buses, 278
Balances, 145, 148
Bandpass filter, 39, 202, 270, 271, 319
Barometers, 137, 360
Basic input/output system (BIOS), 154
Battery(ies), 240
    alkaline manganese dioxide, 242
    capacity, 241
    carbon-zinc, 241
    energy density of, 241
    lead-acid, 242
    lithium, 243
    nickel-cadmium, 242
    shelf life, 241
    silver-oxide, 243
BCD, *see* Binary-coded decimal
BCS, *see* British Calibration Services
BDLC, *see* Byte data link controller
Beam, 67, 89, 118, 127, 136, 331
Bessel–Tomson filter, 271

EOB, *see* Electrical oscillator-based
EPA, *see* Environmental Protection Agency
EPROM, *see* Erasable programmable ROM
Equivalent circuits, 82
Equivalent-time-sampling technique (ETS), 316
Erasable programmable ROM (EPROM), 156
Error(s), 12, 19, 23
    budget, 24
    calibration, 26
    dynamic, 19
    gross, 27
    random, 26, 27, 128
    reduction, 27
    systematic, 24
Ethernet data acquisition systems (EDAS), 357
ETS, *see* Equivalent-time-sampling technique
Examples and applications of portable
        instruments, 311–362
    biomedical instruments, 334–343
        animal biotelemetry, 342–343
        human biotelemetry, 338–342
    chemical instruments, 329–334
        gas detectors, 333–334
        portable chemical laboratories, 332–333
    computer-based and intelligent
        instruments, 357–359
    domestic, personal, hobby, and leisure
        portable instruments, 359–362
    environmental instruments, 323–329
        air analyzers, 327–329
        soil analyzers, 325–327
        water analyzers, 323–325
    GPS and telemetry-based instruments, 354–357
    laboratory instruments, 312–323
        multimeters, 312–314
        oscilloscopes, 315–318
        power and energy meters, 314–315
        spectrum analyzers, 318–323
    mechanical, sound, and vibration
        measurement instruments, 343–350
        mechanical measurement
            instruments, 343–346
        sound measurement instruments, 348–350
        vibration measurement instruments, 346–347
    radiation, light, and temperature
        instruments, 350–354
        dosimeters, 353–354
        long-wave radiation instruments, 350–351
        short-wave radiation sensing, 352–353
        temperature measurements, 354
Excess noise, 37, 38
External bus interface (EBI), 165

F

Failure mode analysis, 234, 297
Fast Fourier transform (FFT), 13, 170, 320
Fault, 27, 66, 307
FCC, *see* Federal Communications Commission
FDA, *see* Food and Drug Administration
FDM, *see* Frequency division multiplexing
Federal Communications Commission
        (FCC), 130, 339
Feedback, 7, 28, 83, 172
Ferromagnetic material, 51, 54, 113
FET, *see* Field-effect transistor
FFT, *see* Fast Fourier transform
Fiber optics, 54, 72
Fieldbuses, 33, 213, 215
Field-effect transistor (FET), 39
Filter(s), 9, 11
    anti-aliasing, 16, 184, 210, 320
    bandpass, 39, 202, 270, 271, 319
    bandstop, 270, 271
    Bessel–Tomson, 271
    Butterworth, 62, 271, 349
    Chebyshev, 62, 271
    digital, 35, 170, 174, 184, 196, 205
    FIR, 173, 205, 206
    high-pass, 270, 272
    low-pass, 199, 270
    notch, 270, 271
    telecommunication, 256
Finite impulse response (FIR), 170
FIR, *see* Finite impulse response
Fire sensors, 118
Flame ionization detector, 102
Flame photometric sensors, 102
Flash converters, 182, 280
Fluid, 73, 100
    dynamics, 141
    measurements, 74, 131
Fluxgate magnetometer, 113
Foil resistors, 249, 250
Food and Drug Administration (FDA), 335
Force
    measurement, 57
    sensors, 139